Construction
Final Presentation
U0411388

Space, Tectonics and Design

空间、建构与设计

顾大庆 · 柏庭卫 著

Gu Daqing · Vito Bertin

中国建筑工业出版社

China Architecture & Building Press

前言一 | preface 1

顾大庆　　Gu Daqing

本书的内容是笔者和同事于2001至2009年这段时间在香港中文大学建筑学系（2009年更名为“建筑学院”）的一个设计课程。这个名为“建构实验”的课程是“建构工作室”设计教学架构的一个部分。借助于“建构实验”这一教学平台，我们试图探讨一个关于空间和建构的设计方法。为什么要进行这样的研究是有多方面原因的，而其中的一个原因很“中国”。鉴于本书主要面向的是国内的读者，故而这里需要就此作特别的说明。

中国的建筑设计在过去的十年间发生了巨大的变化，延续了几十年的巴黎美术学院（“布杂”）的形式主义设计传统好像一下子衰退了，一种对空间、材料和建造的兴趣正逐渐兴起。这个变化最先于1990年代末通过一批青年中国建筑师的设计作品体现出来，他们的作品引起国内外建筑界的普遍关注。他们或有着海外留学的背景，吸收了国外的新设计思想；或通过自我反省和艰苦的探索，实现了设计思想的更新。但是，这些零星的设计实验得以在一个不长的时间内形成一个流行趋势，在很大程度上则要归功于书籍、展览和互联网等现代传媒的作用。我们现在比以往任何时候都能在第一时间了解到那些一流建筑师的最新动态。从某种意义上来说，我们现在也更容易地去“追随”那些新的形式。那些源自模仿的新形式不免显得有点表面化。我也不得不忧虑这会不会只是以前的形式主义设计方法的一种延续呢?

这些新的建筑形式的出现除了突显现代传媒的巨大潜能外，似乎并没有太多的学术支持。这里特别是指建筑院校通过它的设计教育和学术研究对建筑实践的支持。相对而言，巴黎美术学院的形式主义方法就有着深厚的学术基础。事实上我们的整个的建筑教育体系就是为了传授和发展这一形式主义的方法而建立的，不但通过教育培养形式主义的设计人才，还通过学术研究来阐述和发展形式主义的理论和方法。但是，如今的建筑教育在这一新趋势中则要明显滞后于建筑实践。现在的情形是实践影响教育，而不是相反。建筑教育的滞后尤其反应在基础研究方面的缺失。基础研究，不是指建筑设计入门课程，而是建筑设计基本问题和基本方法的研究。当然，这类研究往往与基础教育有关，不过我们不可以将两者混为一谈。纵观中国建筑教育过去三十年的历程，教育的规模呈现飞速扩张，国际化程度也逐渐提高，学术研究也不能说不活跃，但是在建筑设计的基本问题和方法的研究方面则几乎是空白。这大概就是当前的对空间、材料和建构的兴趣显得后继乏力的根本原因。基础研究的缺乏导致我们只能模仿，难以创新。这个就是我们所面对的一个挑战。

在建筑学校中的建筑设计基本问题和方法的研究主要借助于设计课程来展开。在“建构实验”这个课程中我们试图解决这样几个问题：界定一套空间、建构和设计的语汇；开展一个以体块、板片和杆件为线索的空间研究；重新梳理模型和图在设计过程中的作用；最后也是最重要的，发展一个建构设计方法。这一建构设计方法的特点是以模型作为设计发展的主要手段。首先，对模型材料的操作产生一个建构的概念，而后运用这一概念来组织空间和形式，再经过多种模型材料来丰富空间和形式的表达，最后通过从模型材料到建筑材料的转换成为可建造的形式。本书将通过教案、要点、练习、研究、设计等几个主要篇章来全面介绍这个课程。我们希望这个建构实验课程以及本书的出版对于中国的建筑设计基础研究能够起到一个抛砖引玉的作用。

This book is about a design course that my colleagues and I developed at the Department of Architecture (renamed "School of Architecture" in 2009) of the Chinese University of Hong Kong from 2001 to 2009. The course under the name "Tectonic Lab" was part of a teaching programme provided by the Tectonic Studio. By means of this platform, we tried to develop a design method in relation to space and tectonics. There are many reasons behind this endeavour, one of them appearing to be very "Chinese". Since this book is intended for readers mainly in mainland China, I should elaborate this point specially here.

Architectural design in China has been undergoing a dramatic transformation in the last decade. The long lasting tradition of the Beaux-Arts formalistic design seemed to decline gradually, while a new interest in space, material, and construction has begun to emerge. This change was first signalled by the buildings designed by a group of young Chinese architects. However, its popularisation into a national phenomenon within a short period was largely due to the power of modern media. It is true that it has never been so easy to obtain the latest news on leading architects' work. In other words, it also allows us to conveniently "follow" these new forms without any delay. As a consequence we can sense a kind of superficiality appearing in such works. What I have begun to worry about is that this new trend might be merely the continuation of the Beaux-Arts formalism in another form.

What we can observe behind this phenomenon is that, except demonstrating the power of modern media, it actually gains little support from architectural education through its design teaching and academic research. On the contrary, the Beaux-Arts formalism has rooted itself deeply in architectural education. As a matter of fact, our whole system of architectural education was built for the purpose of promoting Beaux-Arts formalistic design. Trapped by its formalistic tradition, today's architectural education falls behind architectural practice. This is obvious particularly in the research on basic design issues and methods. What I mean here is not the foundation course although both are often related. In view of the development of architectural education in China in the last 30 years, there has been a tremendous expansion in the scale of education, an increasing level of globalisation and richness of academic research. However, there is very little achievement in basic design research. And, we take this as a challenge.

The design studio is the best venue for conducting research on basic design issues and methods. Through the Tectonic Lab, we have defined a set of terminologies in relation to space, tectonics and design; conducted an experiment on space based on three space defining elements: block, slab and stick; reemphasised the critical role of design media – models and graphics – in the design process; and the most important, developed a tectonic design method. The method is based on the use of models as the primary design tool. First, the operation of model material generates a tectonic concept; then it is applied to the organisation of space and form, next the use of multiple model materials further enhances the expression of space and form, and finally a built form is achieved through the transformation from model materials to building materials. All these matters will be presented thoroughly in this book in the following chapters: programme, essentials, exercise, study and design. We hope that the experiment of the Tectonic Lab and the publication of this book can be a catalyst for the flourishing of basic design research in China.

前言二 | preface 2

柏庭卫　Vito Bertin

我们在这本书中所讲述的内容没有多少是我在做学生的时候就学会的。如果要说有的话，那么就是“空间限定要素”这个术语。在苏黎世联邦理工学院建筑系，由勃那德·赫伊斯利教授主持的一年级设计课程的练习中，空间限定要素主要是板片，单片的，也可以是“L”形和“U”形的。这个术语强调在可知觉空间和空间限定要素之间存在一定的联系，通过后者我们才能知觉到前者。这种板片的空间限定要素对应于连续空间的概念，这是现代运动为对抗巴黎美术学院的设计传统而提出的主张。在“布杂”的设计中，空间并不是通过独立的板片，而是通过一个连续的包裹界面来界定的。

很久以后，我在欧洲第一次看到这种旧的、“布杂”的空间类型在当代建筑中被非常清晰和纯粹地来处理是由吉耿和癸耶在达沃斯于1992年设计的库奇纳博物馆。在以后的研究中我陆续发现有很多的建筑师运用和发展我称之为包裹空间的概念。这种旧的空间类型的复活，其中的一个主要的原因可能是对建筑的隔热层的要求，这需要将建筑的体积包裹起来，而包裹空间恰恰与之相配合。

我在选修课中开始与学生一起研究符合包裹空间类型的建筑。后来，我和布鲁斯·隆曼在二年级的设计课中引入这两种空间的概念。在建构工作室中，受到马库斯·卢契尔的启发，我们开始研究三种空间限定的要素，即体块、板片和杆件。最终，有关空间限定要素的类型和空间的类型的研究两者结合在一起，我们确定了要研究空间限定要素和空间类型之间的关系。

连续空间和包裹空间这两种空间类型均有深厚的历史渊源。我们将第三种用杆件限定的空间称之为调节空间。调节空间也有历史先例，但是却没有太多的研究。因而，关于这第三种类型的空间的研究也显得比较困难，不但是对学生而言，也同样对我们。但是，其中也充满了令人着迷的可能性。

观察在我们的教学中具有非常重要的作用。它让我们可以去发现一个建筑概念。我们认为建筑概念是设计的一个部分，因而是可以从设计中辨识出来。尽管有时候一个设计会有许多与建筑不太相关的概念，但是我们还是应该能够识别其中的建筑概念。这个观点与另外一种设计态度形成对比，即设计的想法来自于想像，正如我们常常听到这样的表达：“我要这样做”。而观察首先需要有一个观察的对象。这就是为什么在我们的教学中我们强调不需要经过一个漫长的准备阶段就直接进入设计的操作。在这个课程中，我们的直接行动就是对模型材料的操作。

从操作入手使得每个学生要直接面对手头的任务，而不需要依赖在设计还未开始之前就有的“先入为主”之见。观察不是生来具有，而是需要学习和练习。我们在观察一个对象时的所见受到我们先前知识的影响。但是，我们也应该有一种开放的态度来发现未知的知识。在我们的教案中，我们提供了许多制作、观察和纪录的循环。通过这些过程，我们来学习如何观察。我们在每个环节对观察的内容、方法和媒介均加以界定，如此使得这些练习的目的更加明确。

强调空间及观察是贯穿整个教案的核心，也是这本书的主线。从这两点出发，我们可以达到一个新的设计境界。

Not much of what we describe here in this book did I learn as a student. But one term – space defining element – stems from that time. In the exercises of the first year programme by Prof. Bernhard Hoesli at the ETH-Zurich, the space defining elements were mainly slabs which could also become L-shaped or U-shaped. The term emphasises that there is a relationship between the perceived space and the built element through which space becomes perceivable. This type of element supports the idea of continuous space, which in the modern movement is a polemic reaction to the space preferred by the Beaux-Art tradition, in which the space is not defined by discrete elements, but by a continuous envelope.

The first time I saw this older type of space clearly and almost purely articulated in contemporary architecture in Europe was in the Kirchner Museum Davos by Gigon & Guyer in 1992. In subsequent studies I found that quite a number of architects work with and have developed what I call enveloped space. One reason for the renewed interest in this type of space could be the need for thermal insulation, which requires a continuous layer wrapping a building, which could easily be supported within this concept.

I studied buildings which were, in my view, based on the idea of enveloped space with students in elective courses and later introduced the two space types in a second year studio together with Bruce Lonnman. In the tectonic studio, based on ideas by Markus Lüscher, we started to work with the space defining elements block, slab and stick. Finally we merged the two in the sense that we tried to clarify the relationship between element type and space type.

The two types have a clear historical background. We call the third type, space defined by sticks, modulated space. It has historical precedents, but seems not to have been articulated and discussed that much. It is therefore more difficult, for students too, but also has fascinating possibilities.

Observation plays an important role in our teaching. It provides an occasion to discover an architectural idea. It is based on the view that an architectural idea is part of the design, and therefore recognisable in the design. Even if there are links to ideas outside the design, there should be traces of it in the design. This contrasts with the attitude of a general idea, something imagined, or even wilful as expressed in the words, "I want to do". Observation requires something to look at. That is why we start with actions which produce results without a lengthy preparatory phase. Our initial action is the direct manipulation of model material.

Starting with actions provides each student with something to work with, removing the dependence on an idea before the design process can begin. Observation has to be learned and practised. What we see when looking at something is influenced by what we already know. But we should also be open to discover something unexpected. Our programme provides many cycles of making, observing and recording. This provides occasions for practice. And we try to support the learning by identifying in each step what to look for and suggest methods and media, which might make it easier to focus.

The emphasis on space and the concentration on what we can observe are central to all phases of the programme and can be traced throughout the book. Through this focal point a bigger richer picture is projected.

前言三 | preface 3

Markus Lüscher

当许多年前我最初与顾大庆和柏庭卫相识时，我们都在苏黎世联邦理工学院建筑系克莱默教授的工作室教书。若干年后我们又在香港相遇，我向他们介绍了一个建筑教育新课程的展览，这是我作为客座教授在苏黎世联邦理工学院教授的一个关于建造和建筑的一年级课程。基于我早年与克莱默教授一起教学的经历，我想通过强调以下几个问题，从而将建筑教育推进到一个新的高度。

一个关于建筑教育的课程应该将重点放在建筑学的基本问题上：空间、结构和体积。最基本的生成活动必须是能够在一个抽象的层面上被发现、观察和探索的，以保证有足够的实验的自由。

学习活动必须是可以像用模型来做游戏那样的容易。首先你用模型做些东西，然后你仔细观察，你就会在一个抽象的层面上发现空间。学生应该要享受操纵第三维度的乐趣，归根结底，游戏是最佳的学习方式。

教学不是为了发展一个设计方案，而是为了发展建筑设计的能力，是为了探索建筑，是为了发展概念性思考。对建筑教育的最基本的认识是我们完全没有必要一下子就投入到竞赛中去。每个足球或网球运动员都知道在参与到正式的比赛之前都需要经过基本的训练课程。团队合作和交流比起竞赛来要更为重要。学校与现实的最大区别在于，学校可以提供探索和研究的自由。而在实践的环境中，只有那些最优秀的建筑设计事务所才会享有同样的创作自由。

这些主题也深得顾大庆和柏庭卫的认同，他们决定也在香港中文大学建筑学系的建构工作室中尝试类似的方法，我们称这个设计课程为“建构实验”。

这是一个四周的课程，三周的训练和一周的评图。第一周的任务是通过对三种不同尺寸的模型材料——体块、板片和杆件——的操作来生成和发现空间。第二周的任务是设计一个体积，该体积的不同摆放或成为一个楼房，或一个板房，或一个平房。第三周的任务是用设计的体积来形成邻里空间，即体积之间的空间。这就是2001年我们最初开始这个课程的情形。

以后经过多年的不断发展，该实验逐渐成为一个横跨一个学期的课程。概念的方法、结构的影响，以及空间的生成，这些问题在发展一个设计方案的过程中被有系统地检验。建造、功能活动，以及作为城市环境之一部分的场地，这些实际的问题不断地与空间、结构，以及体积这些抽象的问题相互碰撞，最终形成一个新的设计方法和相关的技能。

这个课程的最新发展是在中国用它来培训实践建筑师以及大学的设计老师，以更新他们的设计和教学能力。这更加证明了由顾大庆、柏庭卫，以及其他同事所协同发展的这个设计课程的重要性。我衷心希望这个团队今后能够在教学、研究和发展中取得更多的成果。

I first met Gu Daqing and Vito Bertin as teachers at the ETH-Zürich, where we were all working with Professor Herbert Kramel. Some years later we met again and I presented to them an exhibition about a new programme for architectural education that I had developed as a guest professor in the first year course for construction and architecture at the ETH-Zürich. Starting from my former experiences in teaching together with professor Kramel, I wanted to put architectural education on a new base, focusing on the following theses:

A course for architectural education should focus on the main aspects in architecture: space, structure and mass. The basic generators should be discovered, observed and explored on an abstract level, assuring the freedom of action for experiments.

Learning should become a game on a playground with models which can easily be handled. First you manipulate, then you observe and you will discover space on an abstract level. Students should enjoy exploring the third dimension. Finally, to play is the best way of learning.

Teaching is not only about developing a project, but it is about training abilities in architecture. It is about exploring architecture; it is about conceptual thinking. The idea is not to first of all participate in competitions. Every football team or tennis player knows that he has to attend training programmes before entering a competition. Co-operation and communication are more important than competition. The difference between school and praxis is the freedom of exploration and research which can be granted at school; in practice you find it only amongst the best architectural teams in the market.

These theses have also been the essence of a new teaching experiment which Gu Daqing and Vito Bertin decided to test at the Department of Architecture of the Chinese University of Hong Kong. We called it the Tectonics Lab.

The first run of the programme lasted four weeks. The first week for creating and discovering spaces was based on material in the three main dimensions sticks, slabs and blocks. The second week was for developing a volume which can be considered a tall, a thin or a flat object. The third week was for testing arrangements and neighbourhoods, the space in-between objects. That's how the programme started in 2001 in Hong Kong.

Since then it has been continuously further developed into a one-term course for students. The conceptual approach, the structural impact and space as the main subjects have been tested step by step in the process of developing a project. Materialisation, a programme of activities and a site as part of a city context have been confronted with the abstract approach of space, structure and mass and new practices and skills to develop a project have been discovered.

The latest development of the studio is targeted at the direct interest of architectural practice exposing leading staff to a fresh design approach, and is also used in teacher training in China. This attests to the notable quality of the course which was successfully further developed by the team of Gu Daqing, Vito Bertin and a number of persons supporting the exceptional teaching experiment. I wish the team good luck in the next steps of teaching, research and development.

目录 | table of contents

中大
建築

1 教案 | programme

建构工作室 the tectonic studio
建构实验 the tectonic lab
关于本书 about this book

1 建构工作室 | the tectonic studio

设计教学的基本组织单位是设计工作室，一个建筑学校如何来组织它的设计工作室，取决于特定的建筑教育理念。本书所介绍的“建构实验”是“建构工作室”之一部分，而“建构工作室”又与学系的设计教学体系密切相关。欲了解建构实验课程，就必须要先了解建构工作室，以及它所依存的学系的工作室组织架构。

1.1 香港中文大学建筑学系的设计工作室制度

香港中文大学建筑学系的建筑学专业教育包含三年的本科非职业学位和两年的职业硕士学位两部分。在完成三年的本科教育后，学生必须在设计事务所实习一年，再回来修读两年的硕士课程。2001年，中文大学建筑学系开始实行新的设计工作室组织。一反按照年级来组织设计工作室的传统做法，新的教学结构将全系的设计教学分为五个工作室。一年级的基础课程作为一个独立的“基础工作室”外，一年级以上的设计训练都在四个“主题工作室”中进行。一个主题工作室由数位具有相同或相近设计观念和学术兴趣的教师所组成。主题的界定来自于建筑设计不同的立场和方法，即人居、城市、技术和建构。一种是自内向外的，从人们如何使用空间的角度来决定建筑的形式，是为“人居”；一种是自外向内的，从建筑所处的场地和环境的角度来研究建筑的形式，是为“城市”；一种是从建筑与自然力的关系，从建筑如何解决重力问题的结构和材料的角度以及从建筑如何解决气候环境的角度来研究建筑形式，是为“技术”；最后一种是从建筑物本身，从建造的目的和手段来研究建筑的形式，是为“建构”。

相对于以年级来水平划分的设计教学组织，主题工作室是一垂直性的组织结构，每个工作室同时有除一年级以外的各年级的学生在一起工作，故而又称其为“垂直工作室”。在本科的阶段，二和三年级的学生在两个学年的四个学期中必须通过所有四个工作室的学习，以了解建筑设计的不同态度和方法，而修读这四个工作室的先后秩序则由学生自行决定。在硕士班阶段，学生在第一年可以自由选择工作室，以鼓励学生在某个设计方向的深入发展，最后一年的毕业论文也是在工作室中完成。

一个学期的教学周期（大约13周）划分为工作室主题研究（自选课题）和学系建筑设计（规定课题）两个部分。前者是一个短期作业，约四到五周，由各工作室自行命题，可以是分析性研究，也可以是特定的设计练习，其目的在于提出一个与工作室的目标相一致的设计观念和方法。后者是一个长期的作业，约八周，设计任务由几个工作室共同协商决定，先确定一

学系《黑皮书》中的图解，以白色的光经过三棱镜折射出七色色带来说明四个主题工作室的概念：人居、城市、技术和建构。

The illustration in the Department's "Black Book" uses the analogy of a prism refracting light rays to explain the concept of four "primary" thematic studios: habitation, urbanisation, technics and tectonics.

The design studio is the basic unit of design teaching. The way a department of architecture organises its studio is determined by its particular approach to architectural education. The Tectonic Lab, which this book describes, is one component of the Tectonic Studio. And the Tectonic Studio is one component of the Department's studio system. Therefore, we need first to understand the organisation of the Department's studio system and then that of the Tectonic Studio before describing the Tectonic Lab.

1.1 The studio system at CUHK

The professional education of architects at the Chinese University of Hong Kong (CUHK) consists of two parts: the three-year undergraduate non-professional bachelor's degree plus the two-year graduate professional master's degree. Students are required to take a one-year internship before applying for the master's course. In 2001, design teaching at the Department of Architecture was restructured. Instead of the traditional way of organising studios according to progression by years, the new structure grouped studio teaching into different themes. The first year studio is the Foundation Studio. After the first year, all of the studio courses are conducted by thematic studios, which are defined in terms of design approach: habitation, urbanisation, technics and tectonics. Habitation is concerned with designing buildings according to the way people inhabit a space. Urbanisation covers how the land and the urban context inspires the form of a building. Technics is about how the forces of nature and building technology shape a building. Finally, tectonics focuses on the organisation and expression of a building in terms of space and the materials that define it.

According to the initial design of this studio system, within each thematic studio, teaching should be organised vertically, whereby students from undergraduate years 2 and 3, and graduate years 1 and 2 work at different levels on projects with similar issues. Undergraduate students can choose studios each term, but must pass through all four studios before graduation. Graduate students are free to choose any studio for their first year studio courses and second year theses, according to their interests.

The teaching within a term of approximately 13 weeks consists of two parts: the studio project and the school project. The studio project is formulated by each studio to explore its particular position, approach and issues. The school project is defined on the basis of one of

个设计的主题，再在一个主题中就本科二、三年级和硕士班一年级各年级的任务在规模和复杂性方面作合理区分。设计主题有以下的关键词：居住、祈祷、工作、学习、表演、移动等。每个主题可以进一步衍生出不同的设计课题，如“居住”这个主题可以分为独立住宅、单栋集合住宅及居住区三个设计任务。每个学期的设计主题是变换的，如此可以让学生接触到不同的建筑类型。

每个学期的结束，学系安排以年级来划分的大评图。在一周的时间内每天是一个年级的评图。这个教学架构的特点是既鼓励各工作室发展各自的设计态度和方法，又通过期末大评图等环节提供了一个在学系的层面上各工作室之间可以相互交流和横向比较的平台。设计工作室的运作由学系的教学经费支持，包括工作室的各项活动、旅行参观、邀请学者来讲课及评图等。

1.2 设计工作室的本质：垂直抑或主题

这一新的设计教学体系同时具有“垂直”和“主题”两个特点，如何来理解和定位工作室的角色，就导致各设计工作室之间在基本的教学观念上的不同侧重。

垂直的设计教学组织可以追溯到“布杂”的“图房”。图房的特别之处在于不同年级的学生共处同一个空间，高年级的学生可以辅导低年级学生的作业，反之，低年级的学生成为高年级学生的助手。垂直组织概念还受到设计事务所的设计项目运作方式的启发，即每个设计项目团队由不同设计经验的成员组成，一个主要的建筑师同时可以负责几个设计团队。把这一垂直组织的概念运用到设计工作室的教学就是每个教师负责的教学小组内的成员包含了各个年级的学生。垂直组织方式的特点显然在于不同背景和经验的学生之间的相互交流的可能性。而真正隐藏在垂直组织概念背后的是经验式设计教学，即希望设计教学在教师和学生，学生与学生之间的互动中自然发生。

主题组织理念来自于对建筑师之间存在不同的设计态度和方法这一事实的认识。同样道理，一个学系的教师对于建筑设计的态度和方法也会由于教育背景和设计经验等因素而不一致。这种不一致往往成为设计教学中矛盾的根源。比如以年级来组织设计教学，一个教学组内的设计教师的设计态度各不相同，必然不利于达至共同的教学目标。而且，在这样的情况下，教师之间不同的设计态度和方法的差异成为一个消极的因素被放大了。如果我们换一个角度来思考这个问题，把具有相同设计态度和方法的教师组织起来，那么原先消极的因素就可能转化为积极的因素。当然，教师的组合不是以友谊和利益为

the following fundamental architectural programmes – places for living, working, learning, worshipping, travelling and performing. Historically, buildings of fundamentally similar programmes evolve from one another. Hence, it is possible, for example, to see a room, a house and a village as buildings nested within a line of development of places for living. This course structure makes it possible for students to work on the same type of projects as their peers but according to a particular approach as defined by the selected studio. Students develop particular design attitudes and methods in the studio projects and then apply them in the school project. The different approaches developed in the four thematic studios should become visible at the end of each term through the final review system with external examiners.

1.2 The nature of the studio: thematic or vertical?

This new studio system has the dual characteristics of being thematic and vertical. The position that a studio takes towards the two leads to a very different studio operation.

The origin of the vertical organisation of design studios can be traced back to the atelier at the Ecole des Beaux-Arts in Paris, where junior students worked as assistants to senior students, and in return seniors provided instruction to juniors. This vertical idea is also inspired by the common practice of design firms where a chief architect leads a team of architects with different levels of experience. Sometimes, a chief architect may run several design teams parallel on different projects. To adapt this idea to today's studio context, a studio teacher can have a studio with mixed students from various years, either working on one single task or several tasks. Deep behind this vertical idea of studio organisation is the hope that teaching and learning can happen empirically through the interaction between the teacher and students, and among students themselves.

The idea of the thematic studio at CUHK derives from the recognition of the fact that an architect approaches a design problem in a particular way. Naturally, design teachers within a department should represent different attitudes and preferences toward design due to differences in their educational background and practical experience. It can lead to quite a negative situation of confusion, where teachers with different design attitudes have to teach together in the same class, which can easily happen when studios are taught according to the year of study. The difference in attitude can however become a

建构工作室的设计教学体系由三个层次所组成：本科的建构实验课程、硕士班一年级的建构专题研究，以及硕士班的毕业设计。

The studio teaching system of the Tectonic Studio has 3 levels: the Tectonic Lab for undergraduate students, the Advanced Tectonic Study for graduate year 1 and the thesis.

基础，而是以共同的设计态度和方法为基础。人居、城市、技术和建构就是对不同的设计态度和方法所作的概括，是对建筑设计的知识体系的一种理解，以此来作为工作室组织的依据就是主题工作室。更进一步地说，主题工作室把每个设计方向作为一个独立的学术研究领域，四个工作室确立了建筑设计研究的四个领域。这个关于建筑设计教学的本质的新认识成为建构工作室开展其设计教学和研究工作的基本出发点。

1.3 建构工作室：一个协作模式的主题工作室

一个设计工作室以何种方式运作取决于工作室的指导方针以及具体的成员组成。建构工作室采取的是一种以教案为核心、以研究为基础的团队合作运作方式。这并不是一个随意的选择。

前文指出设计工作室的组织特点有“垂直”和“主题”两种选择，这是就工作室的整体结构而言。进一步来探讨工作室内部教师之间的相互关系，通常有两种选择，即“梯队制”和“协作制”。“梯队制”具有明确的从上至下的组织结构，一位高年资的教师挂帅，由数位不同年资的教师辅佐。梯队制组织具有的一个明显的优势就是可以举一个团队的力量（资源和人力）来进行学术研究工作。至于“协作制”组织结构，顾名思义，就是组成一个教学团队的成员是一种平等的同事关系，团队的负责人名为“协调人”，负责协调教学事务。既然这种工作室的运作是建立于平等的同事关系之上，也就不存在所谓的学术领导的问题，也不大可能集中资源来从事共同的研究。协作制的组织，往往是参与的教师各有不同的观点，各自为阵，除了大家在共同完成特定的教学任务外，没有任何学术意义上的合作。

关于设计工作室的组织和教师的工作关系就有“垂直”和“主题”，以及“梯队”和“协作”四种选择。那么，这四者之间又有什么样的内在关系呢？很显然，垂直式的设计工作室组织和协作制的教师工作关系好像是一种组合，其特点是组内教师的独立性以及松散的组织运作。而强调主题性设计工作室则最好是采取梯队制的组织模式。但是，中文大学的教师组织结构本质上是协作制，也就没有实行梯队制的可能。那么，如何才能在协作制的组织模式条件下实现设计工作室的主题性和研究性呢？

毫无疑问，建构工作室的成员对建筑设计和设计教学的共同兴趣是一个必要的条件。此外，我们把设计教案的研究作为工作室的核心，这才是关键。所谓的“以教案为核心”意味着把教学的重点放在建立一个设计的知识体系。这一体系是独立

positive force if teachers with a similar attitude are working in the same group. The thematic studio provides a structure for this, reflected in the names of the four studios – habitation, urbanisation, technics and tectonics – each emphasising a different design approach. In addition, each theme defines a particular field of academic research and a particular body of design knowledge. The four thematic studios, therefore, define four fields of design studies. This understanding of studio organisation is critical for us to structure the Tectonic Studio.

1.3 The Tectonic Studio: thematic and coordinative

How a studio operates depends on the guiding principles of the studio and the dynamics between its members. The operation of the Tectonic Studio can be described as programme-centred and research-based team teaching. This is certainly not a casual choice.

In the above text we have articulated two ideas of studio organisation: thematic and vertical. Similarly, there are also two different ideas as to how teachers work together: hierarchical and coordinative. Normally in a hierarchical working organisation, a senior teacher heads the studio formed by a group of junior members with different levels of experience. One of the most obvious advantages of the hierarchical structure is that it can allocate all the studio resources to accomplish a common task such as design research. For a coordinative working organisation, members of a teaching team collaborate on an equal basis. Members work together based on consensus rather than a single command. The role of the team leader is to coordinate different interests rather than to lead the direction. Therefore, this type of organisation has little advantage for conducting research in terms of resources.

Now we are confronted with four ideas or two pairs of ideas: thematic versus vertical and hierarchical versus coordinative. What is the relationship between these two pairs of ideas? It seems that the idea of a vertical studio is more suitable for a coordinative working organisation, in which members of a teaching team work independently at the same time connected loosely through coordination. It seems that a thematic studio should best adopt a hierarchical working organisation for the purpose of design research. However, the basic operation mode at CUHK is based on the idea of coordination. This means that a hierarchical organisation is not an option. How can a coordinative organisation realise the educational objective of a thematic studio? This was one of the challenges when organising the Tectonic Studio.

The first condition for a thematic studio based on coordination is

于个别教师的、一种可以描述的、可以讨论和批评、可以运用的、以及可以以系统有序的方法来传授的设计工作方法。在同一个教案下工作的教师以一个共享的观念、知识和方法体系为教学的依据，如此来保证在整体上达到一定的设计教学质量。参与教学的教师以一种合作的方式来共同发展一个教案，各自在这一过程中贡献自己的专长。

1.4 建构工作室：课程体系和运作方式

在建构工作室内学习的学生包含本科的二、三年级和硕士班两年（其中一年为毕业设计），相应地我们把整个工作室的教学分成三个层次，每个层次各有侧重，形成一个完整的建构设计方法的研究体系。

在这个研究体系的最底部是本科二、三年级的设计教学，它的目的是帮助学生建立一个建筑设计方法的基础。这也是本书所描述的内容。建构实验课程完全是团队式的教学，参加的教师共同去发展一个教案。我们将这部分的教学称为“建构实验”。“实验室”一词突出了教学的探索和研究性。最后的成果就是一个包含了四个设计练习的课程。我们将在下一章来重点阐述建构实验课程教案的设计理念和特点。

硕士班一年级的设计教学是建构实验课程的进一步发展。这里我们强调以教师的特长为主导的专题研究，如有关结构和空间的研究，有关参数化设计方法的研究等。这类专题性研究帮助学生开扩与建构和空间问题有关的设计视野。

硕士班二年级的毕业设计是这个设计研究体系的最后一个环节，强调学生的独立研究能力。学生需要自己选定一个与建构和空间问题相关的研究课题。与论文指导教师的研究兴趣相关，毕业设计所涉及的问题也呈现不同的方向，如公共住宅设计、工业建筑的改造、活动房屋的设计和建造、利用可循环使用材料的建筑、工业化预制建造、参数化设计等。

总之，这三个层次的设计研究形成一个从共同的设计基础到以教师的研究为主导的专题研究，再到以学生为主导的设计研究的设计教学体系。如何设计一个工作室的运作方式使得这三个层次的教学研究可以形成一个整体，而不是各自为政，互不关联？

首先是通过工作室内部的评图制度来加强三个层次之间的交流，即在一个学期内设定几次公共评图，教师有机会了解相互的教学和提供各自的意见，这对于课程的发展至关重要。同时，不同课程之间的学生也能增进相互的了解。工作室还特别邀请一位海外的建筑师或学者参加一周的期末评图，这样该位访客有机会了解从建构实验课程、硕士班的专题研究课程，以

the shared common interest in design. However, what is really critical is a defined common goal for teaching and research. The term "programme-centred" implies that all the studio members contribute to the development of a common programme, to a body of design knowledge which is independent from the individual teacher's experience, which is describable, debatable, applicable and teachable. The collective teaching based on such a common programme ensures a higher standard of quality. The way each team member contributes to the development of the programme is through research, which implies a particular attitude toward design teaching.

1.4 The Tectonic Studio: courses and operation

Students in the Tectonic Studio are from undergraduate years 2 and 3, and graduate years 1 and 2, i.e. the thesis year. Accordingly, the courses offered by the Tectonic Studio consist of three different levels, each has particular objectives, and together they form a coherent system of tectonic study.

At the bottom of the system is the programme for undergraduate year 2 and 3 students. Its objective is to build a theoretic and methodic foundation for tectonic design, which forms the content of this book. Teachers who teach this course work collectively to build a common programme. We call this programme Tectonic Lab. The term "lab" implies the experimental nature of teaching. The actual programme is a sequence of four design exercises to explore ideas of space, material and tectonics. We will deal with this topic in the next chapter and in other parts of this book.

The graduate year 1 studio is organised differently from the undergraduate years. Here, we emphasise topical studies based on the studio teacher's interests and strengths such as structure and space, or parametric design, etc. These topical studies expand students' knowledge of tectonic design and further develop their skills for dealing with more complex design problems.

The mode of teaching changes again in the thesis year as students exercise more initiative in defining a topic and conducting research. However, we emphasise that there should be a correlation between the student's initiative and the teacher's interest.

These three levels of studio courses form a progression of design development from a common basis taught by a team of teachers to a diversity of topical studies led by different teachers, and finally to a student motivated mode of design thesis. We need to devise a mecha-

及毕业设计的整个过程，并在评图后给予意见。

其次是公共的讲座系列。除了不同的课程根据各自的教学安排相应的讲座外，工作室还有一个公共讲座系列。讲座由工作室的教师主讲，还邀请外面的建筑师和学者作客座讲座。这些讲座面向工作室的所有学生和教师，有些还面向全系的师生。经过若干年的不断发展，工作室教师主讲的讲座形成几个固定的题目。这些讲座的目的在于描述与课程有关的建构研究的基本范畴、问题和方法。有关题目详见本书的附录。

再次是工作室的联合教学活动。工作室多次组织去深圳等地的建筑参观，增进学生对实际建筑的体验和认识。在工作室运作最后的几年，工作室还组织了计算机辅助设计的工作坊，结合设计练习训练学生计算机作图和建模的能力。我们可以在本书的学生作业中清楚看到作业表现方式的转变。

最后是设计工作室的研究项目，如关于香港现代主义建筑的研究，香港公共住宅的研究等。设计教学的研究（指建构实验课程）和香港现代建筑的研究形成一个互补关系。教学研究为香港本地建筑的研究提供了方向和方法的支持，反之，后者也为前者提供了本地的参照系。

总而言之，我们在这里对建构工作室的教学体系和运作方式，以及它背后的指导方针作详细介绍的目的是在于说明本书所描述的建构实验课程是一个特殊的设计教学研究架构中的一个组成部分。它的重要性在于为整个课程奠定了一个方法学的基础。同时，它的抽象性需要从整个建构研究的体系角度来理解。

建构实验课程与香港现代建筑的研究形成一个互补的关系，前者为后者提供了一个研究的态度和方法，后者为前者提供了一个本地的参照系。左为苏屋邨，1960年；右为天光道已婚警察宿舍，1962年。

Studies on Hong Kong modern architecture complements the Tectonic Lab as the latter defines a particular point of view for observation and analysis, and the former provides a local reference to what is doing in the studio. The left is So Uk Estate, 1960, and the right is the Police Married Quarters at Tin Kwong Road, 1962.

nism of operation in order to run these three levels of teaching as a coherent package.

The first component is the arrangement of joint reviews in which all the members of the studio take part in one level's review and students of different years can see their peers' work. We also invite an external critic to attend the one-week final review at the end of the term so that the guest can have an overview of the whole studio's teaching and give better comments on how to improve our teaching.

The second component is the studio lecture series to the whole studio delivered by studio teachers and invited guest speakers. These lectures aim at defining a field of tectonic study, to provide a theoretic basis in tectonic thinking, and to relate the studio work to historical, international and local references. The list of lecture topics can be found in the appendix at the end of the book.

The third component consists of common activities organised by the studio. One popular activity is a day trip to Shenzhen to visit buildings and exhibitions. In the later years, we began to organise CAD workshops to teach students how to use digital tools for modelling and drawing. The change from traditional to digital media can be seen in students' work found within this book.

Finally, the last component consists of the studio research projects, like the study of Hong Kong's police housing programme in the 1950s and 1960s, and the study of Hong Kong's public housing, etc. These studies complement the studio teaching. The latter defines a particular point of view for observation and analysis, and the former provides a local reference.

We articulated the programme structure of the Tectonic Studio, its operational mechanism, and its guiding principles to provide a context for discussing the Tectonic Lab – the subject of this book – which in its abstract nature provides the theoretic and methodic foundation for the whole studio.

2 建构实验 | the tectonic lab

教案就是一个设计课程的内容和运作方式的计划。把建构工作室中有关基本设计概念和方法的教学的部分称之为“建构实验”，就意味着相关的教案就不是一般的设计教学计划，而是一个设计研究的计划；就意味着这个课程的目的不是传授已知，而是探索未知；就意味着我们要采取不同的工作方法，即在其他的实验室中普遍采用的工作方法。首先，我们要建立一个关于建构的假说，然后设想去设计一些专门的设计课题来验证假说的正确性。我们有必要在设计教案时排除某些因素，同时强化某些因素，以便能够就特定的问题展开讨论和研究。每一次的教学就是一次设计方法研究的实验。前一次的实验成为修正下一次实验的依据，如此不断地推进我们的研究。

2.1 一个关于建构设计的假说

在开始筹划这个关于建构设计的课程之前，我们首先要解决的问题是如何去界定“建构”？“建构”这个术语在不同的语境中似乎有不同的解释。比如在某些情况下建构等同于建造技术，或者重力和结构力的表达，在另外的一些情况下等同于抽象造型。我们并没有刻意去构筑一个深奥和复杂的理论架构，而是采取了一个更加直接的和朴素的方式，即对建筑设计活动的基本观察和再思考，从而寻求一个独特的“建构”视角。我们观察的要点在于材料如何形成空间，以及使得空间可以被知觉。这个建造材料激发对不可视的空间的体验成为我们研究的出发点。如果我们把建筑设计的本质理解为通过建造的过程用材料来塑造空间，那么，建构就是有关空间和建造的表达。再深一步的思考，在塑造空间的手段和所生成的空间的特性之间应该存在一定的内在关系。如此，我们便找到了一个切入建构研究的独特视角。

就设计的工作方法而言，我们特别关注模型制作对设计本身的影响，因为在它的作业过程中始终包含了对模型材料的操作。这在某种程度上似乎又回到了建造的本意，即通过建造的过程用材料来塑造空间。直接用模型进行设计研究是当今最常用的设计方法之一。一张纸板，可以用切割的方法先切成若干的板块，再用这些板块来围合和限定空间；也可以用折的方法来直接将纸板折成所需要的空间。这个折的操作会给设计一个明确的形式和空间特征。如果这个建筑最后是用钢筋混凝土来实现的，这种材料显然是不能折的，但是这个在构思阶段用纸板折出来的形式特征却成为建筑形式表达的一个重要的方面。如果在构思时用了不同的模型材料，那么最后的结果会不会就不一样呢？也就是说，用不同的模型材料来做设计是不是也和用不同的建造方式来建房屋一样产生不同的空间呢？这是一个

The programme is a working plan for the content and process of a design course. By calling the foundation course of the Tectonic Studio the Tectonic Lab implies that there is a research agenda behind teaching, that its purpose is not to teach something known but to explore something unknown. To do so, we need to adopt a different working attitude, similar to the one commonly used in a scientific lab. First, we have to formulate a hypothesis regarding tectonic design and then design a set of experiments to test it. We might need to ignore certain issues in order to be able to concentrate on the issues that we want to explore. These experiments are carried out in the studio through teaching, evaluated through reviews and developed in cycles over the years.

2.1 Formulating a tectonic hypothesis

Before we started to plan this course about tectonic studies, we needed first to declare what tectonics means for us. The term "tectonics" seems to have different interpretations in different contexts. In one context it could mean construction techniques, in another the expression of gravity and the flow of forces, and in a third plastic form-making. Our approach to a definition of tectonic studies is not based on a profound and complicated theory, but on the observation of primary building activities, from which we articulate a particular point of view of tectonic studies. We focus on the basic architectural interest of how material forms space and makes space visible. This interest in how the perception of built material evokes an experience of its invisible space became the starting point of our study. If we understand the nature of architecture as using materials to form space through a process of construction, then tectonics is the expression of space and construction. This further suggests a connection between the quality of space and the way the space is formed.

Regarding the method of work, we have a particular interest in model-based design methods because they require direct operations on model materials. This relates to the original concept of construction – working directly with building materials. Although making models is common in the design process, it does not mean that we are always conscious of how the working method influences the outcome. But we can easily imagine that cutting a piece of cardboard into smaller pieces and combining them to form space or folding a piece of cardboard to form a space are not only different operations on the same material, but also produce different results. Each strategy gives a clear

关于建构的方法学命题，即建构假说。

我们确定三种基本的空间限定要素即体块、板片和杆件作为研究的对象，进而认识到在要素和空间之间存在一一对应的关系，所谓的体块的空间、板片的空间和杆件的空间。为了充分了解每一种空间限定要素的造型潜力，最好的方法是把它放在一个抽象的环境中来考察，将设计过程中各种力量的冲突推到一个极致。这从实际的设计要求来说也许是偏颇的，但是从实验的角度来看却是必要的。

对块、板和杆这三种空间限定要素的研究，从方法学的角度来说，就是从构思形式到建造形式的发展过程。根据建构命题，建构构思直接来自于对不同要素类型的模型材料的操作，特定的操作方法产生特定的要素和空间组织特征。就模型操作而言，要素是物质的，不同材质材料的运用丰富了建构表达的可能性。最后，建造方式和建筑材料的运用应该是强化，而不是弱化最初的建构构思。如果我们把从构思形式到建造形式的转换作为主轴，就可以将整个设计过程划分为几个典型的发展阶段或练习。每个练习集中解决一个核心问题，几个练习串联在一起组成一个连续的练习序列。这个练习的序列实际上就是一个建构设计的方法学框架。经过几年的发展，最后形成了四个练习。

2.2 建构实验：问题与方法

阶段一：构思——操作与观察

用块、板和杆其中的一种要素来探讨生成空间的可能性。练习的目的有二，一是建立一个建构研究的基本态度和方法，二是寻求一个建构的构思，引出后续的一系列练习。操作和观察的互动是建构设计的基本态度和方法。用模型材料来研究设计，块、板和杆这三种材料分别激发设计者不同的操作。体块的操作方法可以是掏空、切割和位移等，平板可以用围合或折叠等方法，杆件用密度、框架或围栏等方法。这个初步的练习可以用来讨论两个基本的问题：首先是空间的概念。空间设计的最困难之处在于操作是直接作用于要素本身，学生往往容易忽视操作的目的，即空间。其次是操作的方法，是不是可以识别出一些明确的操作方法来。在这个讨论的过程中我们必须要借助于观察空间的方法来知觉空间，从人眼的高度来体验空间。我们要求学生把视点放低，边观察边画素描，研究光线和空间的关系。这个空间体验的环节是十分重要的。操作若没有观察作支持，就失去其目的性。一旦学生对自己所创造的空间有明确的体验，这一体验就会转化为明确的设计意图，使得开始还是比较盲目的操作成为有意识的设计活动。因此，操作和观察是一个互为因果的关系。

formal and spatial character to the model. If, in the case of folding, the building is eventually built in concrete, the adopted building material obviously cannot be folded. However, the initial folding idea becomes the most important expression of the building. Will the use of different types of model material in the design result in different types of space in a similar way as the use of different building materials in construction? This question forms the second part of our tectonic hypothesis.

To test this hypothesis, we make a distinction between three basic types of space defining elements – block, slab and stick. We also assume that, corresponding to these three elements, there are three basic types of space. In order to fully explore the potential of each element type to form space, we designed a set of abstract exercises as design experiments.

Regarding the methodic aspect of the study of the three space defining elements, we focus on the transformation from conceptual form to built form. We expect that the direct operation on the model material will lead to a tectonic concept. This concept formulates the relationship between the operation and the resulting space and form, which we call the conceptual form. This conceptual form has to be transformed into built form and the model material has to be changed to building material. In a studio context the built form does not refer to the actual building construction, but to a consideration of construction to realise the concept. The transformation process guides the articulation of a design process consisting of a series of distinct but related exercises, each with several working steps. After years of development, the exercise sequence finally comprises four phases: concept, abstraction, materiality and construction.

2.2 The Tectonic Lab: issues and methods

Phase 1: Concept – operation and observation

The first phase initiates the process by introducing the basic working method and finding a tectonic concept. The basic attitude and method of tectonic design is a reciprocal process of operation and observation. Different model materials provoke different kinds of operation. Blocks can be cut, carved, punched, shifted, etc. Slabs can be folded, combined, interlocked, etc. Sticks can be bent, framed, repeated, etc. The aim is to find out at which stage an operation on the elements creates space, and which properties of that space are related to the element type and the operation. It is crucial to observe the resulting space at eye level, describing its form, depth and scale under light.

建构实验课程的四个练习：构思、抽象、材料和建造。

Four exercises of the Tectonic Lab: concept, abstraction, materiality and construction.

阶段二：抽象——组织与体验

根据前一个练习的操作方法用单一的模型材料来设计一个矩形建筑单体。该单体的尺寸为6mx12mx24m，根据不同放置方式，有平房、板房和楼房三种类型。练习的重点是抽象要素与空间的关系。我们的假设是体块、板片和杆件这三种要素所形成的空间应该具有不同的特点，其区别是可以从模型中观察得到的。相对于体块是“勾勒空间”，相对于板片是“模棱两可空间”，相对于杆件是“调节空间”。

设计的目标之一是追求形式和空间语言的清晰，其关键在于操作的清晰。一般来说，学生在开始阶段的操作倾向于杂乱和盲目，解决的办法是分辨自己在设计中运用了哪些操作方法，再从中确定一种主要的操作方法，作为进一步发展的基础。设计的另一个目标是通过简单的操作达到丰富的空间体验。简单和清晰的操作并不意味着形式和空间的单一和单调，这是练习的困难之处。一味地简化操作并不能产生一个好的设计。或者说，一个好的操作应该能够产生丰富的空间。丰富的空间体验主要取决于空间序列上空间之间的对比和变化，如空间之大小、形状、比例、方向、视线的位移、射入空间的光线的变化等等因素。

阶段三：材料——区分与诠释

用两至三种模型材料重新制作模型，练习的重点在于探讨材质因素的介入而引起的表达建筑体的可能性。从前一个练习的单一模型材料发展为现在的几种模型材料需要对原先的形式和空间关系作新的诠释。这里，关键词是区分，即在原先单一材料的要素之间作出区分。区分意味着差异，意味着对比，意味着重新诠释原先的设计，意味着在原先的秩序基础之上建立新的要素和空间之间的秩序。当然，区分和诠释并不是随意的，它实际上是在抽象形式和空间的表达基础上寻求更丰富的表达内容，其依据在于对结构、空间和使用等方面的考虑。从视知觉的角度来考虑，我们主要考虑模型材料的三个特性：即材质肌理，如木和金属材料的对比；色彩和明暗，如各种颜色的纸板的对比；材料透明性，如透明材料、半透明材料和不透明材料的对比。在运用材料区分对原先的设计作重新诠释的过程中，一个重要的手段是用透视图来观察材料的区分对空间知觉的影响。通常对结构、围合和使用的诠释存在多种的可能性，设计的基本原则是加强而不是弱化最初的构思。

阶段四：建造——构思与实现

在这个阶段，学生要从建筑材料和建造的角度来研究建构

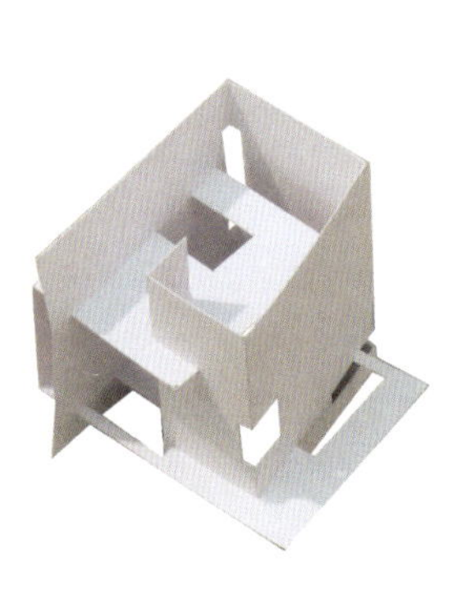

Without direct observation, the exploration will lose its purpose. Once the student can appreciate the quality of the space created, this spatial experience will turn into a clear design concept.

Therefore, the manipulation of the elements and the direct observation of the resulting space should occur simultaneously as a reciprocal process.

Phase 2: Abstraction – organisation and experience

Phase 2 introduces a cuboid object that has the proportion of 1:2:4 and the measurements of 6x12x24m. According to the way the object is placed on the ground, it has three formal expressions: flat, thin and tall. Within the given form, we explore how space can be defined and organised based on the principles of operation from the previous phase. We work with an abstract model made from only one type of material. We put emphasis on the relationship between the physical elements and the perceived space. We use the terms "outlined space", "ambiguous space" and "modulated space" for the space types resulting from working with the respective space defining elements block, slab, and stick. Different operations on the same material will result in variations of these space types.

One of the design objectives in the organisation of the object is to achieve clarity of language through clarity of operation. Usually, students tend to use several operational methods randomly at the beginning and only can reach clarity of operation when a dominant operation can be recognised. Another design objective is to achieve a complexity of experience through the simplicity of operation. This is possible because the experience is based on perception. This again requires the observation of the elements to find corresponding spatial properties.

Phase 3: Materiality – differentiation and interpretation

In this phase we make a step in the direction of materialisation by introducing multiple types of model material, which provoke new possibilities in the expression of the object. The translation from a single material to multiple materials forces students to establish new orders between elements and spaces. In addition, the object must be further articulated by considering the relationship between structure and enclosure and the distinction between inside and outside spaces, which implies a climatic border. In this phase, we emphasise the visual quality of the model material: colour, texture and translucency. Viewing

表达的问题。一般来说，学生在将构思形式转换成建造形式时都对如何根据原先的构思来选择合适的建造方式和建筑材料感到困难。对此，我们引入一个照相拼贴的练习。通过一个透视和立面局部的拼贴练习来研究从抽象材质到建造材料转换的可能性。在此基础上再制作一个局部模型来探讨建造与建构表达的问题。这里所说的建造并非实际去建造的活动，而是从建造的角度来研究设计的构思和实现之间的关系，即从构思形式向建造形式的转换。这里我们要特别区分构件与要素之间的关系。一个要素如板在模型材料中可以是一块硬纸板，而在建造材料中它必须有若干的构件组合而成。但是在构思形式中的板的要素，未必就一定要以板的构件来实现，它很可能是由块或杆的构件组合而成。同样的道理，砖作为建筑材料，可以被认为是块，但是砖不但可以建造出块的实体，也可以建造出作为板的墙和作为杆的柱。具体的练习作业特别要注意将研究的重点放在建构表达方面，而不是建造技术，即研究建造的不同处理方式如何支持建构意图的充分表达，主要体现在构件在横向和纵向的相互关系，即拼接和层次两种关系。

2.3 教案的设计要点

在设计建构实验教案时我们主要要考虑解决以下四个方面的问题，以求达到预设的研究目标，以及学生的最佳学习体验。

首先是保持练习的抽象性。这个练习的系列的本质是抽象性，即这不是一般的包含了具体的建筑类型和场地的设计课题。从教学法和设计研究的角度来说，抽象的目的在于将一些和所要研究的问题不太相关的设计因素排除掉，使得问题更加鲜明和聚焦。抽象性具体体现在以下几个方面。一是将设计研究局限于一个给定的建筑体量，将研究的注意力集中于体量内部的空间问题，而不是体量本身的变化上；二是将功能问题抽象为人在空间中的活动，以及空间的不同特性上，本质上存在一个抽象的功能／空间关系；三是将场地问题抽象为单一体量的三种摆放方式，即平房、板房和楼房三种类型，不同的摆放关系对内部的空间组织是有影响的。从设计训练的角度来说，抽象练习也是非常必要的一种基础训练，学生必须经过抽象的练习才能把握空间和建构这类问题的本质。

其次是如何安排体块、板片和杆件这三种空间限定要素的研究。最理想的情况是每个学生都有机会对这三种要素作分别的研究，但是在给定的教学条件下（最初是4到5周的时间）这是不可能的。我们采取的是一个三种要素的研究平行发展的结构，即每个学生只能选择一个要素作研究，三种要素的研究平

the space and object at eye level is again an important tool for studying the effect of materiality on perception. We use a single view to study variations of material differentiation. These new possibilities to interpret the object in terms of structure, enclosure and use are done in an exploratory way, but with the aim to maintain or even strengthen the original spatial concept, and not to weaken or lose it.

Phase 4: Construction – concept and realisation

In this last phase, students explore aspects of tectonic expression by considering the issue of building materials and construction. In general, it is difficult to transform conceptual forms into built forms with a lack of experience of what kinds of materials and construction methods are appropriate to specific tectonic concepts. Hence, we introduce a collage exercise to help bridge the gap between the two types of form. For this collage, cut-outs of photos of existing buildings are applied to either elevations or perspective views of the object. In contrast to material textures, these images contain information about building material and construction, and therefore can give hints as to how a concept in model materials can be realised with building materials. Following the collage, a partial model which is representative of the whole object and includes part of an elevation and part of a space is built to study the organisation of the building material. One issue of study is to distinguish lateral and depth relationships between components. Laterally, components form joints and in depth they form layers. The role of materials used in this partial model is symbolic, representing or pointing to the chosen building materials. For example, grey cardboard with scored lines can indicate a concrete wall that shows formwork panel joints. When identifying components, their dimensions as building materials must be considered. At this stage we are not interested in the technical aspect of joints as connectors; instead, we focus on the position, proportion, and form of joints and layers. In other words, we are interested in the issue of tectonic expression.

2.3 Programme design considerations

The following four aspects are our main considerations in the design of the Tectonic Lab programme in order to achieve the redefined research goal and the best learning experience for students.

The first is to retain the abstract nature of the design exercises. Unlike in a normal studio project, which should have a specific pro-

第一个建构实验课程借鉴了瑞士建筑师马库斯·卢契尔于1996年在苏黎世联邦理工学院建筑系开设的一年级《建筑与建造教程》。在中大的工作坊包含三个练习：基本、物体和环境。

The first tectonic course was a workshop conducted by the Swiss architect, Mr. Markus Lüscher, which is based on his first year building construction and design course in the Swiss Federal Institute of Technology Zurich in 1996. The workshop at CUHK consisted of three exercises: basic, object and context.

行进行。在一个教学小组内，三种要素平均分配，这样每个学生除了自己的研究外，也能从其他同学的研究中了解另外两种要素的相关问题。如此，同学之间的相互学习也有了实际的意义。

其次是如何通过教案的设置来达到练习成果的多样化，也就是说多样化的可能性存在于教案的结构之中。将建构研究分为体块、板片和杆件三种要素就是多样化的一个重要的手段，而平房、板房和楼房这三种建筑体量类型的引入是达到多样化的另一个重要的手段。此外，也就是最重要的一个因素是每个学生对模型材料的操作的独特性，一些细微的操作变化就可能产生完全不同的形式，而多种模型材料的区分，建筑材料的研究等也不断强化每个学生的设计的独特性。从结果来看，最后就呈现丰富多样的成果。多样性的呈现来自于练习的设计。

再次是结构有序的工作方法的问题。我们以从构思形式到建造形式的转换过程作为建构设计方法的主轴，将设计过程划分为几个典型的发展阶段及工作步骤。每个阶段及步骤针对什么问题，采用什么方法，都有明确的规定。这一结构有序的工作方法是实现严谨和有效的设计教育目标的基本前提。在每个阶段，学生可以掌握某个具体的操作方法，研究某个具体的设计问题。通过一个完整的练习过程，学生便可明了各种方法之间的相互关联。这种严谨的设计训练对于提高整体教育水平的作用是毋庸置疑的，对它的批评往往在于担心缺乏设计的创造性和多样性。从我们实际教学的经验来看，严谨的教学结构更有助于学生专注于对问题的探讨，从而产生往往令人意想不到的空间和形式，设计可以产生的变化之多，也是我们在规划教学时所无法预料的。

2.4 课程的发展演变过程

2001年，第一个建构实验课程

建构工作室的第一个建构实验课程借鉴了瑞士建筑师马库斯·卢契尔于1996年在苏黎世联邦理工学院建筑系开设的一年级《建筑与建造教程》。该教程通过体块、板片和杆件这三个概念将空间和建造问题有机结合在一起。教程由四个设计课题所组成，分别是砌块（体块）、混凝土板（板片）、木结构（杆件），以及钢结构（综合）。我们认为这样的一个课程对于我们所确定的空间和建构的研究目标将会是一个很好的开始，于是我们邀请卢契尔来主持一个四周的工作室主题课程，并起名为“建构实验”。因为卢契尔在苏黎世联邦理工学院的课程和我们的建构工作室的课程在目的和条件方面有很大的区别，这也决定了我们不可能直接照搬原来的课程内容和结构，而必

gramme and a site, we want the students to design a cuboid without a specific programme and a site. The simplicity of a cuboid keeps design efforts within a predefined boundary. In addition, the issue of use is presented in an abstract way as a series of spaces of various qualities and that of site as the placement of the cuboid on the ground in three variations as flat, thin, and tall. All these make it possible to focus on what we are really interested in: the formation of space and its expression through materialisation. We believe that these abstract exercises are essential in design training. It helps the students to grasp the essence of space, construction and tectonics.

The second is to work within a limited time frame – four weeks for the studio project under the Department's thematic studio system. Ideally, every student should have the opportunity to study each of the three space defining elements, which is not possible. Therefore, we devised a parallel structure of individual and group study, in which a single student studies one element type, but within the studio group all three types are studied. In this way, each student can attain a deeper understanding of the element type he or she is working on, but also learns about the other two types from his or her peers.

The third is how to achieve a wide range of results with a strictly structured programme. Different means to allow this diversity are embedded in each phase of the programme. The distinction of the three space defining elements, combined with the three object types forms already a range of starting points, which is further differentiated by the exploratory study of possible operations. With this, each student has from the beginning a unique study configuration. The formation and organisation of space, the interpretation with multiple model materials, and the transformation from conceptual form to built form will further differentiate results. Ultimately, it is rare to find two similar designs in the same class.

The last is our insistence on structured learning and the development of good working habits. Taking the transition from conceptual form to built form as the main axis, we divide the design process into four stages of development. This structural and orderly working method is rigorous as issues and methods are clearly declared at each stage, and it is effective because we can focus on specific design problems using particular methods. The complete process can reveal links between different issues and various methods. It also makes the work discussable. It further helps students to practise how to work on a design, so that the result is generated through a process and not copied

“练习加方案”的课程安排先是建构实验的抽象练习，而后是一个全系统一的设计方案。从这个例子可以看出学生将在抽象练习中发展的空间概念直接运用到设计方案中。以后，我们又尝试了几种不同的模式，如“练习即方案”等。

According to the "exercise+design" pattern, the Tectonic Lab exercises are followed by a Department decided project. As we can see in the illustrations, the students directly applied the same space concept to the design project. Later on, we also tried several other operational patterns, such as "exercise=design", etc.

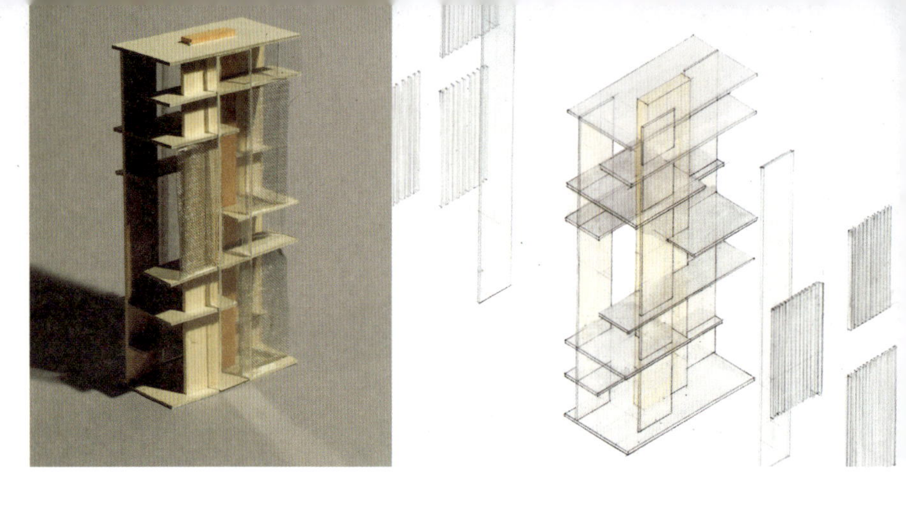

内容和结构，而必须在吸收原课程的要点的基础上建立一个新的体系。在卢契尔的课程中，体块、板片和杆件这三个空间限定要素是延着一个线性关系来展开，现在变成三个要素同时进行的平行关系，如此才可能将一个一年的课程浓缩到短短的四周时间内。整个课程由“基本”、“物体”和“场地”三个练习所组成，即先通过对给定模型材料的操作和对所生成空间的观察来学习一个建构研究的基本态度和方法；再将这个特定的操作方法作为建构构思运用于一个建筑体的设计，该建筑体的几何量度相同，根据摆放的位置产生平房、板房和楼房三种类型；最后，一个小组的同学用各自的建筑体组成一个场地，即建筑体也可以作为空间界定的要素来生成外部空间。

这个建构实验课程的“原型”在以后的教学过程中的演变基本沿两个线索发展，即练习系列的完善，以及练习与设计的关系。

建构实验练习系列的形成

练习系列的完善主要体现在不断推进有关空间限定要素的研究。首先，我们在“物体”练习之后加入“建构”练习，同时取消了“场地”练习。在“基本”练习中，模型材料的不同物理特性激发不同的操作，进而生成不同特性的空间来。而在“物体”练习中，我们则强调用单一的模型材料来做模型，以此来强调抽象的要素和空间之间的关系。在“建构”练习中，我们则引入了多种模型材料的研究，通过多种模型材料的区分来建立更加丰富的形式和空间关系，如不同空间的对比，结构和围合的区分，以及室内外的区分等。所以，“建构”练习将“物体”练习的结果向前推了进一步。再后来，在“建构”练习后又增加了一个从模型材料到建筑材料的练习，研究的重点是如何将前面几个阶段发展的空间和建构概念用建造的手段来实现。拼贴练习在这个过程中是一个重要的转换环节。至此，这四个练习最终形成了一个所谓的建构设计方法。根据设计从构思到实现的逻辑发展过程，我们对四个练习或设计发展的四个阶段进行了重新命名，即：构思，抽象，材料，以及建造。

处理练习与设计关系一些尝试

在建构实验课程的发展过程中，另一个重要的发展线索是工作室自选课题和学系规定设计课题之间的关系。在一个相当长的时期内，我们都是将一个学期的时间分为两个部分，一个是建构实验，接着是一个设计方案。前者时间短，后者时间长；前者抽象，后者具体。这是一个“练习加方案”的课程结

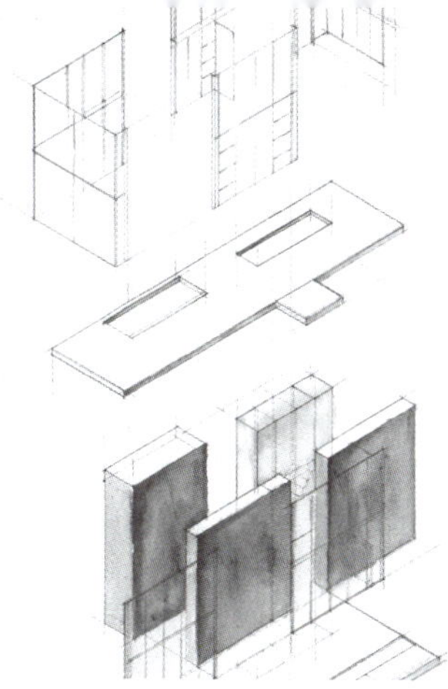

from precedents. The results also demonstrate that our structured approach allows for unexpected discoveries.

2.4 The historic transformations of the programme

The first Tectonic Lab in 2001

The first tectonic study course was a workshop conducted by the Swiss architect Markus Lüscher, related to his first year building construction and design course at the Swiss Federal Institute of Technology in Zurich in 1996. Based on the three distinct elements block, slab, and stick, his course combined the issues of space and construction into one package. It consisted of four small projects or exercises, each dealing with one type of building material: masonry, concrete, timber and steel. As we thought that his could be an appropriate starting point for the Tectonic Studio we invited him to run a four-week workshop which we named "Tectonic Lab". However, the original programme structure needed to be modified. The linear structure of the subsequent studies of the three elements was changed into a parallel structure to fit into a drastically shorter time frame. And the programme was reformulated as three exercises: basic, object and context. The students first learned the basic working method through manipulating model material, then to apply it to the design of a cuboid object and finally to form an urban complex with the objects contributed by the individual students.

This prototype of the Tectonic Lab underwent continuous transformations in the following years along two main lines. One is the perfection of the tectonic study, and the other is to resolve the relationship between abstract exercises and its application to concrete design projects.

The formation of the Tectonic Lab exercise series

The perfection of the tectonic study is shown in the effort to extend the exploration of the three space defining elements. First, we introduced an exercise called "tectonics" after the exercise "object" and at the same time cancelled the exercise "context". This formed a new exercise sequence in three phases: basic, object and tectonics. The students first learned the basic working method through manipulating model material, then to apply it to the design of a cuboid object and finally to reinterpret the object with multiple model materials. At this time, the distinction between a single model material and multiple model materials became apparent. And this new phase "tectonics" ex-

构。我们感到前后两个部分的发展过程重复，且没有充裕的时间来发展前面的练习的部分。于是，我们尝试将建构实验扩展为一个学期，并加入具体的功能内容，以及重新引入场地的因素，把建构实验包装成设计方案，即形成一个“练习即方案”的结构。不过，在作了若干尝试之后，我们感觉到具体的功能内容和场地的因素的介入未必有利于对空间和建构问题的深入研究，故又重新回到建构实验的本质，即“练习即练习”的结构。现在，我们终于有相对充裕的时间来发展这四个练习的细节内容。然而，抽象练习的最终目的是要将其运用到实际的设计方案中，这个教学研究是否成功也将取决于它是否可以转化成一个具可操作性的设计方法。所以，我们最后尝试的就是如何将这四个练习或设计发展的四个阶段自然地融合进解决实际设计问题的过程中，我们称之为“方案练习化”。回顾整个的研究过程，我们在一开始是把空间生成问题从设计过程中抽象出来，形成抽象的练习，最后又通过若干尝试将练习融入到设计过程中去。如此，形成一个设计方法研究的轮回。

tended the study one step further. Later, another exercise "construction" was introduced, which addressed the issue of transformation from model materials to building materials through a collage exercise and a partial model study. Eventually, we arrived at a sequence of four exercises as described in this book. The titles of each exercise were also reformulated as concept, abstraction, materiality and construction.

Exercise versus project

Another major issue in the development of the programme was to search for a proper relationship between the studio project, which is this Tectonic Lab exercise series, and the school project, which is the design project adopted by all thematic studios. Over a fairly long period of time, we followed the Department's thematic studio system by dividing the teaching term into two parts: the first four weeks for the Tectonic Lab and the rest of the term for a design project. The former is characterised as abstract and shorter, and the latter concrete and longer. As a pattern we call this "exercise + project". Later, we felt that these two parts repeated each other in terms of design process while not leaving enough time for in-depth development of either one. To solve this problem, we tried to expand the Tectonic Lab exercise to the whole term and introduced a programme of use as well as a site to make the Tectonic lab more like a design project. We call this pattern "exercise = project". Later on, we eliminated the programme of use and the site again to return to the original exercises but develop them over the whole term. This was an important step which allowed us to further articulate the development steps and clarify the design media and methods. We call this pattern "exercise = exercise". What is presented in this book is largely from this period of development. In the last year of the Tectonic Studio, we explored a new pattern of course structure. We tried to find out how to integrate the tectonic design method we had developed so far into a normal design project. Unlike in the pattern "exercise = project", in which a programme of use and a site is rather artificially and arbitrarily imposed onto the abstract exercises, here our aim was to search for a way to balance the problem solving of a design project with the space-making of abstract exercises in a natural way. We called it "exercise ⊂ project", meaning that the exercises are contained in the project.

3 关于本书 | about this book

这本书的写作历史几乎和课程的历史一样的长，也和课程的演变过程一样是一个不断发展的过程。我们把文本整理作为设计研究的一个不可或缺的环节，通过对学生作业的文本整理以及教学文件的梳理来检讨教学的效果和反思研究中的问题，从而不断推进研究的发展。本书通过教案、要点、练习、研究和设计五个主要篇章来描述建构实验课程的五个不同的方面。

“教案”一章描述建构实验课程的教案设计。本书的主要内容是建构实验课程，该课程是建构工作室建构教学体系的一个组成部分，而建构工作室又是学系的主题工作室制度下的一个组成部分，所以要了解建构实验课程，我们首先要了解该课程所属的教学体系。建构实验课程的教案设计首先是要建立一个有关建构的假说，然后去设计一些抽象的练习来就三个空间限定要素的空间潜能进行探索。本章就建构实验课程教案设计的基本原则，内容和历史发展作一回顾。

“要点”一章介绍贯穿于设计练习序列中的一些最基本的想法，这些想法呈现为一组用来描述设计问题、手段和方法的术语。在进入到具体的设计练习以及其他的几个篇章之前，我们将这些基本术语和想法抽取出来、按照它们的本质进行分类。这些想法对于整个建构实验来说非常的重要，没有这些术语我们将无法来讨论我们的设计，故而有必要作专门的讨论。当然，很多罗列出的想法并不是在这个课程的一开始就已经形成的，而是这些年来不断地反思和澄清的结果。我们将有关的术语分为四类。“命题”重点阐述我们就建构研究所做的基本假设、理论定位、基本观点以及方法；“问题”试图描述建构研究的基本范畴；“方法”将练习系列中的建构工作方法的要点作一梳理；“媒介”则就模型、透视和建筑图这三种主要的设计媒介在这个制作、观察和纪录过程中的作用作一总结。

“练习”一章呈现建构实验的练习系列。这个设计练习系列是基于一个假说，即由不同类型的空间生成要素所产生的空间也应该具有不同的品质。为了要验证这个假说，我们研究三种要素类型：体块、板片和杆件。我们的设计研究方法是以模型为基础的。设计研究的过程分四个阶段：概念、抽象、材料和建造。本章就每个练习的目的、内容和方法逐一详细介绍，以呈现这些练习在实际教学中的情况。在练习的介绍之后是学生的习作，以展示设计练习的发展过程。

“研究”一章通过一些实际的学生作业来讨论建构实验的结果：生成空间的研究。在这个实验中，我们把体块生成的空间定义为勾勒空间、把板片生成的空间定义为模棱两可空间，以及把杆件生成的空间定义为调节空间。这三种空间特征应该是三种要素最纯粹的空间形式。然而，体块、板片和杆件只是

The history of writing this book is almost as long as the history of the course itself. We consider the documentation of students' work an integral part of the research, as a form of reflection, evaluation and development. The book describes the Tectonic lab in the five chapters: programme, essentials, exercise, study and design, each of which addresses one aspect of the course.

The chapter "Programme" describes the design of the programme of the Tectonic Lab. The Tectonic Lab is one component of the Tectonic Studio's teaching programme and the Tectonic Studio is one component of the Department's thematic studio structure. Therefore, we need to first describe the context in which the Tectonic Lab was conceived and developed before explaining the actual content of the course. The Tectonic Lab is based on the hypothesis that directly working with model material can lead to spatial concepts. It consists of a sequence of abstract exercises which explore the potential of three types of elements and the space they can define in a transformation from conceptual to built form. The chapter provides an overview of the concepts, content and historic development of the Tectonic Lab.

The chapter "Essentials" introduces a set of terminologies which we use to describe, discuss and reflect what we are doing in the studio. We collect and discuss the essential ideas, taking them out of the context of the sequence of exercises, articulating them separately from the studies conducted through the exercises and isolating them from the design development process. These terminologies make the tectonic method "teachable". We present the terminologies in four parts: in "Propositions" we state our basic assumptions, position, point of views and approach; in "Issues" we outline the spectrum of the tectonic study through several distinctions of design interests; in "Method" we list different aspects relating to the way of working for both study and design development; and in "Media" we describe the role of models, views, and exact drawings in this process of making, observing, and documenting.

The chapter "Exercise" describes the actual content of the Tectonic Lab. This series of design exercises is based on the hypothesis that there is a fundamental difference in the properties of a space depending on the type of element used to define it. To test this hypothesis we work with the distinction of three types of elements: block, slab and stick. Our design research method is model-based. There are four exercises: concept, abstraction, materiality and construction. In this chapter we explain each exercise in terms of its

抽象的概念，落实到具体的模型材料，结果就变的非常不可预测。我们在研究中发现，除了这三个最极端、最明确的空间特征外，与所采用的模型材料的物理特性以及操作方法相关，所产生的空间有很多的变化形式。只有通过不断地实验研究，我们才能了解空间限定要素的操作和生成的空间之间的丰富内涵。这一章的目的不是罗列体块、板片和杆件三种空间限定要素生成空间的所有可能性，就一个开放的和无止境的研究而言，这几乎是不可能的。但是，我们希望通过每个要素10个案例的介绍来展示如何展开研究的可能性。

“设计”一章讨论抽象练习和综合性建筑设计的问题。根据全系的设计专题工作室的基本设置，建构实验课程最初包含两个部分，即先是抽象的练习，然后是一个通常的设计方案。这是“练习加方案”的模式。从2006－2007学年开始，我们对练习和方案的关系作了三种不同的探索，主要是想解决过往教学中出现的一些问题。如“练习即方案”将功能和场地问题结合进练习；“练习即练习”是将方案设计排除在外，专门集中于练习的教学；最后是“方案练习化”则是把练习的设计过程融入方案的设计过程。关于“练习即练习”的模式已经在前面的“练习”章节中专门讨论过，本章就只介绍练习加方案，练习即方案，以及方案练习化三种模式。学生作业的选取尽量体现体块、板片和杆件三种基本要素的研究和运用。

本书最后的“结语”是对整个建构设计研究的总结，同时也交代建构工作室的结束以及展望后续可能的发展。

本书以中英文双语来写作，但是中文并不是英文的翻译，反之亦然。就某个具体的内容而言，中英文的基本意思应该是一致的，但是文字上未必一一对应。而中英文各自略有侧重的情况也时有发生，或许中英文的综合才最完整地表达出作者的意图。

objectives, content and its operational procedure in detail as it was conducted in the studio. Finally, we use three examples of student work to illustrate the complete process.

The chapter "Study" discusses the outcome of the exercise series: studies of space formation. In this course we focus on three element and space types: block with outlined space, slab with ambiguous space and stick with modulated space. These three types of elements cause three clearly different types of space. After the recognition of the difference in type, the study can reveal many variations in the implementation of each type. These variations are caused by the possibilities of operation on the elements and need experimentation and observation to be discovered. And if we can articulate how the operation of the elements influences the properties of the perceived space, we can communicate and discuss the findings, and use them for purposeful design. As this is an open experiment, we have no intention to provide a complete list but show a wide range of variations within each space type through ten selected examples of student work.

The chapter "Design" deals with the exercise-project relationship and shows different examples of student work. In the exercises, we work with a certain degree of abstraction and with a reduced set of issues. In the project, we deal with less abstraction, with an expanded set of issues and closer to a real project. Over the years, we have experimented with different exercise-project relationships. Initially, we followed the system of studio and school project like the other thematic studios (exercise + project). Later other relationships were tested – to treat the exercises more like an abstract project (exercise = project), to expand the exercises to the whole term (exercise = exercise) or to embed the exercises into a project (exercise ⊂ project). The chapter explains three relationships, except exercise = exercise, using student work.

After these five chapters, the "Epilogue" reflects the whole tectonic study and explains the end of the Tectonic Lab and its possible future developments.

The book is bilingual. However, it is not a simple translation from English to Chinese or vice versa. Generally speaking, the content in both languages is consistent, but does not necessarily correspond word by word. In some cases, the actual content might vary in the two languages, with a different focus in each, so that the complete meaning is expressed by the combination of the two languages.

2 要点 | essentials

命题 proposition
问题 issues
方法 methods
媒介 media

一个术语和想法的目录 | a catalogue of terminologies and ideas

本书的每一个篇章描述这个建构实验课程的一个侧面。在这一章中我们着重介绍贯穿于设计练习序列中的一些最基本的想法，这些想法呈现为一组用来描述设计问题、手段和方法的术语。在进入到具体的设计练习以及其他的几个篇章之前，我们将这些基本术语和想法抽取出来、按照它们的本质进行分类。这些想法对于整个建构实验来说非常的重要，没有这些术语我们将无法来讨论我们的设计，故而有必要作专门的讨论。当然，很多罗列出的想法并不是在这个课程的一开始就已经形成的，而是这些年来不断地反思和澄清的结果。其中的大部分在课程中以讲座的形式向学生作深入的介绍，在此则只能作简要的概括。

在“命题”一节中我们重点阐述我们就建构研究所做的基本假设，我们的理论定位和基本观点，以及方法。我们通过引用一些建筑先例来阐述这些想法。其中最重要的一点是如何来给一个以模型手段为核心的设计过程来进行理论定位。模型对于我们来说不仅仅是一个建筑按比例缩小的抽象表现。通过对模型材料的操作我们可以获取空间的概念，因而模型成为设计的发生器。

在“问题”一节中我们试图描述建构研究的基本范畴，并对建筑形式问题作一些必要的区分，以决定哪些建筑在建构研究的范畴内。我们引入了抽象的要素、空间类型，以及可见的物质要素和不可见但可以知觉的空间之间的关系的概念，我们还提出如何借助多种模型材料来区分空间和丰富表达，以及如何通过建造的原则来组织建筑材料等问题。

在“方法”一节中我们罗列了与学习和设计发展有关的工作方法的几个主要的环节。其中贯穿整个过程的、最核心的方法是一个制作和观察循环往复的工作过程。我们根据设计发展的不同阶段具体的制作任务以及相应的知觉和体验内容对这一核心方法作进一步的区分。

最后，在“媒介”一节中我们系统阐述模型、透视和建筑图在这个制作、观察和纪录过程中的作用。尽管模型是主要的设计手段，我们实际采用了各种不同的设计媒介，从不同种类的物质模型到电脑模型，从草图到精心手绘或电脑制作的图。我们试图对各种模型和图在设计的不同阶段如何表现整个设计以及某个设计问题的功能作明确的区分。

Each chapter describes this teaching experiment in a different way. In this chapter we collect and discuss the essential ideas presented as a set of terminologies, taking them out from the context of the sequence of exercises, articulating them separately from the studies conducted through the exercises, and isolating them from the design development process. Without these terminologies, we wouldn't be able to describe, discuss, and reflect what we are doing in the studio. Naturally, we could not have done this at the beginning. This is a result of a process of reflection and clarification over the years. Many parts have been presented to students in the form of lectures in much greater detail than on these pages.

In propositions, we state our basic assumptions, our position and point of view, and our approach. And we relate these to a few references at other times by other people. We construct an argument for a model-based design process, in which the model is more than a scaled abstraction of the building, by becoming a generator for spatial ideas through the direct manipulation of the model material.

In issues, we outline our spectrum of tectonic study and make several distinctions which help us to decide which types of buildings are of specific interest to us. We introduce our definition of the abstract element and space types, and their relationship as visible physical forms and invisible but perceived space. We look at how material properties differentiate space and expression, and how building material is organised through construction.

In methods we list different aspects relating to the way of working for both study and design development. The core method can be described as a process of working in cycles of making and observing. This process is supported by a number of distinctions relating what is done to how this is perceived and experienced.

And in media we describe the role of models, views, and exact drawings in this process of making, observing, and documenting. Although the physical model is the main generator, we use a wide range of other media, from various types of physical models to computer models, from rough hand sketches to precisely constructed hand and computer drawings. We try to be specific in how each model and each drawing can at the same time represent the whole design and articulate distinct aspects.

1 命题 | proposition

在这一节中我们重点阐述我们就建构研究所做的基本假设、理论定位、基本观点，以及方法。其中最重要的一点是如何来给一个以模型手段为核心的设计过程来进行理论定位。模型的作用在于通过对模型材料的操作我们可以获取空间的概念，因而模型成为设计的空间概念的发生器。

那么，如何来界定“建构”呢？我们并没有刻意去构筑一个深奥和复杂的理论架构，而是采取了一个直接的和朴素的方式，即对建筑设计活动的基本观察和再思考。通过对本能的和设计的两种空间生成方式的考察，我们认识到设计的空间生成方式把设计从建造过程中分离出来，从而失去了用材料建造空间的直接性；通过对作图和模型两种设计方法的考察，我们认识到以模型作为设计的手段，在某种程度上与直接用材料来建造是相通的；更进一步来考察那些主要依靠模型来产生设计的情况，我们发现对模型材料的操作决定了最后的形式。在这些观察的基础上，我们提出了一个有关模型工作方法的假设、界定了基本的要素类型、提出了从模型材料到建筑材料的转换问题、以及设计过程的模型。

In this section, we state our basic assumptions, our position and point of view, and our approach. We construct an argument for a model-based design process, in which the model is more than a scaled abstraction of the building, by becoming a generator for spatial ideas through the direct manipulation of the model material.

Our argument is not based on theoretical considerations but on direct observations and reflections on primary building activities. We distinguish between the creation of space as an instinctive act versus a design act, and between a drawing-based versus a model-based design process. Originating from the observation that the working process influences the resulting form, and from the observation of strong spaces using distinct space defining elements, we argue for a model-based design process in which the main spatial idea is developed through the direct manipulation of model material and a subsequent transformation into building material.

空间生成作为一种本能行为
making space as an instinctive act

这三张照片纪录了一位母亲如何在田边为她的孩子搭建一个临时的蔽所。她选择了田埂边一凹地，再用几根树枝搭在上面，便给了孩子一个既安全又遮阳的栖身之处。这个创造空间的过程不是概念的和抽象的，而是具体的和物质的。用材料来搭建以创造空间，这就是建筑活动的本质。在这个最原始的层面上来看，这位母亲完全是凭借本能来行事，一切的思考皆融合到了建造的过程中。我们可以在其他的乡土建筑形式中观察到相类似的空间生成的直接操作。

In the three photos, we see a mother building – a shelter for her baby near the field she is going to work in. She first finds a suitable place where the ground already forms some space which can provide protection. She than uses some twigs to further articulate the space, increasing the protection and providing shade. She creates the space intuitively, working directly with the material that she finds and providing for the use. In one person, she is the client, the designer and the builder at the same time. Similar direct acts of creating space can also be found in forms of vernacular architecture where people choose a suitable site, adjust and form it, and use local materials to form a structure and enclosure. The need for space, enclosure, and protection is met by directly working with the land and available material. Is it possible not only to learn from this as a possible origin of architecture, but also by reintroducing a directness of working with material into the design process, after design and building have become the responsibility of different groups of people for quite some time?

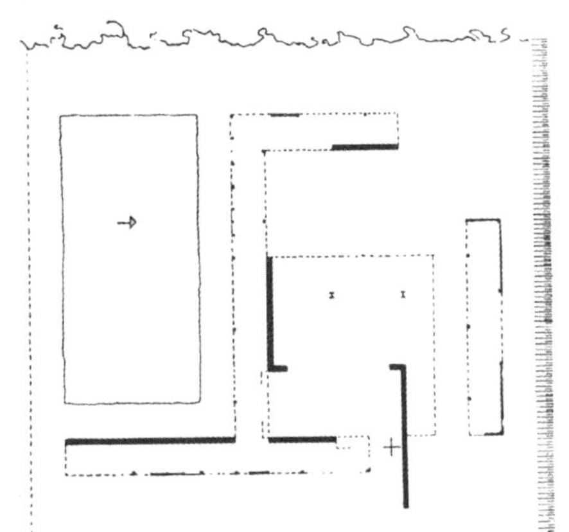

空间生成作为一种设计行为
making space as a design act

而我们现在通常面对的空间生成问题则是一种设计行为。与前述的空间生成的本能行为不同，作为设计行为的空间生成将设计的环节从建造的过程中分离出来，这是现代专业分工的结果。建筑师的主要任务之一就是通过作图和模型等设计的手段来构想空间，而实际的建造则是由专门的建造者依据图纸来完成。也就是说，在一个设计过程中，建筑师并不需要直接操作建造的材料就能完成他那部分的工作。

An architect is a specialist who works together with others to realise a building. Among the specialists, he is responsible to for creating the space apart from his other duties. Over time, architects have developed design processes which might lead from an initial idea, through a worked out project, to a finished building. For the architect, the making of space has become an act of design. During the design process he has to anticipate the construction of the building and its use. For this he has developed various media to develop the design, to communicate it to others and to describe the future building in various degrees of detail. Among these professional media are scaled models, plans, sections and elevations, and perspective views. Compared to the act of building, the act of designing is indirect.

Sculpture pavilion in the Sonsbeek Park in Arnhem by Gerrit Rietveld, 1954

两种设计方式
two modes of design

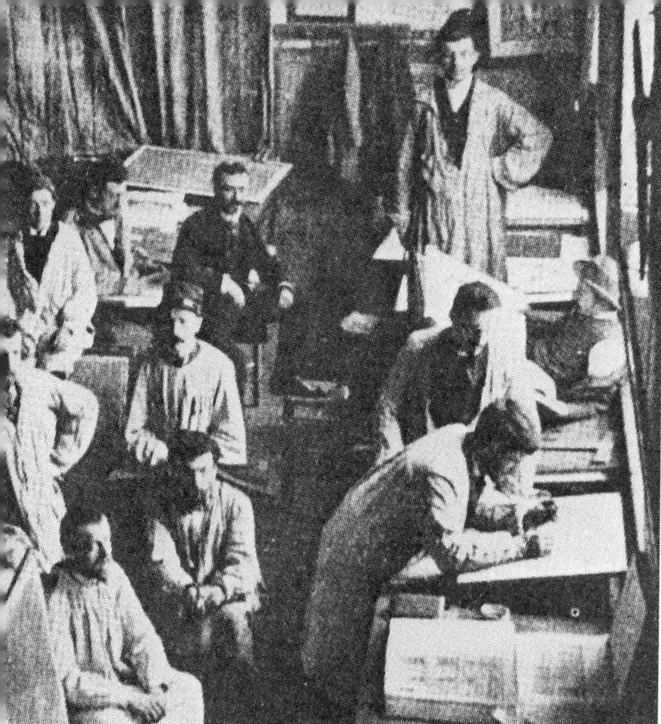

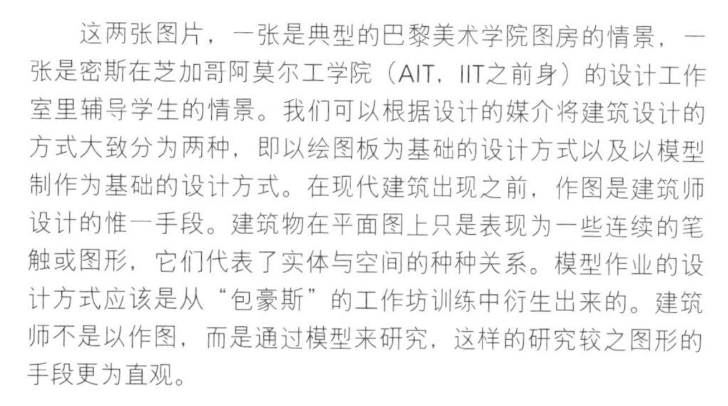

这两张图片，一张是典型的巴黎美术学院图房的情景，一张是密斯在芝加哥阿莫尔工学院（AIT，IIT之前身）的设计工作室里辅导学生的情景。我们可以根据设计的媒介将建筑设计的方式大致分为两种，即以绘图板为基础的设计方式以及以模型制作为基础的设计方式。在现代建筑出现之前，作图是建筑师设计的惟一手段。建筑物在平面图上只是表现为一些连续的笔触或图形，它们代表了实体与空间的种种关系。模型作业的设计方式应该是从“包豪斯”的工作坊训练中衍生出来的。建筑师不是以作图，而是通过模型来研究，这样的研究较之图形的手段更为直观。

The two images – a view into a design studio at MIT, which offered at that time a Beaux-arts-based education, and a view of Ludwig Mies van der Rohe, who shaped modern architecture, working with students at the Illinois Institute of Technology – illustrate and represent two very different design and teaching methods, independent from the formal preferences and ideological differences. They differ in the primary media used for working to learn design. In the first, the complete design process is in the form of drawings, which take various forms in the development of the design from a first quick sketch to very elaborate drawings. In the second, the design is mainly developed and discussed through models, and documented by drawings. The first is a process very far from building. The second can be seen as a building and construction process in an abstract form.

An atelier at the Ecole des Beaux-Arts in Paris at the turn of last century
Mies' teaching at the Armour Institute of Technology in Chicago

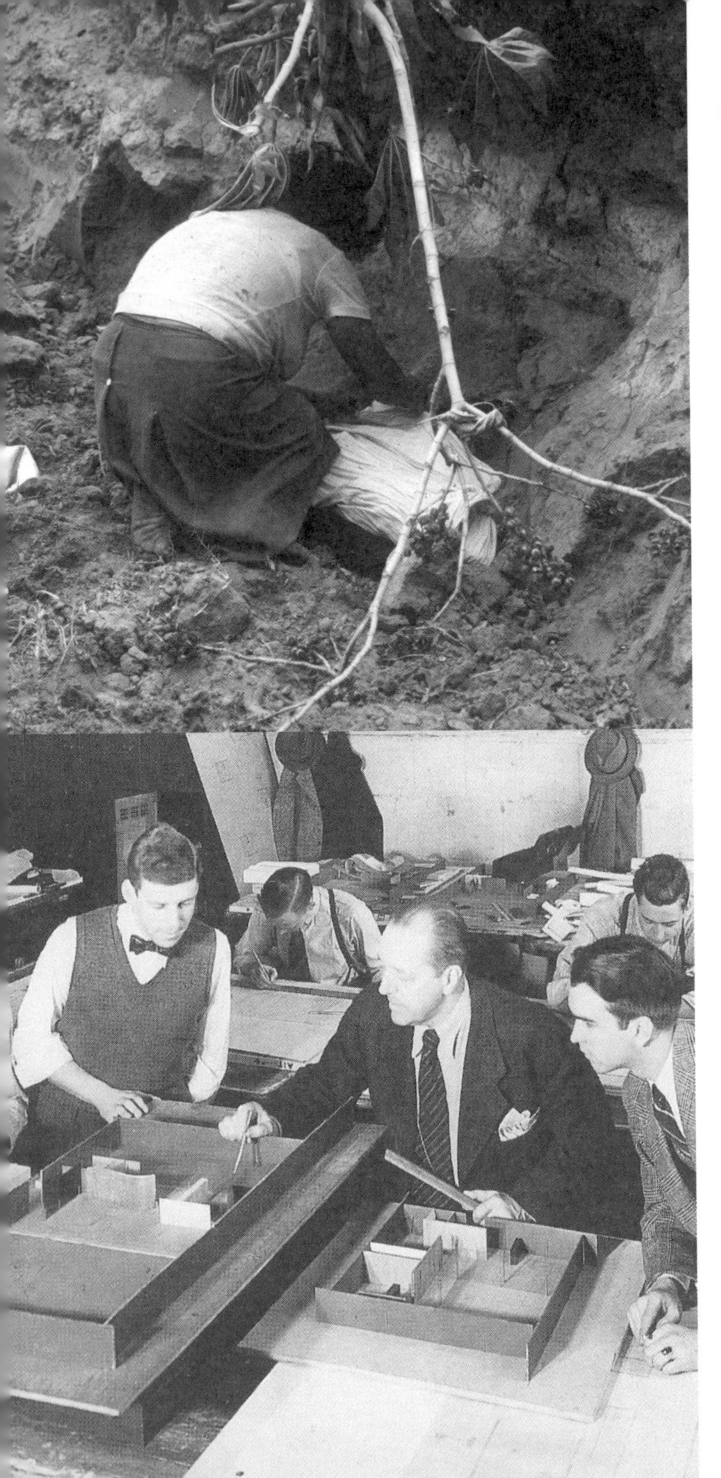

空间：本能行为与设计行为
space: instinctive act versus design act

在民间的和本能的建筑活动中，建造的意图与实现意图的手段之间并不存在一个明确的“设计”环节。人们是直接通过建造，即对建筑材料的操作来实现建造的意图，这也许是我们能够在这类建筑中找到建构的最直接和最丰富的表达的一个根本原因。建筑师的出现，使得建筑意图和建筑手段之间的直接联系中断，建筑师将业主的建筑意图以图纸和模型的方式转化为设计，再以设计来指导建造的活动。在这样的情况下，建构的表达就不大可能是建造的过程中自然产生的结果，而是建筑师需要刻意去追求的设计目标。我们之所以特别强调模型制作的设计方法，是因为在它的作业过程中始终包含了对材料（模型材料）的操作，这在某种程度上似乎又回到了建造的最初定义。

The instinctive act of making space seems very powerful. One person plays many different roles and there is a directness in all the aspects involved. In a modern design process the roles are distributed among many persons and the directness has been replaced by many mediations. We ask ourselves if certain aspects of the intuitive act might be reintroduced into the act of design. Could the directness of working with building material have an equivalent in working directly with model materials? This would require a shift in the view of the model as being a small abstract version of the building to the model as an object on its own, built with model material and not representing something else. It would also introduce a new challenge, that of translating the object into a building and the model material into building material at some stage.

工作方法决定形式
method of work determines form

很多当代的建筑师习惯于直接用模型进行设计研究。制作模型需要材料，不同的材料提供了不同的操作的可能性。一张纸板，你可以用切割的方法先切成若干的板片，再用这些板片来围合和限定空间；你也可以用折的方法来直接将纸板折成所需要的空间。这个折的操作会给你的设计一个明确的特征，因为纸板这个材料提供了这种创造空间的可能性。如果你的这个建筑最后是用钢筋混凝土来实现的，这种材料显然是不能折的，但是这个在构思阶段用纸板折出来的形式特征却成为你的形式表达的一个重要的内容。也就是说，工作方法决定形式。

The illustrations show three stages of a design process for a chapel by the architect Juan Carlos Sancho Osinaga. A first idea is developed by folding a piece of paper into an expressive form defining a space. Another model shows the space for the church based on this idea, but already is no more directly folded. The form of the finished building in concrete reflects the idea discovered by manipulating a sheet of paper, developed in further models, made from model material and finally translated into a building, constructed with building material. This is clearly an instance where the model-based working method of the architect had a direct influence on the space and form of the building. We adopt this attitude as our basic proposition. And we are interested in finding out more about this relationship and how it can support not only a design process, but also a learning process.

Chapel in Valleacerón , Spain, Sol Madridejos and Juan Carlos Snacho Osinaga, 2000

三种空间限定要素
three space defining elements

体块 Block

Kirchner Museum, Davos, Gigon and Guyer, 1992

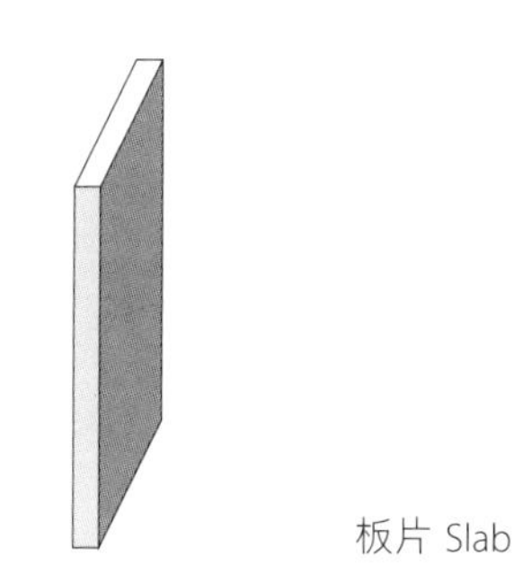

板片 Slab

Barcelona Pavillion, Ludwig Mies van der Rohe, 1929

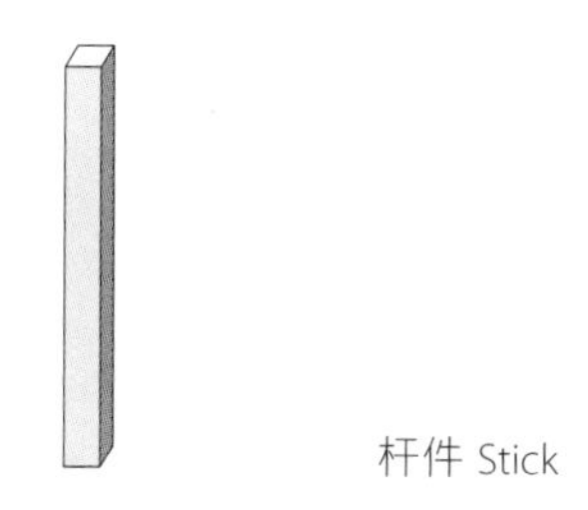

杆件 Stick

Cathedral–Mosque of Córdoba, 10th century

对现实建筑实例的考察证实了从要素着手研究建构的可行性。吉贡和古耶设计的达沃斯基希纳博物馆是体块造型的一例，空间存在于体块之内。而密斯设计的巴塞罗那世博会德国馆则是平板造型的典型，空间被平板由室外引导到室内，或围合或放开。西班牙的清真寺则体现了杆件造型的特征，石柱像树林一样占据了整个空间，不同间距的柱子调节着空间的疏密程度。由此可见，不同的要素所产生的空间也是不同的。再深一步思考，每一个空间限定要素均存在不同空间特征和形式表达的可能性。在这个教学实验里，我们把它归结为操作的问题。即使是同样的一种要素，因为操作方法的不同，产生的空间也应该是不同的。

It is through the observation of buildings, that we realised that there could be a distinct relationship between the type of space defining elements and the resulting type of perceived space. By looking at the three examples, the Kirchner Museum in Davos by Annette Gigon and Mike Guyer, the Barcelona pavilion by Ludwig Mies van der Rohe and the Al Azhar mosque in Cairo, we can easily see a difference in the character of the space. Each of these examples uses one element type in a rather pure or dominant way. Analysing this issue further, by looking at more buildings and making experimental models and drawings, we can articulate the differences of the perceived spaces based on the used elements of block, slab, and stick. In models, the space defining element can be identical to the construction element. In buildings, the relationship between space defining elements and construction elements is more complex, as also construction elements can be in the form of blocks, slabs and sticks.

从模型材料到建筑材料
from model material to building material

我们把一个建筑设计的过程理解为从构思到建造的一系列操作，它的一端是构思形式，另一端是建造形式。所谓的构思形式，就是指一个设计构思，它的形式和空间的特征与设计者所采用的特定模型材料以及由此而衍生出的操作方法密切相关。而所谓的建造形式，就是指这一设计构思的物质化，与建筑材料密切相关，涉及采用何种建筑材料来实现设计构思以及具体的建造问题。建造形式并非建造完成的建筑物，而是指建筑研究的深度考虑到建造材料和建造方法对设计表达的影响。根据建构命题的推演，在设计的过程中构思形式和建造形式之间应该存在一个内在的逻辑联系。从构思形式向建造形式的转化过程中如何把握空间，以及实现建构表达，这正是建构作为一种工作方法的核心之所在。

In a model, we can manipulate the model material directly. For example we can fold a slab-shaped piece of stiff paper or cardboard to explore ways to define space and find form. Building materials have very different properties and are manipulated in other ways than model materials. A building can inherit its space definition and form from the model. For many other aspects, we have to consider the consequences of the transition from model material to building material. The construction of initial models in this respect only relates to the model itself, so the model material does not point at building material. But later models have to be built, in which the model material becomes symbolic, indicating a building material, and also the construction of such a model points to the construction of a building.

Chapel in Valleacerón , Spain, Sol Madridejos and Juan Carlos Snacho Osinaga, 2000

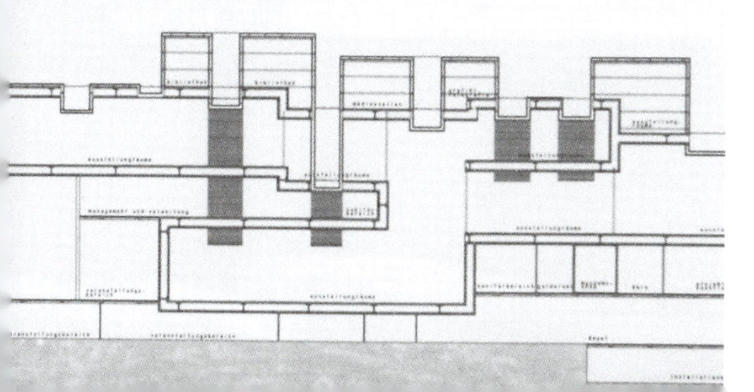

一个设计过程的模型
a model based design process

以模型作为基本的设计手段，设计的阶段也应该可以以模型材料的规定性作为依据。从构思形式到建造形式，我们可以划分四个相对独立的研究阶段。首先是通过对特定的模型材料的操作来寻找一个建构的构思，包括材料的操作方法和相应的空间；然后是根据建构的构思来继续研究如何满足基本的功能和空间的要求；再次是通过引入不同模型材料来研究气候边界等问题；最后将模型材料转化为建筑材料，将构思形式转化为建造形式。在这个过程的不同阶段，模型材料扮演了不同的角色。据此，我们可以建立一个包含四个阶段的设计过程模型作为教学的基础。

The transition from model material to building material can be used to structure the design process into several phases, forming a model-based design process. In the first, the potential of the manipulation of the model material for forming space is explored. In the second, one or a few principles of manipulation are used to define the required spaces, establish a basic structure and give form to the building. In the third, the possibilities offered by using more than one model material are studied to articulate a differentiated expression. The fourth is an investigation into the translation of model material into building material, finding out how the treatment of building material can support the ideas developed through working with model material. In our context of a series of exercises, this process might seem artificial, but it can also be observed in architectural practice.

Bolzano Museum of Modern Art, Italy, Federico Soriano Y Dolores Palacios, 2001

2 问题 | issues

把这个关于建构的设计课程定义为有关空间和建造（即生成空间的手段）的研究仍然显得有点宽泛，因为每个建筑物的设计似乎都自觉或不自觉地要涉及空间和建造的问题，因此我们有必要对研究的范畴作进一步的界定。所谓的建构，其本质是形式表达的问题。通过对三组不同的建筑参照物的观察，我们区别出象形的、抽象的、材料的、建造的，以及结构的表达几种类型。如果我们把这三组建筑归纳为一个图表，就可以发现一个很清晰的图景，即这个课程的研究范畴涵盖抽象的、材料的和建造的三个方面。我们把象形的表达和结构的表达排除在这个课程的研究范畴之外。

更进一步，我们还发现建构研究的三个基本问题也同时构成了设计方法的三个基本的阶段，即抽象的、材料的、建造的表达分别对应于设计方法的抽象、材料和建造三个阶段。如此，我们便可以从三个层次或设计阶段来考察空间和生成空间的手段之间的相互关系。抽象的层面聚焦于要素与空间的基本关系，材料的层面关注材料的区分对空间表达的强化，建造的层面关注从模型材料到建筑材料的转换。在下一节中我们将讨论这三个阶段的具体操作方法，这里则着重讨论这三个阶段所涉及的不同设计问题。

The first set of issues tries to clarify the scope of our tectonic approach and to determine the range of buildings which can serve as references. We looked in built examples at the role of expression, form, material, construction, and structure. These were initially studies related to the issue of a tectonic approach, in order to articulate various differentiations within this approach, but they were not explicitly connected with each other. When these studies are put next to each other, quite a clear pattern seems to emerge: a central area to be included, and a periphery we keep in mind but leave to another context, like the habitation and technics studios. This formulation of observations also helps us to strengthen links between our design learning process and built examples we can use for study and as a reference.

The second set of issues deals with the core issues in the three different phases of abstraction, materialisation and construction. In the abstraction phase, we are interested in the relationship between the built elements and the perceived space. In the materialisation phase, we are interested in how the main properties of material help to differentiate and articulate space and expression. And in the construction phase we are interested in the organisation of the building material into components, articulated through joints and layers.

Shanghai Museum, China, Xing Tonghe, 1996
Rietveld-Schröder House, Utrecht, Gerrit Rietveld, 1924
Residential Home for the Elderly, Chur, Peter Zumthor, 1993

象形、抽象、建造
symbolic, abstract and constructive

我们知道建构并不是一个关于建造技术的研究，它所关心的是建筑形式的表达问题。大致来说我们可以归纳出三种基本的建筑表达形式，即形象或象征的、抽象或塑型的，以及建构的。所谓象征的形式和表达是指建筑的形式所表现的，是与建筑本身没有直接关系的内容，如上海博物馆的基本体量所体现的“天圆地方”的概念，这与建筑本身的结构和功能都没有直接的关系，设计者是在借助建筑的形式来表述一个建筑以外的概念，尽管这种表达也是必须借助于建造的手段才能实现的。风格派的施罗德住宅，它所表达的是一种板的构件的抽象构成，构件表面的涂料掩盖了具体建造的材料和它的结构，因此它是一种抽象的和塑型的表达。彼得·卒姆托设计的养老院，建筑的形式通过材料的运用清楚表达了结构体系关系，它的建造方式是直接可读的，这是建造的表达。

The main expression of the Shanghai history museum is based on the reference to traditional Chinese symbols: heaven represented by a circle and earth by a square. The overall form is that of an old ritual vessel with its distinct handles. In Gerrit Rietveld's Schröder house, Utrecht 1924, very clear abstract elements are related to each other in a composition, without revealing the material they are made of or how they are connected. In Peter Zumthor's home for the elderly, Masans 1993, the material used and the components formed determine the composition and the perception directly. In these three examples, the form of the building is expressed in three very different ways. We try to articulate this difference with the terms: symbolic form, abstract form and constructive form respectively.

抽象的、材料的、建造的
abstract, material and constructive

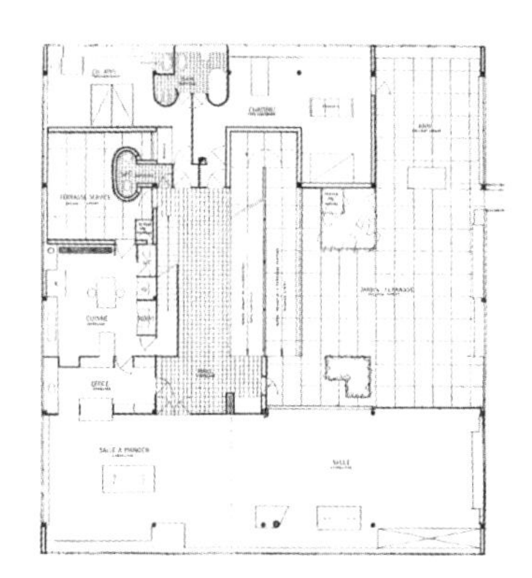

Villa Savoye, Paris, Le Corbusier, 1931

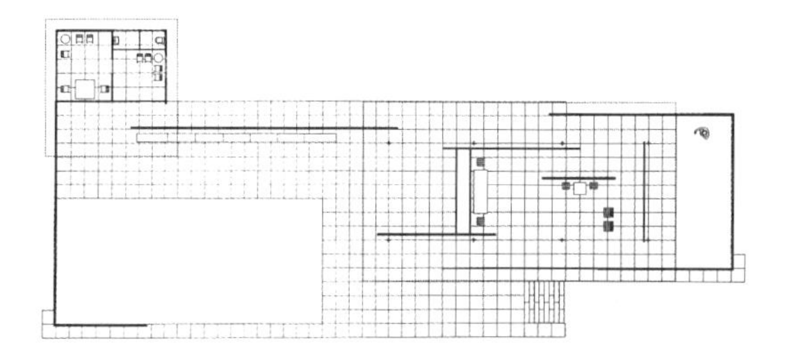

The Barcelona Pavilion, Barcelona, Ludwig Mies van der Rohe, 1929

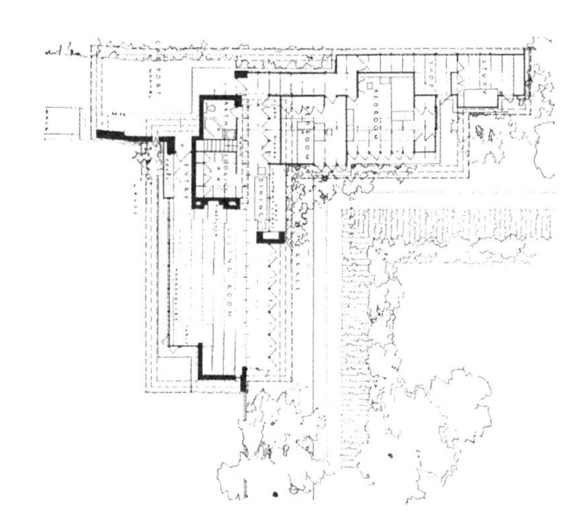

The Herbert Jacobs House, Madison, Frank Lloyd Wright, 1937

这三座建筑，勒·柯布西耶1929年的萨伏依别墅，密斯1929年的巴塞罗那世博会德国馆，以及赖特1936年的雅各布斯住宅均体现了自由平面和连续空间的设计理念，但是三者的具体处理方式则很不相同。其中的区别之一在于空间界定要素的表面建筑材料的作用。萨伏依别墅的表面施于粉刷，掩盖了真实的结构，表达抽象的形式要素；巴塞罗那世博会德国馆则以丰富的材料种类来表达不同的板片要素，而非显示结构关系；雅各布斯住宅显示如何由小的部件构成墙身的结构。总之，即使是相同的空间概念，在材料的表达上可以区分为三种表达类型，即抽象形式表达、表面材料的表达，以及建造方式的表达。

These three buildings, Villa Savoye, Paris 1929 by Le Corbusier, the Barcelona pavilion 1929 by Ludwig Mies van der Rohe and the Herbert Jacobs House, Madison, Wisconsin 1936–37 by Frank Lloyd Wright, share the idea of free plan or continuous space, but deal with it quite differently. One of the differences lies in the role of the building material which forms the surfaces of the space defining elements. In the Villa Savoye, most of the surfaces are painted white resulting in an expression of abstract form. In the Barcelona pavilion, material is used for expression in the form of differentiated surfaces not exposing any construction. In the Jacobs House, the surfaces are those of building material; they have depth and are articulated exposing smaller elements forming bigger components. In the three examples, expression is based on abstract form, material surfaces and construction components respectively.

材料、建造、结构
material, construction and structure

我们借用这三座瑞士的交通建筑来讨论建筑的材料、建造和结构表达的问题。迈里和彼得于1997年设计的苏黎世火车站月台屋顶体现了材料的表达，直接可见的是由木条组成的巨大表面，结构和建造不仅仅是隐藏在背后。而且，由于混凝土柱子在顶棚处消失在木质表面背后而产生模棱两可的阅读。实际上，柱子支撑顶棚后面的巨大桁架，只是结构概念并非该建筑的表达内容。斯宾登于1992年设计的迪蒂孔的自行车停车棚是建造表达的一个例子，结构、围合、屋盖和楼面的各个构件的相互关系均表达清晰，结构和材料支持建造的概念。布洛西和欧布瑞斯特1992年设计的库尔的巴士站表达的是一个结构的概念，即屋顶的荷重如何传递的结构形式，材料和建造则用来支持结构概念。如此，我们在材料的表达、建造的表达、以及结构的表达之间作进一步的区分。

These three traffic buildings are chosen to discuss the role of material, construction and structure in the expression of a building. The bus station roof in Chur 1992 by Richard Brosi and Robert Obrist is mainly an expression of structure. Material and construction are used to support the structural idea. The bicycle parking in Dietikon 1992 by Ueli Zbinden is expressed as construction, the articulation of elements and how they are put together. Structure and material support this idea. In the platform roofs of Zurich railway station 1997 by Meili & Peter with Fickert & Knapkiewicz, the expression is through the material which forms large-scale elements. Construction and structure are not only hidden, but an ambiguity of reading is created by the massive inclined concrete elements ending at the vast wooden surface, but actually carrying a huge invisible truss. The structural idea is not used for expression.

Platform Roofs Zürich Main Station, Zürich, Fickert & Knapkiewicz and Meili & Peter, 1997
Bicycle Shed on Station Plaza, Dietikon, Ueli Zbinden, 1992
Chur Railway and Bus Station Roof, Chur, Richard Brosi and Robert Obrist, 1992

象征
symbol

抽象
abstract

材料
material

建造
construction

结构
structure

建构的研究范畴
the spectrum of tectonic study

把前述三个讨论进行小结，我们可以区分五种形式表达：象征、抽象、材料、建造和结构。很显然，在这个有关建构的研究中我们首先要排除象征的表达，因为此类表达与空间和建造均没有直接的关系。其次，我们还要排除结构表达的问题，此类问题需要特别的结构知识的介入。象征的表达可以是人居工作室的研究课题。结构表达的问题则在建构工作室的一个硕士课程中有专门的研究。如此，我们把这个建构实验课程的研究范畴进一步定义为三个层次或方面，即抽象的、建造的、结构的表达。在下面的几页，我们将就这三个方面或层次所涉及到的具体问题作深入的讨论。

When we combine the previous three – originally unrelated – discussions together into a table of three columns, we can assign the examples to five rows, distinguishing five levels of possible emphasis for expression: symbolic, abstract, material, constructive and structural. Our interest is in the three middle levels. Both the symbolic and structural expressions lie outside our scope and might be more important to other studios. We are still interested in structure in two ways, just not when it is the dominating expression. One is that when working with any of the three element types, a structure has to be formed as well. The second is the issue of the structure-space relationship, which becomes a separate topic as one of the programmes for graduate advanced tectonic study. On the following pages, we try to articulate the basic issues that we are addressing on these three central levels: abstract, material and construction.

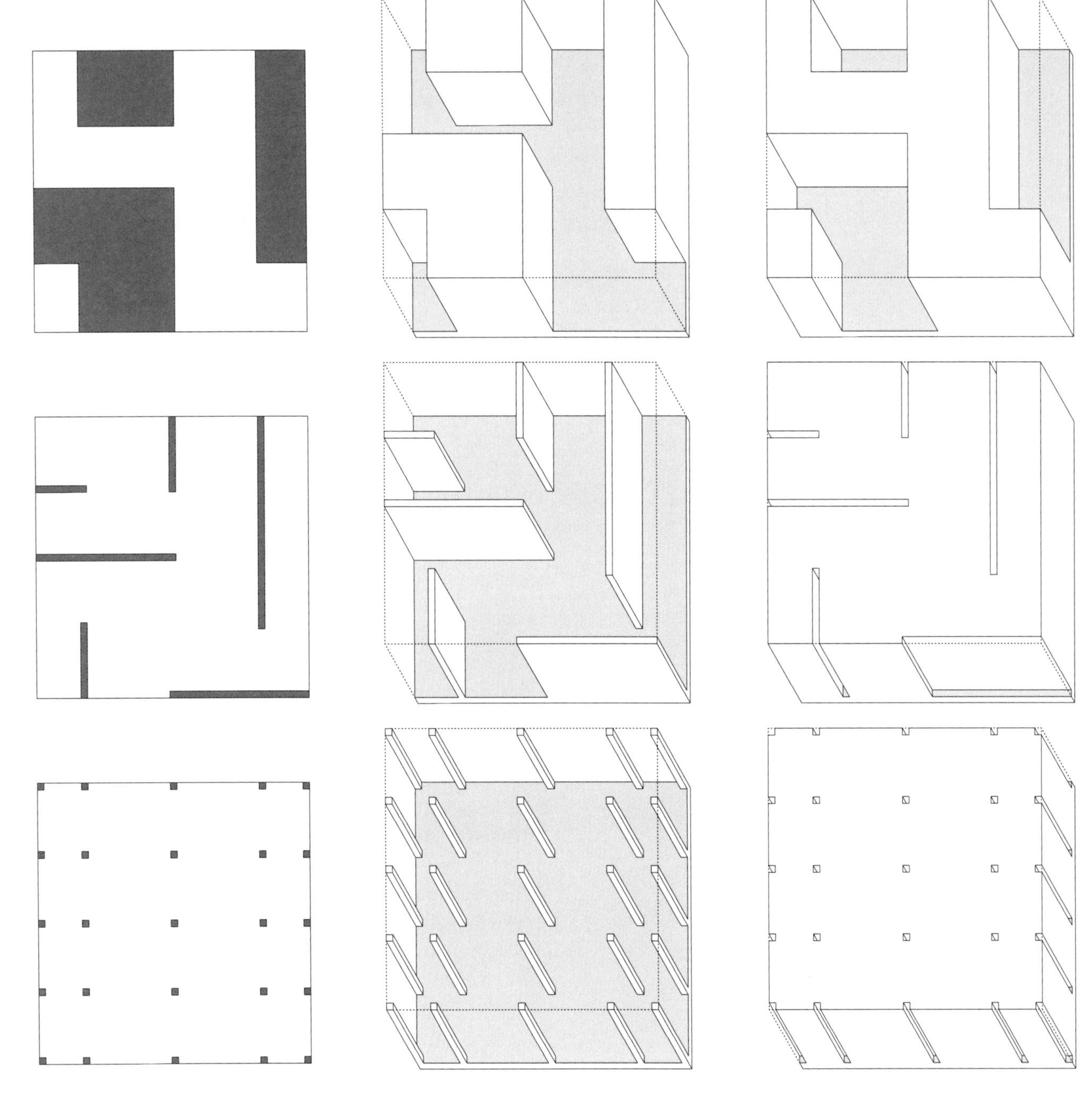

抽象：体块、板片、杆件
abstract: block, slab and stick

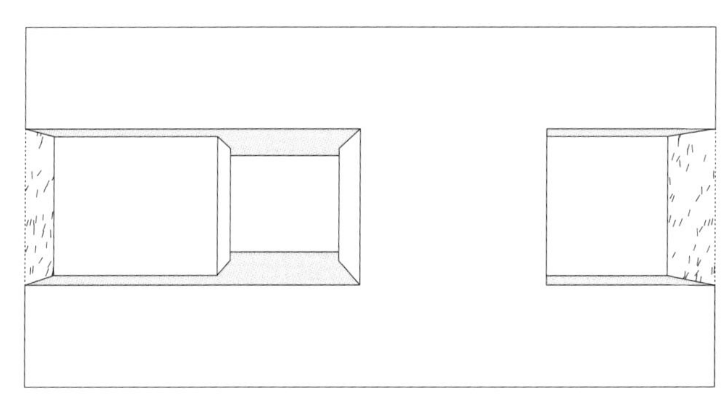

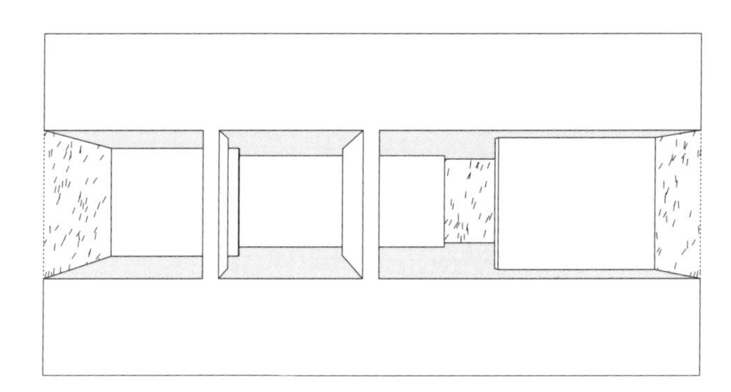

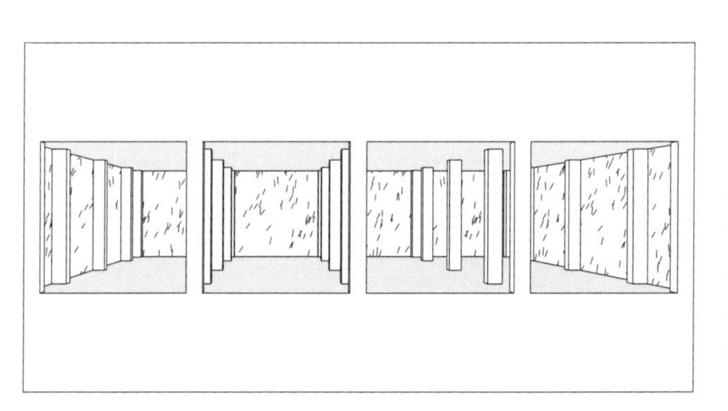

在抽象的层次上，我们区分出三种极端纯粹的空间限定要素：体块、板片和杆件，并且认为不同的要素生成与之相应的空间。我们的研究重点在于这些要素如何生成空间，以及它们在知觉上又是如何的不同。第一栏图示三种基本的要素显示为白色底面上的黑色图形。后面的两栏，要素和空间表达为实体与虚空的互换。最后一栏则是三种空间人眼透视的知觉。很明显，这三种要素类型在纯粹的情形下，它们的空间知觉是完全不同的。在下一页，我们将就这三种要素和对应的空间之特点作进一步的解读和描述。

On the level of abstract form, we distinguish the three extreme space defining elements: block, slab and stick. We are interested in how these elements define space and how the built space definition influences the perceived space. In the first column the elements are drawn as black figures on a white background. In the next two columns, both the elements and the spaces are drawn as complementary solid/void pairs. The last column shows eye-level views of the three spatial configurations. It is obvious that the three element types, when used in pure form, result in very different spaces. The following page shows an attempt to articulate and formulate these differences. On this level, we work with a single model material. The material properties that we consider are limited to its abstract geometrical shape as an element, extending in one, two or three dimensions, and to its malleability, how it can be manipulated to define, form, and differentiate space.

要素
类型：**体块**
element
type: **block**

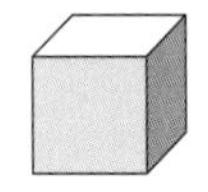

要素 / 空间
实 / 虚：**平衡**
图 / 底：**对等**
element / space
solid / void: **balanced**
figure / ground: **equal**

空间
特点：**勾勒**
关系：**互补**
space
character: **outlined**
relation: **complementary**

要素
类型：**板片**
element
type: **slab**

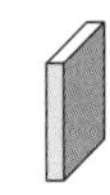

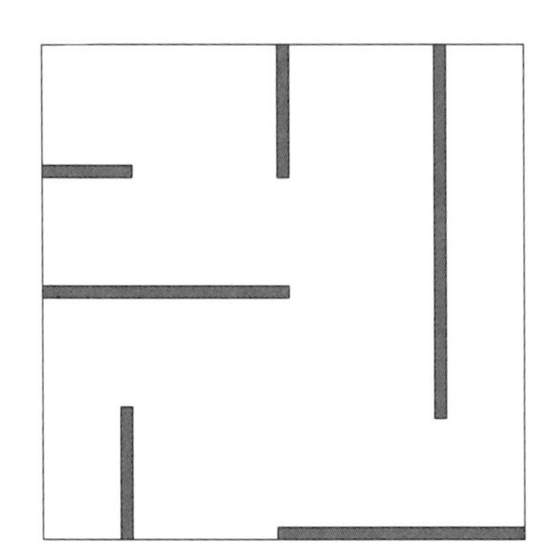

要素 / 空间
实 / 虚：**压缩**
图 / 底：**主导**
element / space
solid / void: **compressed**
figure / ground: **dominant**

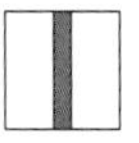

空间
特点：**模棱两可**
关系：**重叠**
space
character: **ambiguous**
relation: **overlapping**

要素
类型：**杆件**
element
type: **stick**

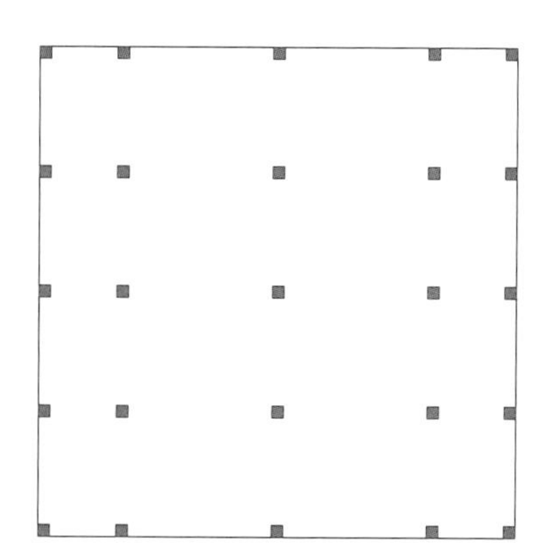

要素／空间
实／虚：**植入**
图／底：**附属**
element / space
solid / void: **embedded**
figure / ground: **subordinate**

空间
特点：**调节**
关系：**相同**
space
character: **modulated**
relation: **uniform**

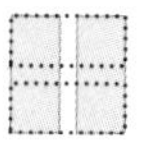

这里我们试图运用一些术语来描述空间和生成空间的要素之间的关系。这个图表从要素、要素和空间，以及空间三个方面（横行）来对体块、板片和杆件三种要素（竖栏）不同空间特点作分析，并以图底关系图解来表达。就体块而言，实体与虚空、图与底处于平衡的关系，而空间具有明确的边界。实体内部的空间，以及体块之间的空间是一种互补的关系。用同样的方法来分析其他两种要素。板片界定出若干相互重叠的空间关系，空间的定位具有模棱两可的特点；杆件在一个空间内作疏密或间隔的区分，调节空间的密度。我们还可以说，体块的空间是在其内（即体块内部包裹的空间），板片空间是在其间（即板片之间的空间），而杆件则是在空间之中。

We attempt to develop a vocabulary which can support the description and discussion of differences between the element and space types. Each column has three small basic diagrams with terms describing them. Between are two configuration diagrams illustrating figure ground relationships and two readings of different spaces. The first column reads: For the element of type block, we can distinguish two element-space relationships: the solid balances the void and the figure is equal to the ground. The character of the space is outlined with each space having a clear boundary. The two types of spaces, inside or between the blocks are complementary. The two other columns can be read in the same way. The slabs define many spatial zones which overlap, so we call the character of the space ambiguous. Sticks differentiate space from within, so we call the space modulated. We can also say that the location of the space defining element regarding the perceived space is outside the space for blocks, between spaces for slabs and inside the space for sticks.

材料：质感、色彩、透明
material: texture, colour and translucency

关于材料，无论是建筑材料还是模型材料，我们可以将其分为三个类别来讨论，即材质、色彩和透明。材质即不同材料的表面特性，如木、石、钢、纸、布等不同的材料具有不同的表面纹理，材质的最重要的感知特性是触觉。而就这里所关注的空间生成问题而言，材质的主要作用是给于不同界定空间的要素表面作出区分。色彩，包括明暗的差别，其作用与材质是相同的。总之，材质和色彩改变和调节我们对空间界面的解读，但是并不能改变空间本身。材料的透明性，即材料的穿透、阻碍、反射光线和视线的特性则具有改变空间知觉的特点，因此也具有特别重要的意义。

We are not just interested in the material itself; rather our interest lies in the role that material plays to articulate mass and space, how its properties of texture, colour, and translucency can be used to support spatial differentiations and the volumetric expression of the building. In the Kirchner Museum, each type of space – exhibition, movement and support – has a specific material articulation, contrasting with the others and the exterior mass. In the Vocational School Baden, the short sides of the building and some walls define an outside space, reflect the directional nature of the prefabricated concrete structure, and contrast with the long bronze-glass facades. In the interior contrasting materials define spaces next to the concrete corridors. In the Hotel Zürichberg, the single material of the exterior envelope contrasts with a differentiated layered interior envelope – wall, corridor and atrium, with the guest rooms in-between.

Kirchner Museum, Davos, Gigon and Guyer, 1992
Baden Vocational School, Baden, Burkard Meyer, 2006
Hotel Zurichberg, Zurich, Marianne Burkhalter and Christian Sumi, 1995

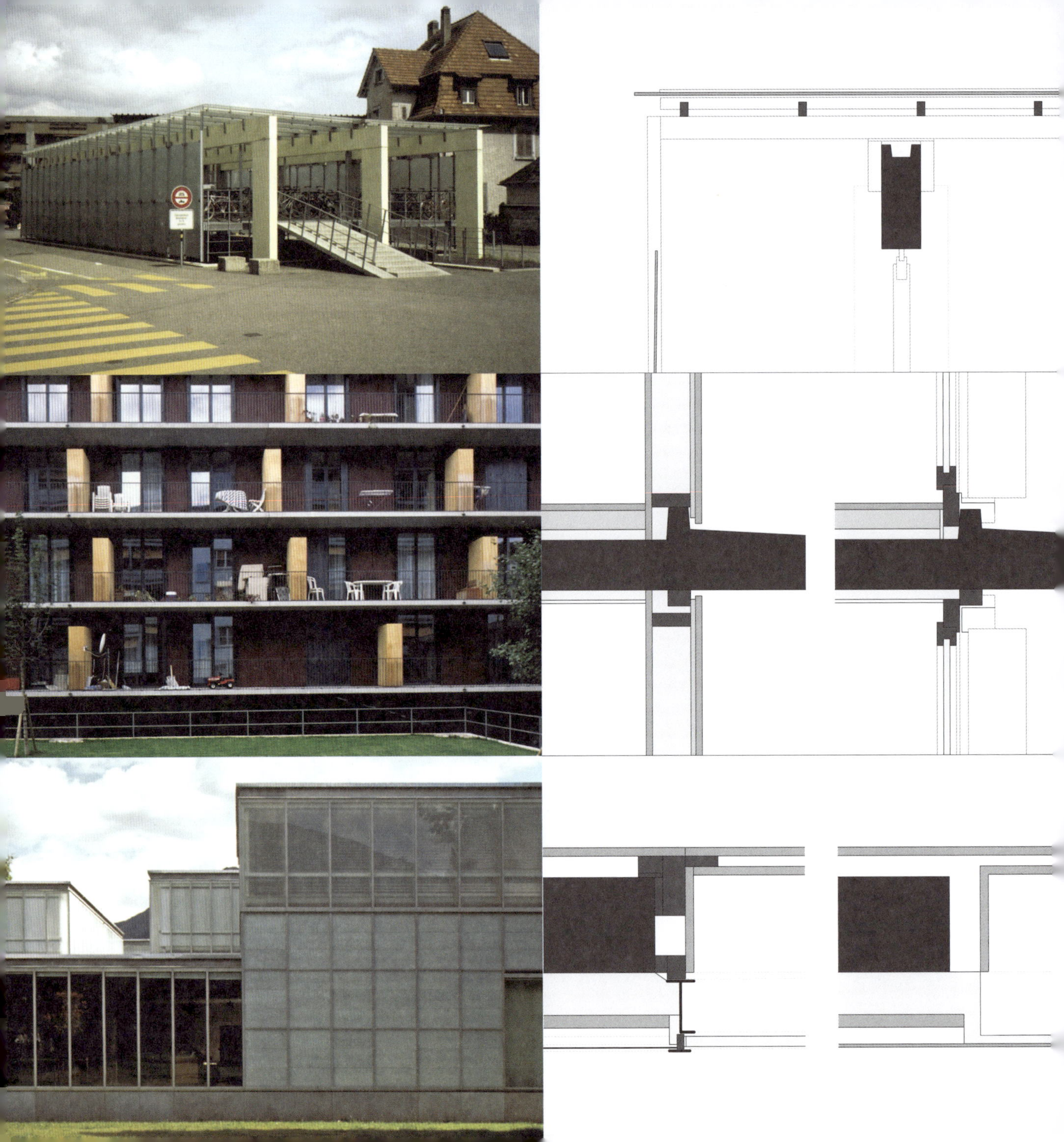

建造：构件，层次与连接
construction: components, layers and joints

从建造的层面来讨论，我们关注如何由建筑材料的构件组合成一座建筑物。比如用砌块砌筑墙体或用木条铺设地面等。建筑材料的尺寸往往是有限制的，需要用小的构件来建造大的墙体、楼面和表皮。我们分别用拼接和层次这两个概念来表述材料构件在平面和深度上的组织。材料构件的拼接所形成的表面图案对于空间的知觉有重要的作用。总结起来，一个清晰的空间概念应该在抽象、材料和建造这三个层次上一以贯之。即在抽象层次的表达在材料的层次得以加强，并最终在建造的层次得以实现。所以，构件的组织，即拼接和层次的问题不是单纯的技术问题，它们决定了建筑物的形式和空间的表达，同时也决定了人们如何来解读建筑的构成。

On the level of construction, we emphasise the order of components from the smallest element to the whole of the building. Within and between components, we distinguish two main expressions of relationships: joints and layers. Joints establish lateral relationships and layers establish relationships of depth. The issues of components, joints and layers are not only technical; they determine the expression of the building and also how its composition can be read. In the first example, a hierarchy of elements and components, related by joints and connectors dominates the expression revealing the load-bearing structure, and layers are secondary. In the last example, components forming layers with the use of translucency are used for the main expression without revealing the load-bearing structure and joints form a secondary order. In the second example components, layers and joints are used to create a complex, partly ambiguous expression.

Bicycle Shed on Station Plaza, Dietikon, Ueli Zbinde, 1992
Public Housing Block Müllheimerstrasse, Basel, Morger & Degelo, 1993
Kirchner Museum, Davos, Annette Gigon and Mike Guyer, 1992

3 方法 | methods

在这一节中我们着重阐述设计发展和学习的工作方法。其中贯穿整个过程的、最核心的方法是一个制作和观察循环往复的工作过程。我们根据设计发展的不同阶段具体的制作任务以及相应的知觉和体验内容对这一核心方法作进一步的区分。

方法是针对建构研究所提出的若干课题而设计的专门实验。我们所提出的课题是从构思形式到建造形式、从模型材料到建筑材料的转换过程中的特定问题，所以这些设计实验构成了一个连续的设计研究过程。具体来说，方法以四个阶段来展开，即阶段一，基本——操作与观察；阶段二，抽象——要素与空间；阶段三，材料——区分与诠释；阶段四，建造——构思与实现。通过对这些设计实验的描述，我们可以基本了解设计的想法是如何生成的，又是如何被推动，以及可以达到怎样的一个程度。每个设计环节包含了模型制作、空间观察和作图纪录等的相互作用，这里只是就核心的方法作简要的介绍，完整的过程描述放在“练习”一章。我们虽然强调从课程的整体架构中来理解各种设计研究方法的目的和操作，但是这里所述的各种实验也具有普适性，即也可以个别地应用到不同的设计环境中。

下面阐述的工作方法，每一个要点包含两个对立面，每两个要点构成一个阶段。

In this section we list different aspects relating to the way of working in relation to design development and the associated studies. The core method can be described as a reciprocal process of working in cycles of making and observing. This process is supported by a number of distinctions relating what is done to how this is perceived and experienced.

In phase 1, material and operation deals with the basic method of making, and space and observation with the associated method of observing, with the intention to arrive at a tectonic idea. In phase 2, clarity of operation and clarity of language and simple operation and rich experience introduce distinctions to help organise elements and space within an object. In phase 3, differentiation and interpretation provides distinctions for further articulation of elements and space, and materiality and perception expands the corresponding scope of observation. In phase 4, two types of transformation process are identified, from model material to building material relating to the role of the models, and concept and realisation relating to design development.

These methods are formulated as pairs of terms and each two refer to each of the four phases of exercises.

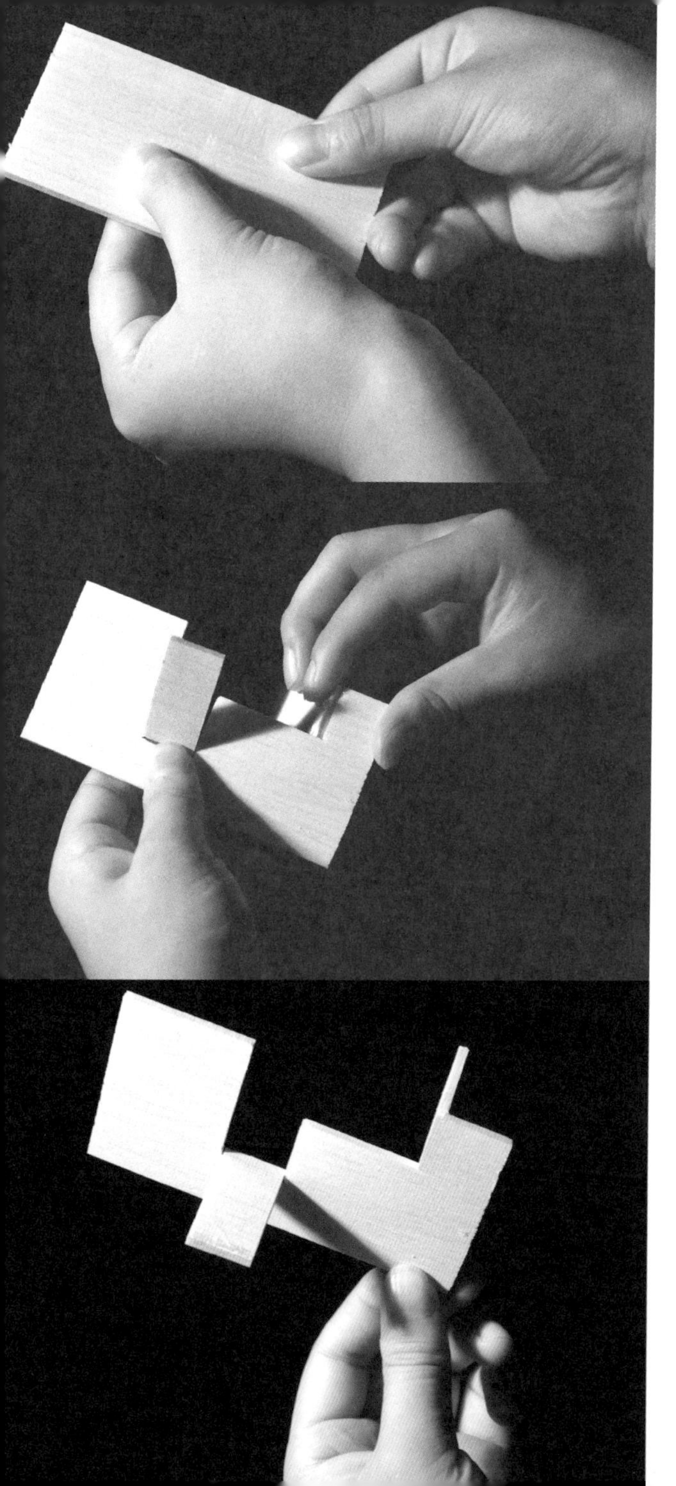

要素与操作
material and operation

建构研究的出发点是如何借助模型的方法生成空间。我们假设体块、板片和杆件每一种要素激发特定的处理方法，进而生成特定的空间。我们把特定的处理要素的方法称为操作。要素是个抽象的概念，模型操作涉及具体的模型材料，所以操作方法既与要素的类别有关，又与材料的特性有关。在模型研究的过程中如果涉及一系列的操作，其中必然有主次之分。有时生成一个空间需要几个操作才能实现，或后者必须在前者的基础上才能完成。根据我们假设的要素和空间之间的基本关系，每种要素决定了相应的空间类型，因此每一种操作就产生了同一类空间的一个特定的例子。

In phase 1, we work with a given model material to define a tectonic concept. Our starting point is the direct operation on the material. The material is model material and the operation is a principle of manipulation creating space. The material is given in the three forms of block, slab and stick, implying that each of the three material types triggers different operations, based on the assumption that they result in three corresponding types of space. The element type determines a distinct type of space and the operation creates an instance with a variation of that type. The material has to be chosen so that this direct manipulation becomes possible with the available tools without adding the need for an additional layer of planning.

对模型材料进行空间生成的研究，一般的规律是我们在开始时往往完全是凭借直觉来行事，经过一段时间的探索才开始系统化的操作。操作意味着动作，不同的要素激发不同的动作，如体块的操作方法可以是掏空、切割和位移；板片的操作可以是围合或折叠；杆件的操作可以是调节、搭接或排列。操作是针对要素而言，但是我们要切记，对材料操作的目的是生成空间。从更深一个层次来说，操作的目的是要寻找一种处理形式和空间的概念。这个概念不是一开始就有的，而是通过对操作方法的不断实验而逐渐清晰的。

It is possible to first instinctively manipulate the material and later explore operations more systematically. If several operations are applied in sequence, we talk of primary, secondary, etc. operations. Some operations are only possible in combinations or are based on a preceding one. The term operation emphasises the element, but it always implies that the operation's purpose is to create space. As an experiment, we first make something and then look at the result. So the intention should not be to try to make something interesting, but to find out how to form space with the material through the operation. Behind this method is also an attitude towards arriving at a concept or finding an idea. It allows work to start immediately without a specific idea or concept and to discover them later in the work itself.

Space

空间与观察
space and observation

操作和观察的互动是建构设计的基本态度和方法。空间设计的最困难之处在于操作是直接作用于要素本身，我们往往容易忽视操作的目的，即空间。其次是操作的方法，是不是可以识别出一些明确的对材料的操作方法来。这两者密切相关。空间不可见，但可以知觉。我们必须要借助于观察空间的方法来知觉空间，从人眼的高度来体验空间。我们要求学生把视点放低，边观察边画素描，研究光线和空间的关系。这个空间体验的环节是十分重要的。操作若没有观察作支持，就失去其目的性。一旦学生对自己所创造的空间有明确的体验，这一体验就会转化为明确的设计意图，使得开始还是比较盲目的操作成为有意识的设计活动。因此，操作和观察是一个互为因果的关系。

Through the direct manipulation of the material, we generate models which can be observed. We can directly see physical elements. To see the space they define is not easy and has to be learned and practised. And we have to find ways to describe the space in sketches and with words. To see the space is a matter of perception. We can look at views and can ask ourselves, what in the view makes us see space. We can also observe what changes when we change the viewing direction and when we move through the space. These issues of observation have to be actively reinforced in discussions and reviews of the work. The operation on the material and the observation of space should become a reciprocal process. Observations of space can lead to modifications of operations and these allow for new observations.

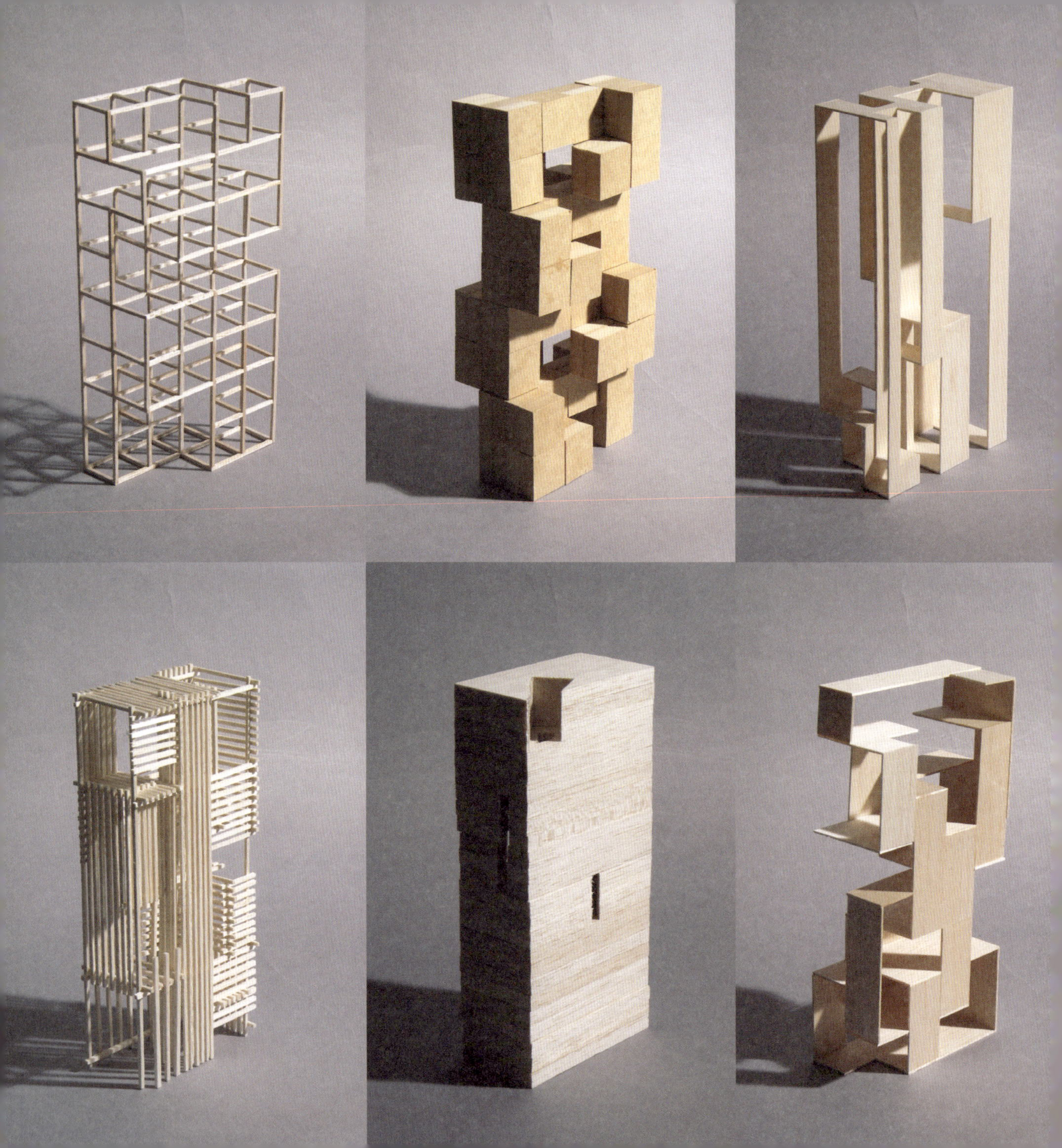

操作的清晰性与形式语言的清晰性
clarity of operation and clarity of language

设计的目标之一是追求形式和空间语言的清晰，其关键在于操作的清晰。一般来说，在设计的开始阶段我们对模型材料的操作倾向于杂乱和盲目，或有意识地追求操作方法的多样化，其结果是一个混乱的形式。克服这个问题的办法是分辨自己在设计中运用了哪些操作方法，再从中确定一种主要的操作方法。对操作方法的辨识的另一个重要的问题是对各种操作生成空间和形成稳定结构方面的潜能的认识。如果在一个设计中能够将清晰可辩的操作一以贯之，那么其结果必然是形式语言的清晰表达。这里所指的形式语言就是对操作方法表达的阅读和理解。

In phase 2, the form is given as a flat, thin or tall object. This suppresses the aspect of form as a shape and allows one to focus on the inner organisation of the form and its expression. What initially was a hypothesis has been confirmed by the evidence of the models: if the operation is clear and consistent, the object will also have a clear formal language, resulting in a strong expression. This issue of clarity of operation and formal language not only applies to both the elements and the space; it also strengthens the complementary nature of element and space. It is also the reason for the wide range of possible expressions. As an experiment, we should not try to design an interesting object, but find out more how the clarity of the operation can actually result in an interesting object.

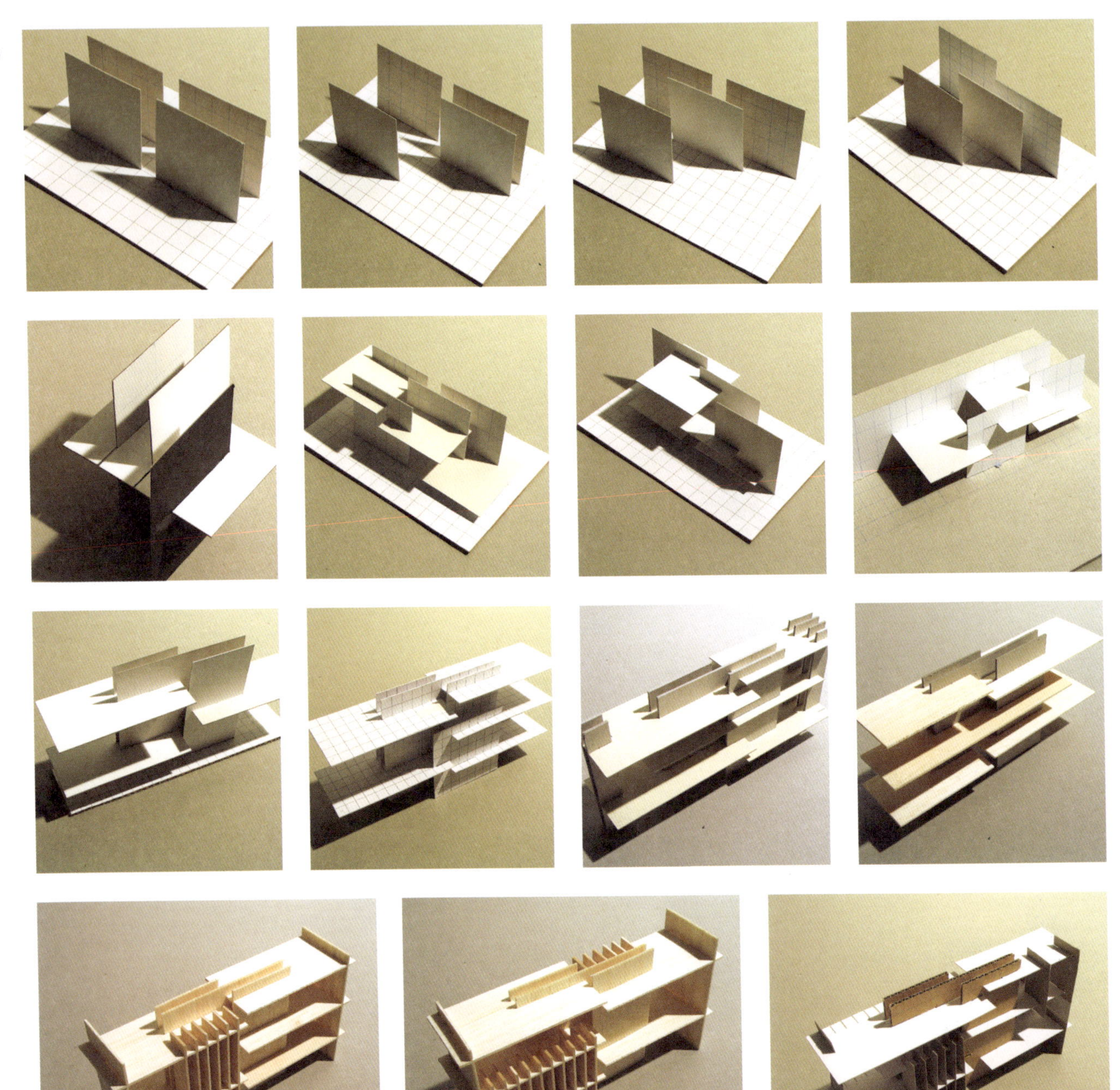

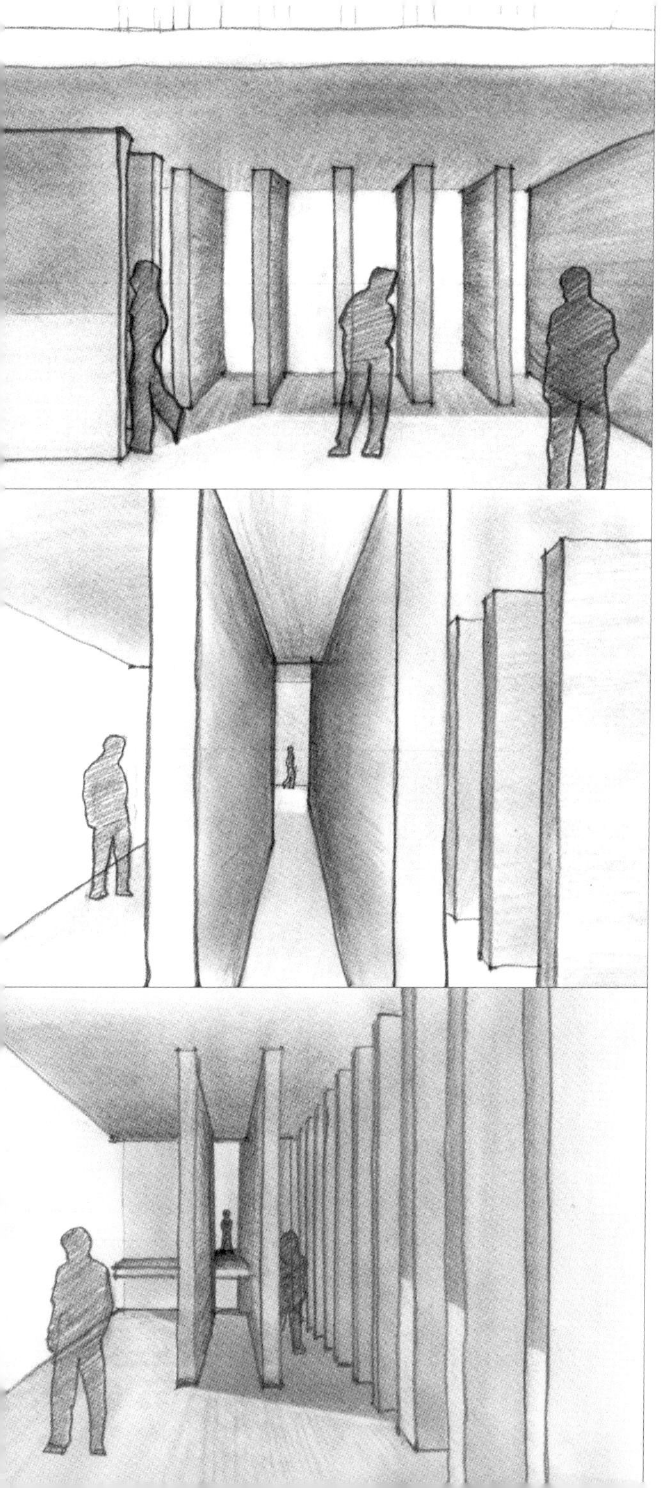

简单操作与丰富体验
simple operation and rich experience

操作的清晰是相对要素和形式的表达而言，就空间而言我们的要求则是简单的操作导致丰富的空间体验。简单和清晰的操作并不意味着形式和空间的单一和单调，这是模型研究的困难之处。一味地简化操作并不能产生一个好的设计。相反地，复杂的操作也未必能产生丰富的空间体验。丰富的空间体验主要取决于空间序列上空间之间的对比和变化，如空间之大小、形状、比例、方向、视线的位移、射入空间的光线的变化等等因素。总之，操作的清晰性和空间的丰富性的共同作用决定了形式空间的复杂性。

The fear of a design being boring often leads to the temptation to make a design artificially complicated. But in our context, the complicated manipulation of the elements does not necessarily lead to a rich experience. It might just lead to many similar views or to views which are difficult to grasp and relate to. The initial hypothesis that a simple operation can lead to a rich experience, can be seen confirmed through the evidence of the results. A simple operation can lead to a rich experience because of the distinct views it can offer to a viewer. Variations in the operation will lead to variations and contrasts of the views. The sequence of these views then leads to the experience.

Zephyr Lam Wing-him

区分与诠释
differentiation and interpretation

在研究中加入多种模型材料的因素，相对于单一模型材料的研究，其重点在于区分。我们借助模型材料的对比所产生的的区分的可能性来深化单一模型材料的设计，比如加入气候边界来区分内部和外部空间。就杆件空间而言，特别需要增加空间的进一步界定。多种模型材料相对于单一模型材料增加了形式表达的可能性，如在主要要素（结构）和次要要素（围合）之间作出区分，在水平要素和垂直要素之间作出区分，在表皮和内部结构之间作出区分，在开放空间和私密空间之间作出区分等。多种模型材料的运用的最重要的作用在于对原先单一材料的模型作出新的解读，如使得原先的操作概念得以更清晰地表达，或导致一个新的概念。

In phase 3, we use the differentiation of model material to study the potential of further development of the model made from one material. Using more than one material allows one to interpret the model in ways which will lead to various articulations with different expressions. For this purpose we distinguish between variations and alternatives of interpretations. An alternative interpretation would give a different reading of the model. In the case of blocks, for example, a different reading of the relationship of mass, space and surfaces would result. A variation only consists of the use of a different material for the same interpretation. Both are important and influence the expression of the space and the form. We ask for at least three models which can also be compared in their representation as drawings.

Alice Wong Chui-kwan

材料与知觉
materiality and perception

多种材料模型的研究必然要导致空间感知和体验的改变，因此在这个过程中我们必须不断通过人眼视点的观察来检验模型，观察原先的概念是否得到加强、或削弱、或产生新的概念，这些观察所得及时反馈到模型研究中去。对于那些不透明的模型材料，如各种颜色的卡纸板或木片，它们的意义在于区分，采用木片并不意味着将来就一定是木材的建筑构件。就空间知觉而言，我们要特别注意那些能够透射、反射视线和光线的材料的运用，因为这些材料的运用可以极大地改变原来空间的知觉。空间知觉的研究除了直接的观察外，其主要的手段是透视图的制作，为了比较不同的材料区分对空间的不同诠释，我们采用相同视点若干透视图并置的方法。若是电脑作图，则要避免直接采用软件附带的建筑材料图库。

The use of multiple materials means the use of multiple model materials. We want to find out how the properties of the model material make us perceive the space and the form. So the careful selection of the material for its properties and the direct observation of the model are crucial. In addition to the direct observation we also use three parallel eye-level perspective views, looking at the same representative space defined by different material interpretations. And we ask for plans, sections and elevations, which can indicate the differences in descriptive drawings. The direct observation of the model at least allows for intuitive reaction. The articulation in drawings makes the differences visible also to others. A next level would be to verbally describe the observed spaces through their differences in materiality.

Alvin Kung Yick Ho

从模型材料到建筑材料
from model material to building material

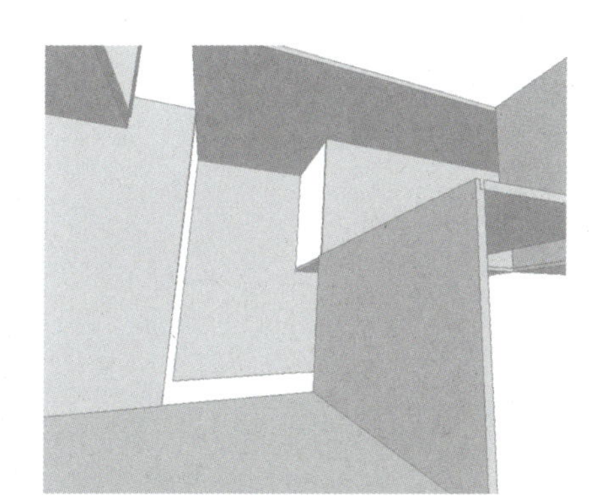

从模型材料到建筑材料的转换并不是一个简单的技术问题。如何决定建筑材料的选择，如何在建筑材料的阶段保持和发展原先的设计概念？其中的一个方法就是照片拼贴，即用建成的建筑实例的照片为素材，来作内部空间和外部立面的拼贴研究。这些照片不仅仅是建筑表面形象，还提供了材料构件和构造方式的信息。就单纯的视觉形式而言，材料构件及其构造方式形成建筑表面的图案，是形式和空间表达的重要手段。拼贴研究的目的就是要找到符合原先设计构思的建筑材料类型、材料构件的形式和尺寸，及其构造方式。以这个设计研究为目的的拼贴研究切忌采用电脑设计软件内附的建筑材料图库，因为这些贴图没有构件和构造的信息。

In phase 4, we use a collage to deal with the transition from model material to building material. We can look for images of existing buildings in publications which have a similar expression as the views of the models made by model material in the previous phase. If we use images from existing buildings, the building material is not just a pattern like the textures in a material library, but contains information about the design and the construction of the building. In a first step we just make use of the expression, without having to worry about construction and in a second step we can look for information as to how that expression was achieved through construction with building material. So we use a visual approach as an initial stage to deal with a technical issue.

概念与实现
concept and realisation

在设计发展中通常会碰到的一个问题是：在前一个阶段得到的一个重要的设计构思在下一个阶段遇到新的设计问题时就很容易地被否定或减弱，经过几个阶段的发展，到了最后原先的设计构思已经荡然无存。前述的照片拼贴研究是在模型材料向建筑材料的转换中保持原先设计构思的重要一步。具体的建造研究则是借助于局部放大的模型来完成。这个模型研究的重点不是建筑材料构件的构造方式，而是建筑材料构件组织的基本关系，即材料平面上的拼接关系和材料的墙体纵深的层叠关系。因为研究的重点是建筑材料构件的形状、尺寸及其组织方式，模型的制作并不要求在材料的材质上与真实的建筑材料一致。总之，作为这个设计研究的结束，建造的研究应该有助于原先设计构思的实现，甚至强化原先的概念，而不是削弱它。

One of the big difficulties in the development of a design is that an important idea in one stage can easily be eliminated when confronted with a new requirement in another stage of development. This is especially crucial in what we call the transition from model material to building material, as the new issues of construction pose a lot of difficulties. After starting with a visual approach, we continue with a model. In this second step we work with a partial model of 1:20. We increase the level of resolution beyond what we intend to handle, to reveal the new aspects provoked by construction issues. The gained insights will help to make a better final model of 1:50 with an again reduced level of resolution. Also the final descriptive drawings should benefit from this change of level of resolution.

Vanessa Chik Tsui-yan

4 媒介 | media

　　在这一节中我们系统阐述模型、透视和建筑图这三种主要的工作媒介在这个制作、观察和纪录过程中的作用。

　　媒介的作用不仅仅在于表现和传达设计，更是发展设计的重要手段。实际上，在一座建筑建成之前，设计只是以媒介的形式存在。在一定的程度上可以这样说，设计媒介就如建造者手中的砖石，构筑想像中的建筑。在这个课程中，我们主要借助于几种媒介来发展和表现设计。模型是最重要的设计手段，模型材料是最主要的设计媒介，在各个不同的设计阶段，模型材料的意义各不相同。透视图是另一个主要的设计手段，用来研究空间的体验和知觉，在各个设计阶段，我们对透视图的应用也有不同的要求。建筑图是纪录设计的手段，不同比例尺的作图要求反映各个阶段的设计内容。简言之，模型的作用在于生成，透视图的作用在于观察，建筑图的作用在于纪录。各种不同的设计媒介在设计的过程中相互作用，推动设计的发展。因此，在可见的成果上，设计的发展就是模型、透视图和建筑图不断积累和演变的过程。而计算机辅助设计工具的运用则为设计研究带来全新的可能性。

In this section we describe the role of the three primary design media – models, views, and exact drawings in this process of making, observing, and documenting.

Media are not only important for the presentation and communication of a design, but also for developing it. Until a building is built, the design exists only as media. There is a danger that models and drawings, instead of being the design, only point to the possibility of a design. During the process, the models and drawings should accumulate in a way that at any stage the work can be put up for review. We try to differentiate the media like model and drawing types to emphasise that specific media can support the design process by making the explored issues visible, and therefore increase precision, encourage observation, and allow discussion. In our programme there is also a pattern to use specific media for distinct purposes: models for generating, views for observing and drawings for documenting the design. We also try actively to deal with the use of the computer.

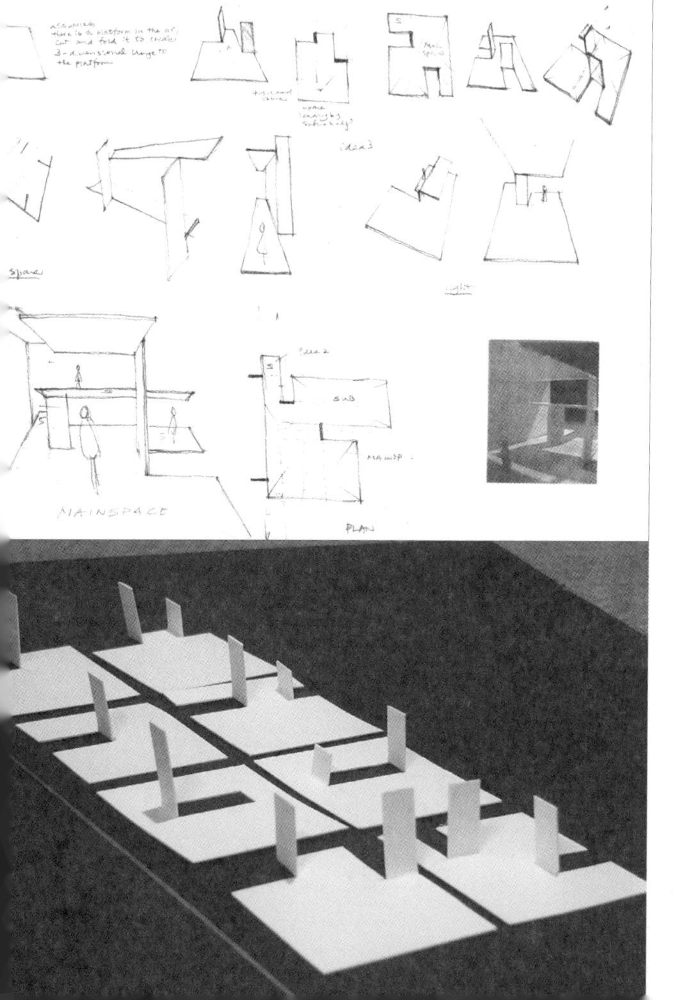

制作、观察、记录
make, observe and document

设计媒介与设计的方法紧密相关。我们鼓励学生制作、观察和纪录循环往复式地推进设计，即在模型、透视图和建筑图三种主要的媒介之间的不断转换。相比较在后面讨论各种设计媒介时所列出的那些正式的练习，设计思维习惯的养成更直接体现在学生为发展设计所作的、以笔记的方式存在的各种草图或草模。整个的设计过程包含了四个阶段，每个阶段的模型有其特定的要求和意义，相应地也有不同的透视图和建筑图的要求。其中建筑图的演变随着设计深度的增加而愈加复杂。每个阶段所产生的各种模型，透视图和建筑图等，一方面反映了特定阶段的设计，另一方面它们的集合体现了一个完整的设计发展过程。

We encourage students to work in cycles of making, observing and documenting to develop the design. With each stage and round of such cycle, specific media are involved. They vary over the iterations of the cycles. In their basic form, we associate making with physical models, observing with perspective views and documenting with descriptive drawings. As the models evolve over the phases, the role of the model material changes corresponding to the explored issues. Each phase has its own type of main view, supporting the observation at that stage. And the drawings will become increasingly denser with the accumulation of new considerations in each phase. The media can be read in two different ways: once as a set consisting of different media, representing the design and its study at each phase and again as three sequences of different media types showing the evolution of the design and study.

Jonas Tang Chin-hong

模型材料
model material

物质性 physical

抽象性 abstract

视觉性 visual

象征性 symbolic

模型材料作为主要的设计研究媒介，它在各个不同的阶段的角色定义各不相同，即材料的物质性、抽象性、视觉性，以及象征性。在方法阶段，对材料的操作取决于材料的物理特性，即不同的材料物理特性激发不同的操作。在这个意义上，依据材料物理特性的操作是最接近于建构本质的一种角色定义。在抽象阶段，规定所有的模型必须用木质材料来制作，其用意是压抑材料的表达，强调抽象要素的几何特性。在材料阶段，多种材料的运用，一方面利用材料的区分来深化设计，另一方面研究不同材料阻挡、穿透或反射光线和视线的特性对空间知觉的影响。在建造阶段，模型材料的作用实际是模拟真实建筑材料，只是一种象征性的表达，比如用纸板来模拟砖砌墙体，重点是构件及其组织。

We use models as the primary generator of the design. They take the form of testing models, working models and presentation models. Over the span of the programme the model material changes from an easily workable material in the phase concept to an arbitrarily unified material in the phase abstraction, to differentiated model materials in the phase materiality and finally to model material which symbolises building material in the phase construction. This conscious articulation of the role of the model and the type of model material should support the issues explored at different stages of the design process, and hopefully lead to more precise models – precise in the sense that the model contains and expresses the design.

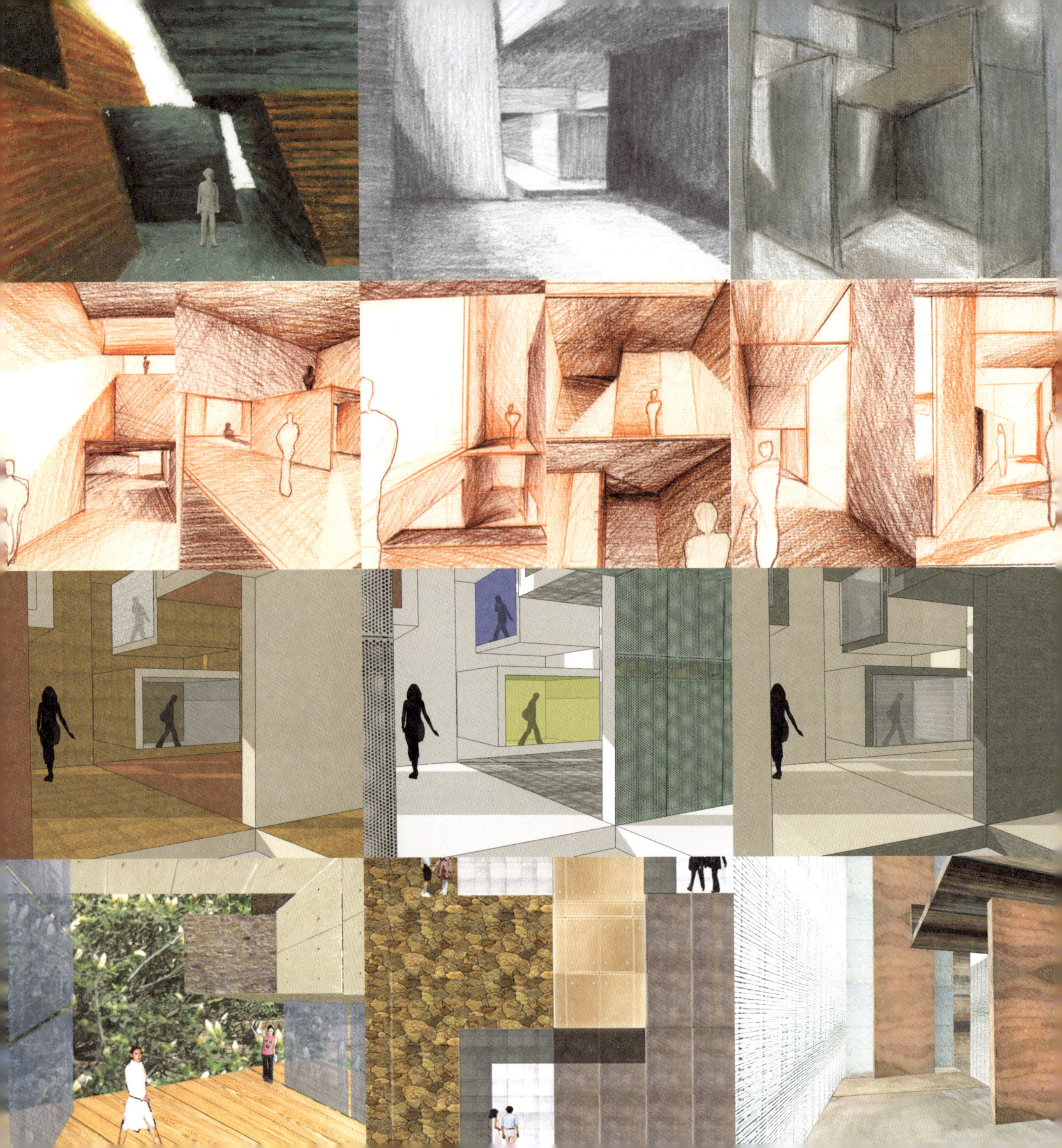

透视图
perspective views

基本 basic

序列 sequence

比较 comparison

转译 translation

对空间知觉的研究主要借助于透视图的手段。在研究的各个不同的阶段，透视图的作用各不相同。在方法阶段，我们强调基本观察方法的养成，体现在对模型的直接观察，通过摄影和素描的方式来强化对空间的知觉；在抽象的阶段，我们强调穿越空间的连续体验，以配合简单要素操作和丰富空间体验的设计研究；在材料阶段，我们强调同一视点若干透视的并置，来研究不同材料区分对空间知觉的影响；在建造阶段，我们则通过照片剪贴的方式研究建筑材料的组织对空间知觉的影响。根据不同的研究目的，透视图的制作要求也各不相同。总之，如何选择视点、调控光线、描述材质及表现场景，这些技术问题也同时反映了作者对空间的感知和表达能力。

We use perspective views as the primary support for observing space. They take the form of sketches, photos, videos, constructed perspectives, collages and rendered key drawing. Over the span of the programme the main type of view changes from multiple hand sketches in the phase concept to a sequence of computer generated views in the phase abstraction, to three parallel versions of a single view in the phase materiality and finally to a hand or computer collage of a representative single view in the phase construction. In each phase the specific type of view emphasises the newly introduced issues: element and space, spatial organisation, material differentiation, and construction with building material.

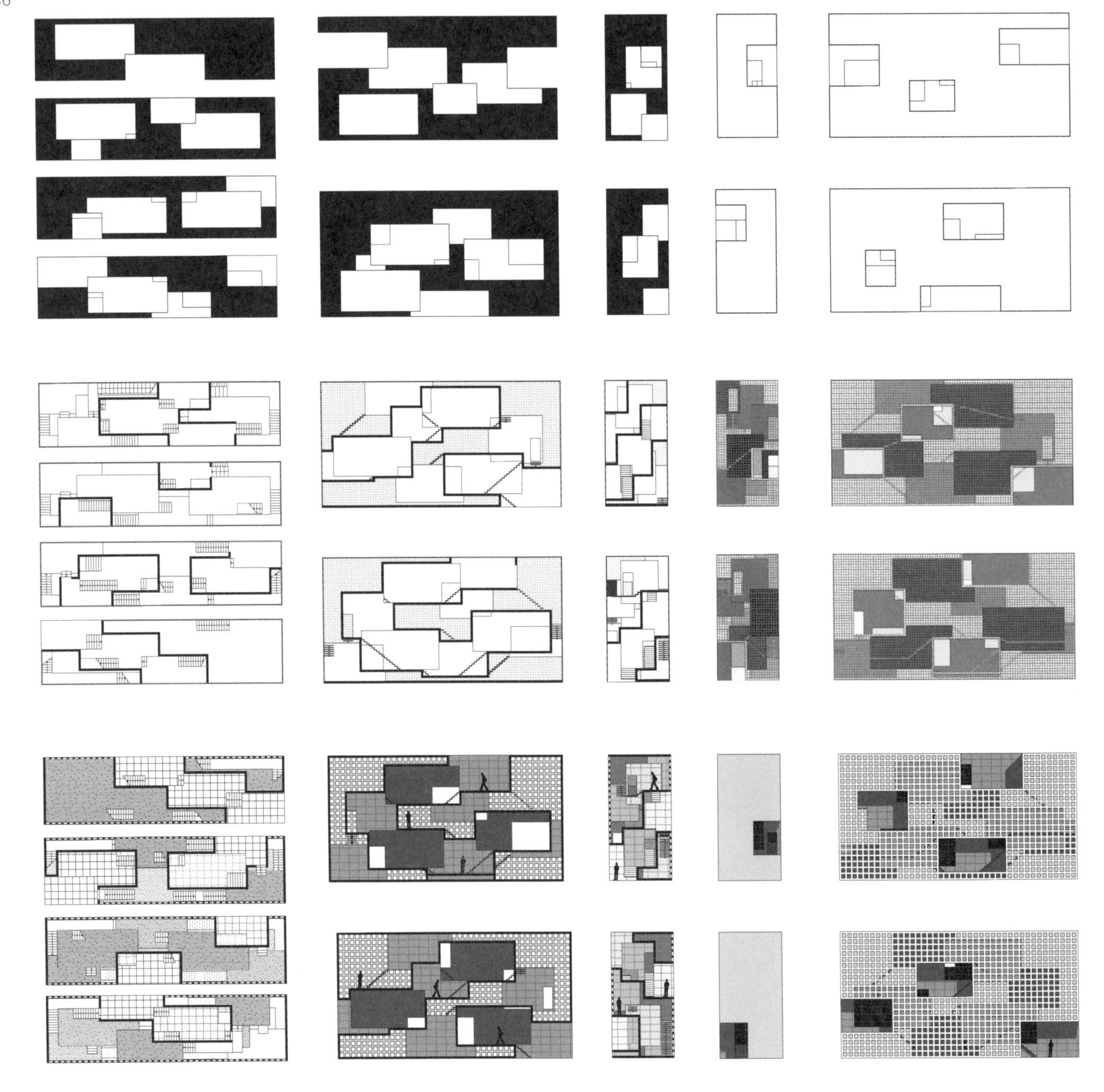

建筑图
descriptive drawings

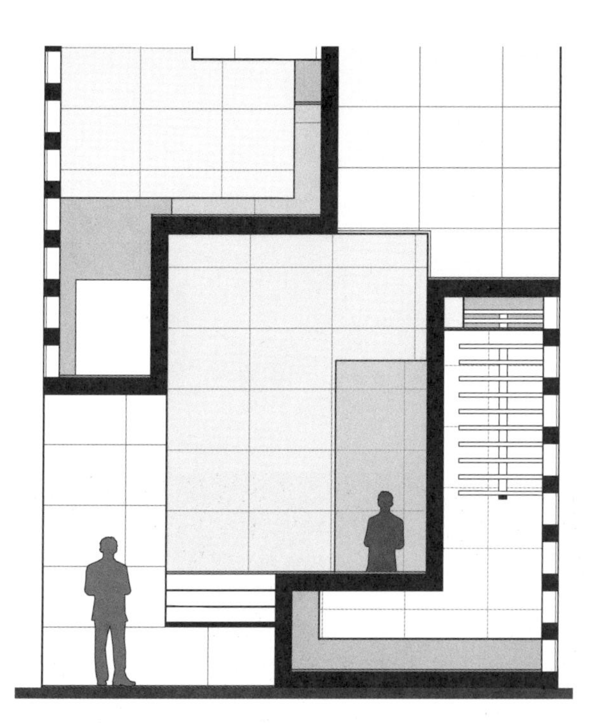

Daniel Ho Hao-yen

我们对用来描述设计对象的平、立和剖面图也有与几个设计阶段相应的要求，以求图面表达的准确性。在抽象阶段，对象的描述强调体块、板片和杆件基本要素和相应的空间的表达，立面和剖面则强调光影的描述；在材料阶段，要求在建筑图中表达材料的区分；在建造阶段，则要求在建筑图中表达建筑材料构件的组织。建筑图制作的另一个重要的原则是比例尺的控制，即不同的比例尺与不同的研究深度相关。我们主要在三个比例尺上作图，即在比较概念的阶段用1:200，局部深入的研究用1:50，建筑材料的表达用1:100。作图的比例尺与模型研究的比例尺相差一级，即相应的模型研究的比例分别为1:100，1:20和1:50。

We use descriptive drawings as the primary documentation of the design. They take the form of plans, sections and elevations. Whereas the models and views are specific and change in type over the phases, the drawings are standard and don't change in type. They only change through the accumulation of information over the phases. The standard set of drawings is in the scale of 1:200. In the final presentation we ask for a layout of these drawings which also allows the reading of individual projections over the phases. We also make use of the change of level of information through the change of level of scale. In the construction phase a partial section and elevation in the scale 1:50 is used to identify issues of building material and construction. They become the basis for more accurate drawings for the final presentations in 1:100.

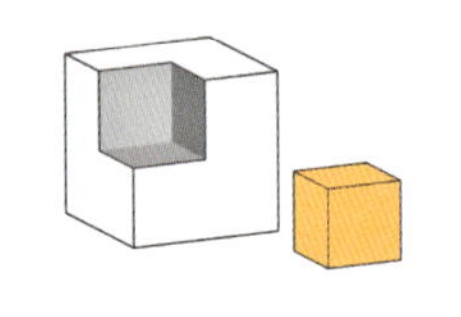

CARVING BLOCK TYPE 1

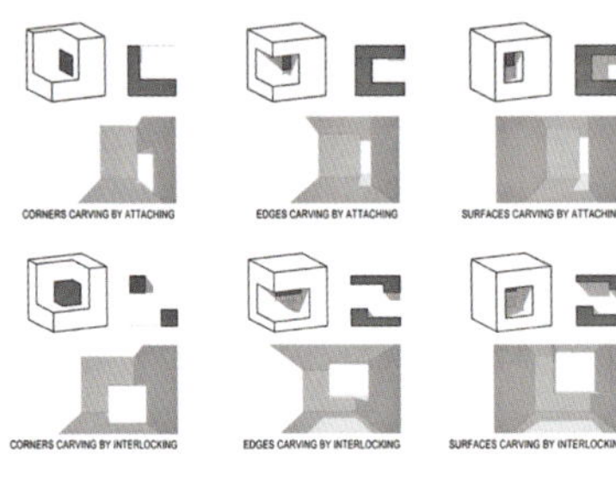

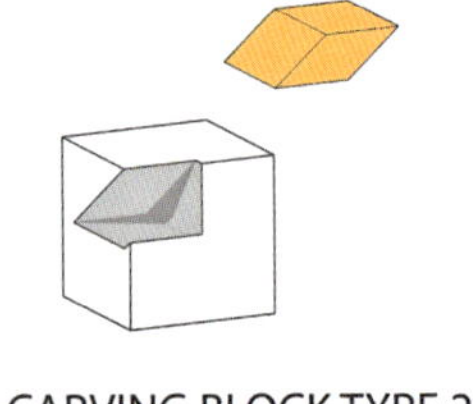

CARVING BLOCK TYPE 2

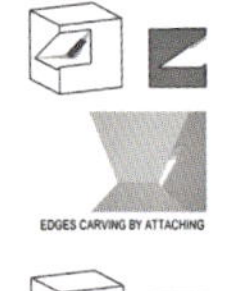
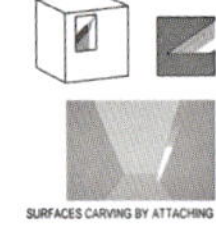
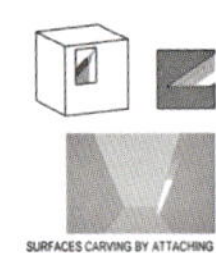
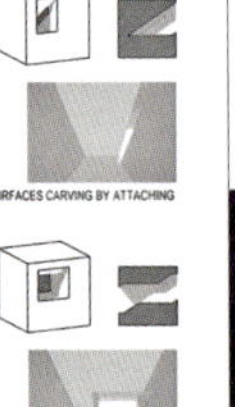

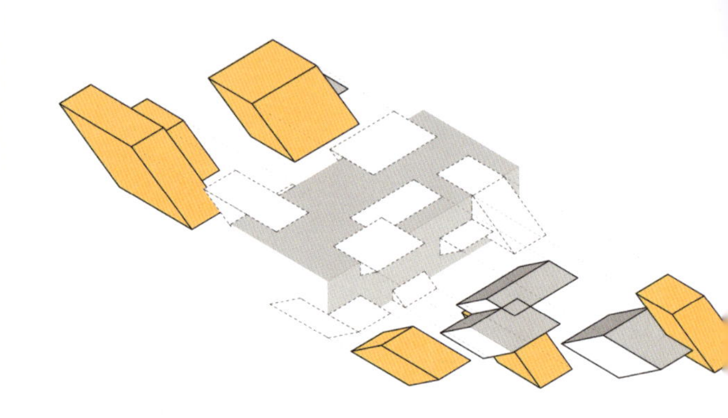

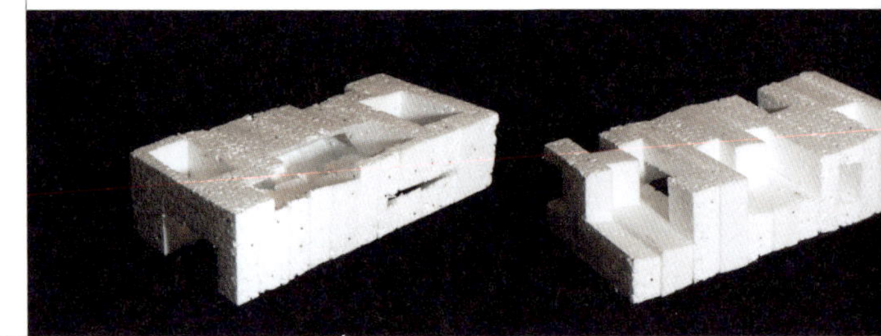

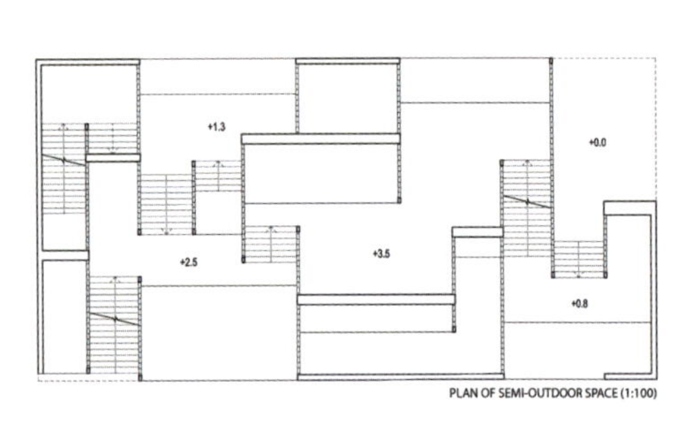

PLAN OF SEMI-OUTDOOR SPACE (1:100)

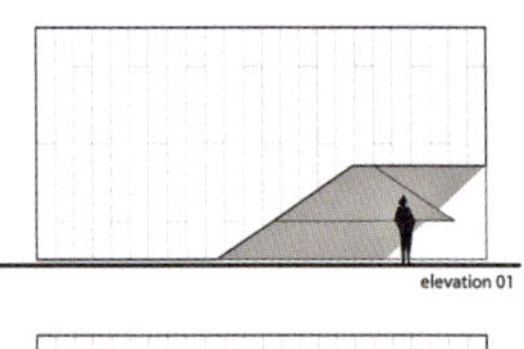

elevation 01

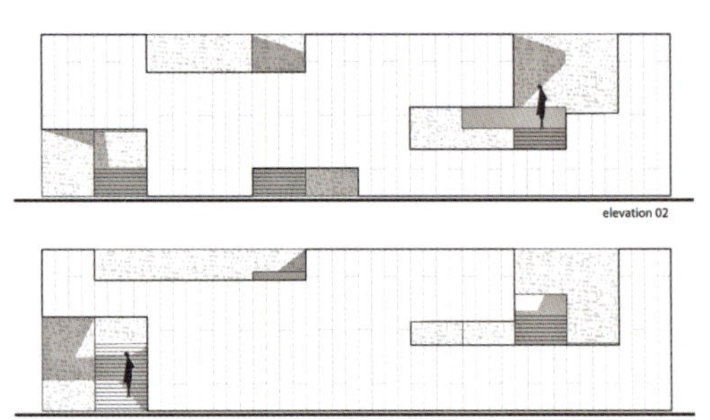

elevation 02

elevation 04

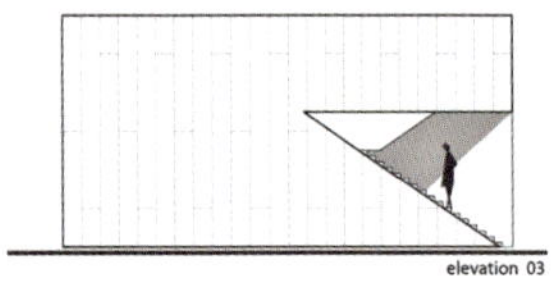

elevation 03

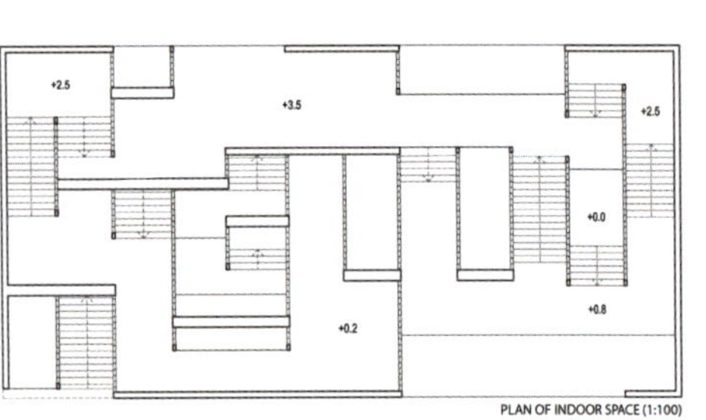

PLAN OF INDOOR SPACE (1:100)

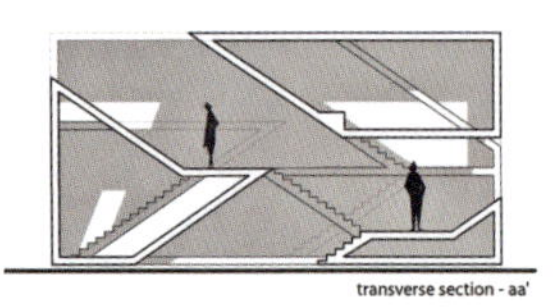

transverse section - aa'

longitutinal section - bb'

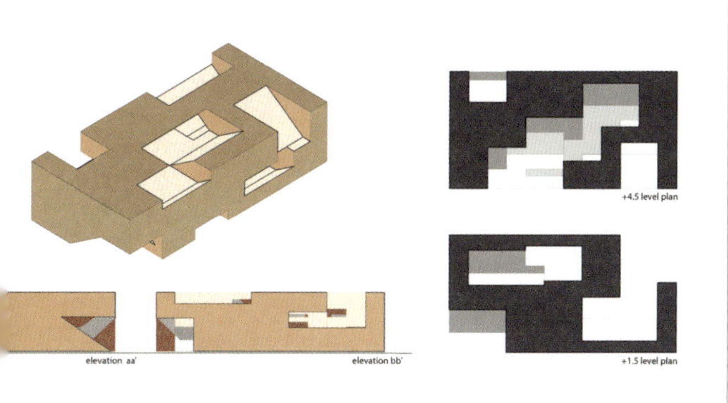

计算机辅助设计
the role of the computer

计算机辅助设计的作用与前面几种设计媒介不同，它将模型研究、空间体验和建筑图制作集于一身。对模型材料的操作，如切割、推拉、弯折等可以在SketchUp或ArchiCAD这类软件中完成。但是，电脑模型的操作不能取代实物模型的操作。实物模型的操作受特定的模型材料的物理特性的影响，而在电脑模型中，这种因素是不存在的。电脑模型在空间体验上具有实物模型不及的优点，即我们可以容易地在空间中走动，从不同的角度来观察空间。而对实物模型的观察则往往受到模型大小的限制，很难深入到空间中去。计算机辅助设计还在建筑图的制作、建筑表现和版面制作等方面给设计研究提供更多的方便和可能。

We use the computer as a complementary to working by hand. When the physical models reach a limit, we also construct computer models, and as we observe the physical models from eye level, we also set up eye level views on the computer to observe the computer model. The computer also helps to generate variations and therefore encourages exploration. And depending on the software used, the drawings can be derived from the models. In this way, the computer supports the development, study and documentation of the design. In addition, we intend to encourage the acquisition of good working habits: working in parallel with hand-made and computer generated models and drawings, encouraging the use of appropriate software for specific tasks and emphasising the importance of actively controlling the graphic quality of the output through careful settings and editing. The results should not show the brand of software used, but the quality of the work.

Cathy Wong Kai-fung

3 练习 | exercises

概念 concept
抽象 abstraction
材料 materiality
建造 construction
习作 work

建构作为一种工作方法 | tectonics as a method of work

这个设计练习系列是基于一个假说，即由不同类型的空间生成要素所产生的空间也应该具有不同的品质。为了要验证这个假说，我们研究三种要素类型：

1. 体块
2. 板片
3. 杆件

我们的设计研究方法是以模型为基础的。直接操作模型材料反映了一种建构设计的态度，即对模型材料的操作类比于用建筑材料来构筑建筑物。除了模型作业外，我们还采用电脑模型、透视图、建筑图，以及分析性图解等手段。

我们的基本工作方法是通过制作和观察的循环作业，将概念问题与感知问题联系在一起来研究。设计研究的过程分四个阶段：

1. 概念—操作与观察
2. 抽象—组织与体验
3. 材料—区分与诠释
4. 建造—概念与实现

在这个过程中，最核心的问题是从模型到建筑，从模型材料到建筑材料的转换。学生应该通过这些练习达到以下教学目的：

1. 如何产生一个建构和空间概念
2. 如何将一个建构和空间概念发展成一个建筑
3. 如何运用不同的设计媒介达到设计的目的

每个学生分配三种基本要素类型之一种做练习，如此一个班级有1/3的学生研究一种要素类型，三种类型的研究平行进行。这样，一个学生既可以从自己的研究中学习有关该要素的相关问题，又可以从其他同一类型的研究中了解不同的操作方法，此外，还可以从其他两类要素的研究中了解不同要素类型的操作和空间问题。

这些练习经过若干年的发展和演变，四个练习的结构基本未变，但是每个练习的内容和操作方法一直不断更新和修正。最近两年的一个重大的改变是将原先4周的练习扩展为一个学期，如此获得更多时间来发展这个练习系列。而且最后增加了一个总结的阶段。

This series of design exercises is based on the hypothesis that there is a fundamental difference in the properties of a space depending on the type of element used to define it. To test this hypothesis we work with the distinction of three types of elements:

1. Block
2. Slab
3. Stick

Our design research method is model-based. The direct manipulation of the model material is a reflection of a tectonic position, an analogy to the making of a building through the manipulation of building material. Besides physical models, we also work with CAD models, views, descriptive drawings and analytical diagrams.

As a working method we use cycles of making and observing, linking conceptual aspects with aspects of perception, over four phases:

1. Concept – operation and observation
2. Abstraction – organisation and experience
3. Materiality – differentiation and interpretation
4. Construction – concept and realisation

This makes the transformation from a model to a building, from model material to building material, a central issue.

The main objectives of the exercises are:

1. How to generate a tectonic concept
2. How to transform a conceptual form into a built form
3. How to use different media at various design stages

Each student is assigned one of the three element types so that each third of the class works with the same type in parallel. This allows students to learn in three different ways: from one's own exploration, from classmates working with the same element type and from classmates working with a different type.

The exercises evolved over years. While the basic structure of four phases is consistent, the content and operational procedure of each exercise has undergone constant modification and clarification. One of the major changes in the last two years was to extend the four-week long exercise to a whole term which makes it possible to deal with each step with more attention. A final conclusion phase is added to summarise the whole process. The presented sequence reflects the most current state of the programme.

阶段 | phase	1 概念 | concept	2 抽象 | abstraction
条件 | given	块体、板片、杆件 block, slab, stick	平房、板房、楼房 flat, thin, tall
主题 | theme	操作与观察 operation-observation	组织与体验 organisation-experience
模型 | model	 不同材质的模型材料 various material type	 1:100单一材料模型 1:100 single model material
透视 | view	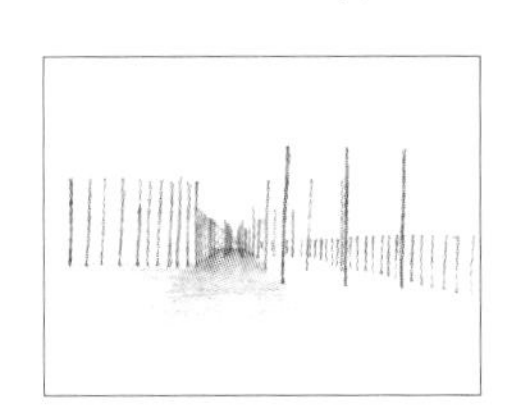 素描，尺度和光影 sketch, scale and light	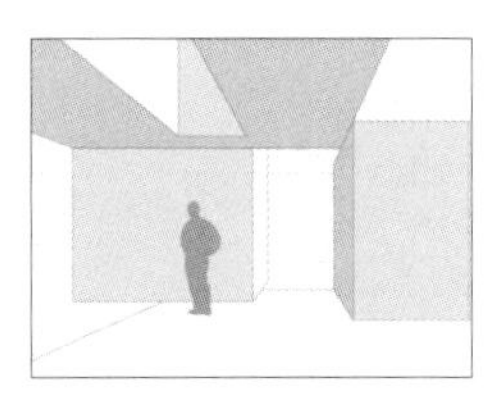 穿越空间的连续透视，尺度和光影 walkthrough views with light
建筑图 | drawing	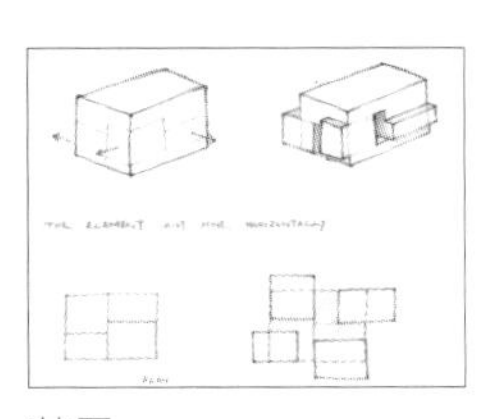 草图 sketch drawings	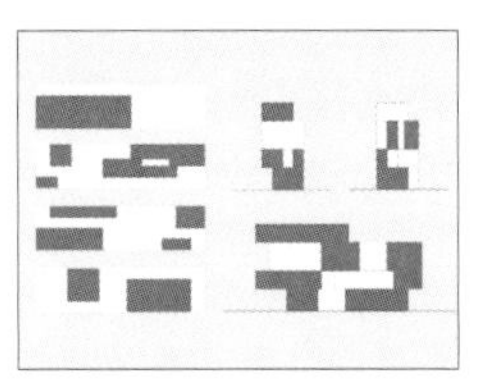 1:200平、立、剖面图， 表达抽象的要素和空间 1:200 plan, section and elevation expressing element and space

3 材料 | materiality

材质、色彩、透明性
texture, color, translucency

区分与诠释
differentiation-interpretation

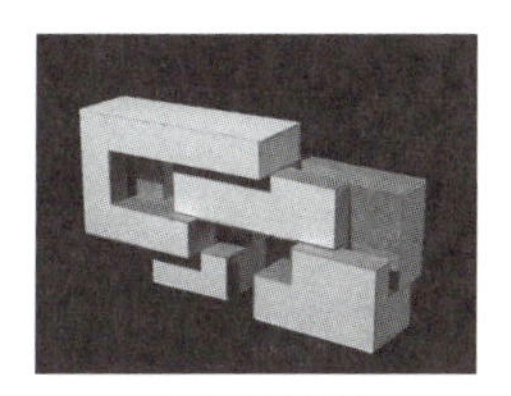

1:100多种材料模型
1:100 multiple model materials

同一视点、不同材料区分的比较研究
comparative views with materiality

1:200平、立、剖面图，
表达材质区分
1:200 plan, section and elevation
with material indication

4 建造 | construction

构件、拼接、层次
component, joint, layer

概念与实现
concept-realisation

1:20局部模型
1:20 partial model

建筑材料、建造和表达的拼贴研究
collage based on built examples

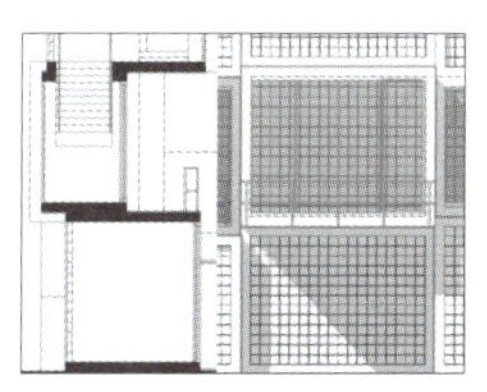

1:50局部立、剖面图，
表达建筑材料的构件和组织
1:50 partial section and elevation
indicating components and construction

5 总结 | summary

1:50 表现模型
1:50 model

内部空间和外部形式的表达
interior and exterior views

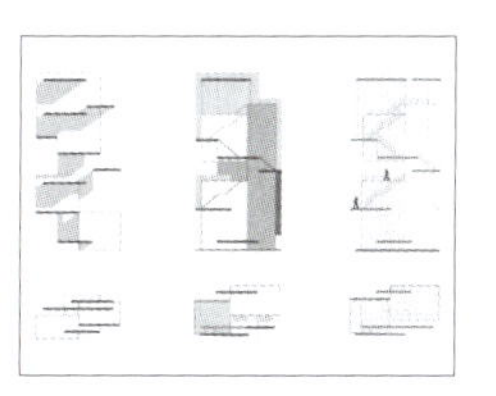

1:100平、立、剖面图，
表达建筑材料
1:100 plan, section and elevation
with building material indication

1 概念 | concept

用体块、板片和杆件其中的一种要素来研究如何通过对材料的操作来形成空间。练习的目的有二：一是建立一个建构设计研究的基本方法，二是探求一个建构的基本概念，引出后续的一系列练习。

根据形成建筑空间的要素的基本形式特征，我们可以定义三种要素类型，即体块、板片和杆件。体块的形式特征是一个大的体积，它的长宽高三边尺寸基本相当。板片的形式特征是一个平板，它的两个方向的尺寸要比另一个方向的尺寸要明显来的大。而杆件则是一个方向的尺寸要明显比其他两个方向的尺寸来的大。这一根据形式特征进行要素分类的方法决定了以下如何来研究空间和空间形成的方式之间的关系。在那些主要运用一种要素类型的建筑案例中我们可以清楚地看到要素类型和空间特征之间内在联系，即体块、板片和杆件这三种要素分别对应相应的空间类型。

根据这一理论假设，在这个练习中每个学生将指定用三种基本的要素类型之一做操作和观察的基本研究。模型研究的工作方法决定了我们不可能在抽象要素的层面上来讨论空间生成的问题。我们必须根据这三种要素类型来选取模型材料。体块材料可以是木块、泡沫塑料或雕塑泥。板片材料可以是纸板、三合板或纸片。杆件材料可以是木杆、铁丝或竹竿。因此，在练习中我们所面对的不是抽象的空间限定要素，而是特定的模型材料。不同的模型材料的物理属性决定了我们如何来加工它们。

建构工作方法的最基本核心是操作和观察两者之间的互动过程。一种由特定的模型材料所激发的操作方法，一方面取决于材料本身的特性，另一方面取决于如何用它来限定和组织空间。对材料的操作和对操作结果的空间的观察应该同时进行。观察的要旨是从人眼的高度和透视的角度观察空间的内部。借助于草图来辅助观察，在作图时特别要注重对空间的基本特征的认识，以及对光和影的描述。

这是一个相对短暂的练习，步骤的划分实际上并没有什么意义。

Kirchner Museum, Davos, Gigon and Guyer, 1992   Barcelona Pavillion, Ludwig Mies van der Rohe, 1929   Cathedral–Mosque of Córdoba, 10th century

体块
block

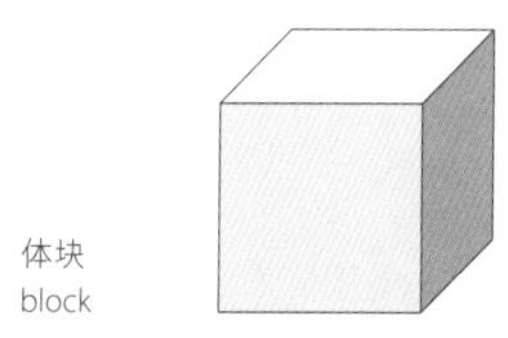

板片
slab

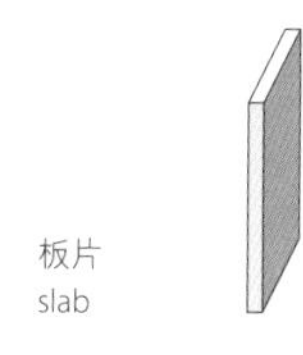

杆件
stick

In this first phase, we study how to generate space through the operation on one of the three given elements – block, slab or stick. The objectives are twofold: to establish a basic working method of tectonic study and to search for a tectonic concept for the following development.

The distinction of three primary elements which define architectural space is based on their maximised difference in occupying space. A block extends in three directions, forming a mass and a continuous surface. A slab extends in two directions, forming two distinct surfaces and a continuous edge. A stick extends in one direction, forming one edge but no surface. This provides us with a point of view from which we can study the relationship between the perceived space, the defining elements and the way of working with them. The uniqueness of the spatial character becomes visible in buildings which use predominantly one type of space defining element.

Based on the hypothesis of a relationship between types of space defining elements and types of space, each student will be assigned to one of the three basic elements for a basic exercise on operation and observation. In this exercise, we adopt model materials which correspond to these three element types. A block can be made of wood, styrofoam or clay. A slab can be made of paper, cardboard or plywood. A stick can be made of wood, bamboo or wire. The physical properties of the model material determine the method of operation.

The proposed fundamental working method is a reciprocal process of operation and observation. The operation is the direct manipulation of the model material. Observed are two different issues. One is how a manipulation creates or differentiates space. The other is the characteristics of this space. Sketching and taking photos can support the observation and at the same time document it. The preferred form for the two are different: axonometric views for the manipulation and eye-level perspective views for the second. The observation of surfaces and edges in light and shadow can help to capture the spatial characteristics.

This is a relatively short exercise. The listed three steps should be treated as the three aspects of one action.

操作与观察 | operation and observation

1.1 要素和操作

每个学生被指定一种要素类型。作业的目的在于尝试对指定的要素类型和选择的材料进行空间生成的操作，同时研究所生成的空间的特点与要素类型相关。

- 模型材料的选择以方便操作的研究为基本原则，如泡沫塑料块、卡纸、纸板、木杆或粗铁丝等。
- 首先要关注基本的工作方法。你先从材料的操作开始，再从人眼的高度对所生成的空间进行观察，用素描方法来描述空间，用相机对操作过程和结果作纪录，绘制平、立、剖面图。
- 在掌握基本的工作方法后，你就要将着眼点放在寻求建构构思上，即一种明确的操作方法及其相应的空间。

1.2 空间和观察

虽然练习从对模型材料的操作开始，但是对空间的观察必须紧随其后，两者是一个过程的两个不可分割的方面。

- 对空间的观察必须要从人眼的高度来进行，模型不能太小，在观察时尽量把视点放底。
- 虽然可以用照相机来观察和纪录空间，但是不能代替素描的研究。素描的过程包含了对空间基本特征的抽象。
- 在素描的过程中着重对光影的描述，加强对空间的体验。
- 空间尺度的问题可以以照片剪贴的方式来研究。

1.3 操作和观察的互动

操作和观察的互动推进研究的进行。我们特别鼓励对要素和空间关系进行系统的和基本的研究，尤其不要太关注最后的结果必须是一个完成品，还要避免先入为主的想法。

- 把模型或系列研究模型固定在硬纸板上。
- 空间观察素描以及平、立、剖面图。
- 有关操作方法的纪录，可以是相片，也可以是图解。
- 根据研究的过程来组织以上的各种成果。

1.1 element and operation

With the given element type, we try to find out how an operation on the element creates space, and what properties that space has as a result of this specific operation.

- The selection of model material should allow and encourage direct manipulation, such as styrofoam > carve, thick paper > cut and fold, wire > bend, etc.
- To practise the method of operation, observation and documentation.
- After an initial intuitive practise of the method, the focus can shift to the search for a tectonic concept – a particular way of operation which creates a particular type of space.

1.2 space and observation

The exercise is initiated by the operation, but the observation should immediately follow. The two are inseparable aspects of one process.

- To observe the model from eye level in order to perceive the space in the model similar to a real space. For this purpose, the model cannot be too small.
- To support the perception of space by taking pictures for the record and drawing sketches for clarification and abstraction.
- To articulate light and shadow in the sketch to express the visual experience.
- To indicate scale through the insertion of figures in photomontages.

1.3 a reciprocal process

Operation and observation work as a reciprocal process. We specially encourage systematic and elementary study to allow discovery, and discourage intentional design.

- Model or series of models fixed to a stiff board.
- Perspective sketches, plans, sections and axonometric views can be drawn in any media freehand.
- The operational method can be recorded either with a digital camera or through sketch diagrams.
- The output should be arranged and presented to show the process of exploration and discovery.

2 抽象 | abstraction

根据前一个练习所得到的建构和空间概念，用单一的模型材料来设计一个非功能性的建筑单体。研究的重点是抽象要素与空间的关系，以及空间的组织与体验。

我们将单体的基本尺寸限定为6mx12mx24m，即比例1:2:4。该矩形体块不同的放置方式产生平房、板房和楼房三种类型。尽管单体的几何尺寸一样，但是它的放置方式产生不同的设计问题。每个学生被指定平房、板房和楼房中的一种。这样，三种要素类型和三种建筑类型的组合形成若干新的设计条件。

该单体并没有特定的使用功能，或者说只有抽象的功能，如应该包含大小不同的空间，这些空间由一条以人的体验和光线控制为线索的路径来组织。垂直交通方式（如楼梯）不一定要在模型中做出来，但是应该是可以说明的。结构的合理性主要体现于完成的模型是否可以独立站立。

单一模型材料的条件限制保证了这一阶段的研究的抽象性，即着重于抽象的要素和空间的关系。接续前一阶段对要素和空间之间的关系的初步感性认识，我们要具体探讨体块、板片和杆件这三种基本的要素及其所生成的空间之间的关系，引入专门的术语来定义不同的空间属性。对要素和空间的问题，不但能够感知，还要能够在讨论中清楚地用语言来描述。

与前一个阶段的要求不同，这里的主要挑战是如何通过对特定模型材料的操作来达到抽象形式的清晰表达。我们提出一个设计的原则，即操作语言的清晰性导致形式语言的清晰性。一般来说，学生在开始阶段的操作倾向于杂乱和盲目，解决的办法是分辨自己在设计中运用了哪些操作方法，再从中确定一种主要的操作方法，并将其一以贯之。

设计的结果如何，需要从空间的体验来检验。设计的另一个目标是简单的要素操作导致丰富的空间体验。简单和清晰的操作并不意味着形式和空间的单一和单调，这是练习的困难之处。一味地简化操作并不能产生一个好的设计。丰富的空间体验主要取决于空间序列上空间之间的对比和变化，如空间之大小、形状、比例、方向、视线的位移、射入空间的光线的变化，等等因素。

Weekend house, Ryue Nishuzawa, 1998

Tokyo Kasai Rinkai Park Visitor Center, Yoshio Taniguchi, 1995

Kalkriese Archaeological Museum, Gigon & Guyer, 2001

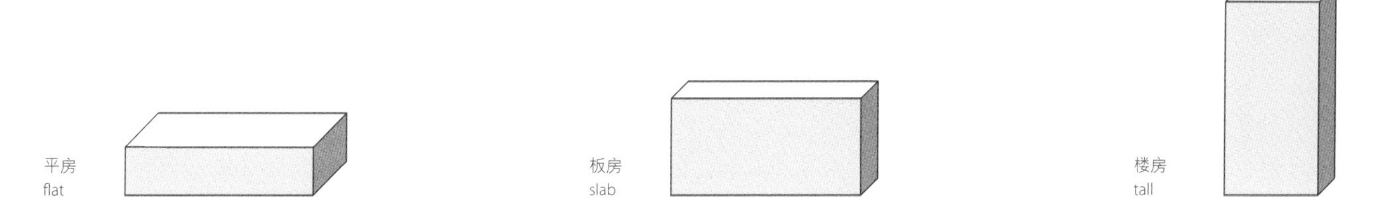

In this phase we design a building-like object with a neutral programme, based on the tectonic concept from the previous phase, using a single model material. The focus of the study is on the relationship between the definition, the organisation and the experience of space.

We work with a given overall form, a cuboid of the proportions 1:2:4 and the measurements of 6x12x24m. We make a further differentiation in how this form is placed in relationship to the ground plane. This differentiation of the volume into flat, thin and tall creates three different spatial conditions which will affect the perception of space even for the same manipulation of the elements. Each student will be assigned to one of these three objects. In the combination of element types with object types, nine different design conditions are generated.

There is no specific programme for these objects, except an abstract programme of spaces of various sizes linked by a path and lit by natural light. There is no requirement to articulate the vertical circulation as long as it can be explained through spatial connections. The requirement of structural stability is satisfied if the model can stand by itself without additional support.

Working with a single model material supports the focus on the abstract relationships between space defining elements and perceived space. Expanding the ability to articulate the perception of space from the previous graphical description, we also try to develop verbal descriptions and include the organisation and experience of the space along a path.

In the making of the object, one of the main objectives is to establish an order in the organisation of elements and space. This is determined by the clarity of the operation. As a principle, we should avoid employing too many different principles of operation in a single object. This order of organisation not only affects the built elements but also the perceived space.

Another challenge in this study is to achieve a rich spatial experience from a simple operation. This spatial experience is presented as a spatial sequence from entering the object to passing through it in both horizontal and vertical directions. The richness of experience depends on the contrast of spaces in size, direction, geometric form, light conditions, and relationship such as separated, adjacent, overlapping, and within.

组织与体验 | organisation and experience

2.1 模型，1:100，单一材料

根据前一阶段的建构概念来制作指定的平房、板房和楼房之一的单体模型。通过系列模型的制作来研究如何借助于对特定材料的操作来建立要素和空间的结构秩序。

- 限定于单一的模型材料。所选择的材料应该能够支持相应的操作方法，比如，用吹塑块来切割，用硬卡纸来折叠。
- 明确材料操作的方法，并将其贯穿模型制作的始终。
- 完成的模型能够体现特定操作方法，可以观察辨识，可以用语言描述。
- 所有的模型为1:100，没有底板固定，能独自竖立。

2.2 连续透视

根据完成的单体模型来建电脑模型，并借助电脑模型来研究空间的感知和体验。在实物模型和电脑模型、制作和观察之间转换，推动设计的发展。重点研究如何在保持简单操作的前提下实现空间体验的丰富性。

- 用ArchiCAD或SketchUp来建模，用相同的操作方法。
- 以系列透视图的方法来研究穿越单体内部时的空间体验，辨识各个空间的基本特点。
- 以观察的结论作为继续实物或电脑模型研究的依据，直至实现简单操作和丰富体验的设计目标。
- 电脑模型不附设材料，利用光影来增强空间知觉。

2.3 建筑图，1:200

根据最终完成的实物模型和电脑模型来绘制建筑的平、立、剖面图。同时，制作这个阶段的最后成果，包括模型和透视图。

- 建筑图比例为1:200。强调基本要素的表达，在立面中表现光影。
- 不论前面的工作模型是用什么材料来完成，最后的实物模型必须统一用木材。比例为1:100，没有底板。
- 一系列的人眼高度透视图，表达穿越空间的体验，线描附加光影，加入人体以显示空间尺度。
- 可以用图解来表达操作的概念或设计的过程。

2.1 model 1:100 single material

Based on the tectonic concept explored in the previous phase, we develop a spatial complex in the form of a flat, thin or tall object. We use working models to explore how a principle of operation can help to establish order for elements and space.

- Limitation to a single model material, which should support the operation, for example carved out from a styrofoam block or folding a stiff thin cardboard.
- Clarification of the principle of operation through the model-making process.
- The method of operation should be recognisable through observation and describable in words.
- Models in the scale of 1:100 as freestanding stable objects.

2.2 sequential views

Based on the working models, we build CAD models to study the perception and experience of space. We develop the design by shifting between physical and CAD models, and alternating between making and observation.

- Build ArchiCAD or SketchUp models using the same operational principles.
- Use the CAD models to identify the basic characteristics of main spaces, to set up a series of views and to study the experience of a spatial sequence.
- Feed the result of observations back into the model studies and try to achieve a rich experience with a simple operation.
- No materials in CAD models. Use light, shade and shadows to articulate space.

2.3 drawings 1:200

Finally, we record the design in plans, sections, elevations and axonometric views. We rework the models and edit the views for the final production.

- Descriptive drawings in the scale of 1:200 emphasise the expression of basic elements and add shadows in elevations.
- The final physical model should be made of wood, in the scale of 1:100, as a freestanding object, stable in itself, with no need for a base.
- A series of eye level views as line drawings with shadows and human figures.
- Use diagrams to demonstrate the tectonic concept and the operational strategy.

3 材料 | materiality

前一阶段的研究限定于单一模型材料，使得我们把研究的焦点放在基本的要素和空间秩序方面，现在我们将引入多种模型材料，通过区分和诠释的方式来建立更加复杂的形式和空间秩序。具体的任务是用两至三种模型材料重新制作1:100的模型，练习的重点在于探讨材质因素的介入而引起的结构、围合和空间的表达问题。

引入多种模型材料的意义在于不同材料的对比和区分，并不暗示将来真实的建筑材料，即采用木材并不意味着将来的建筑材料也是木材。我们主要是从视觉的角度来考虑模型材料的三个特性：即材质肌理，如木和金属材料的对比；色彩和明暗，如各种颜色的纸板的对比；材料透明性，如透明、半透明、不透明材料的对比。材料的透明性对空间知觉有重要影响。

从前一个练习的单一模型材料发展为现在的几种模型材料，需要对原先的形式和空间关系作新的诠释。这里，关键词是区分，即在原先单一材料的要素之间作出区分。区分意味着差异，意味着对比，意味着重新诠释原先的设计，意味着在原先的秩序基础之上建立新的要素和空间之间的秩序。当然，区分和诠释并不是随意的，它实际上是在抽象形式和空间的表达基础上寻求更丰富的表达内容，其依据在于对结构、空间和使用等方面的考虑。比如在室内和室外空间之间增加气候边界，给予空间进一步的界定，在结构要素和空间要素之间作出区分，在主要结构和次要结构之间作出区分，以及在不同用途的空间之间如公共和私密空间作出区分，等等。

此外，三种基本的空间限定要素在这个阶段的问题可能略有不同，但是同样面临如何在引入多种材料后保持或加强原先的构思的问题。

在运用材料区分对原先的设计作重新诠释的过程中，一个重要的手段是用透视图来观察材料的区分对空间知觉的影响。通常对结构、围合和使用的诠释存在多种的可能性，设计的基本原则是加强而不是弱化最初的构思。因此，在设计的过程中就需要对不同的诠释的结果进行检验，即同一视点的不同要素和空间诠释的比较研究。我们可以借助模型照片或电脑模型来做该研究。

多种模型材料的研究最终也反映在建筑图的制作上，我们要求平、立和剖面图也要表达材料的区分。

Kirchner Museum, Gigon & Guyer, 1992

Baden Vocational School, Burkard Meyer, 2006

Hotel Zürichberg, Burkhalter & Sumi, 1995

Based on the clarified operation and spatial organisation from the last phase, we continue to rework and develop the object as a model 1:100 using two or three different model materials. This allows us to differentiate and interpret the order of elements and space. The distinction of structure and enclosure, and the articulation of use can be supported by material differentiation. We further introduce the idea of a climatic border which separates inside from outside space.

We mainly distinguish three material properties of the model material: texture, colour and translucency. The differentiation can range from subtle differences to strong contrasts of one or more properties. The strong material expression of the material as a model material is emphasised. The model material has no meaning regarding actual building material. This allows the use of any kind of suitable material: not only different types of paper, cardboard, wood and metal, but also wax, soap, foams, etc.

The consideration of new aspects always has a danger of weakening existing concepts. We therefore pay attention to the development of the design, trying to support the tectonic concept from the previous phases, differentiating it and hopefully even strengthening it, when considering structure, enclosure, use and climatic border.

Through this further articulation, we explore how material properties and differences affect the perception of space. It also provides an additional chance to find out more about the specific character of the three different space defining elements. We distinguish between interpretation, which adds a layer to the spatial order by material differentiation, and variation, which shows the same interpretation with different materials.

The observation of the expression of the object and the perception of the space uses comparison through parallelism: three models, three axonometric views and three perspective views. The media are physical models, computer models, photographic views and computer views.

Finally we want to express the material differentiation also through the graphic representation of plans, sections and elevations.

区分与诠释 | differentiation and interpretation

3.1 模型，1:100，多种材料

制作若干个多种材料模型作为对前一个单一材料模型的诠释。快速地产生不同的方案来探讨结构、围合和功能表达的种种可能性。这项研究最终导致新的要素和空间秩序的建立。

- 采用各种不同特性的材料如卡纸、木材、金属材料、玻璃、塑料、腊、黏土、丝网等。
- 每一个模型限于两到三种模型材料，作为气候边界的透明玻璃除外。
- 单体范围内60％的部分必须是室内空间，其余为外部空间。内外部以气候边界相隔。
- 在研究时特别注意这些诠释是不是加强了原先的建构构思，还是引出新的构思。

3.2 空间比较观察

电脑模型的研究辅助实物模型的研究。电脑模型不仅可以方便地生成不同的方案，而且还可以用作透视观察的比较研究。

- 电脑模型中的材料保持模型材料的特性，不要采用建模软件中的建筑材料图库。
- 选择一个最能反映建构构思的典型空间角度，从人眼的高度来研究材料变化对空间知觉的影响。
- 生成至少三个相同视点的透视来比较不同诠释。
- 每个诠释的透视辅助以一个轴测图。

3.3 建筑图，1:200

在练习的最后阶段，你需要同时在实物模型、电脑模型、透视观察，以及建筑图几方面工作。建筑图需要重新绘制，以反映有关材料区分的研究。

- 平、立和剖面图的比例为1:200。
- 建筑图要表达气候边界，以及垂直交通方式如楼梯或坡道。
- 在未知具体建筑材料的条件下，墙体厚度以统一的线宽来表示。
- 以线条等级和不同的线型来表示材料的区分。

3.1 model 1:100 multi-material

We make several models, each an interpretation of the previous single material model. Consider the articulation of structure, enclosure and use. Explore how this modifies your concept, differentiates the order, and influences expression and perception.

- Choose from a wide variety of materials such as cardboard, wood, metal, glass, plastic, wax, soap, clay, mesh, foam, etc.
- Use only two or three materials for one interpretation, apart from the transparent material for an additional climatic border.
- About 60% of the spaces within the object should be inside spaces, the rest being outside spaces.
- Pay special attention to whether these interpretations strengthen the original tectonic concept or lead to new ones.

3.2 comparative views

We complement the physical model with computer models. This facilitates the exploration of interpretations and variations, and the use of parallelism for the comparison of axonometric and perspective views. Compare at least three versions.

- In the computer model we also work with model material, not building material. Do not use the building material texture library provided by modelling software.
- For the parallel comparative views, use the same eye-level perspective and the same axonometric view setup.
- Explore the influence of the material on the perception of the space in the perspective views and the change of expression based on material differentiation in the axonometric views.

3.3 drawings 1:200

As in the previous phases, as a last step, you should rework the physical models, the computer models, the views and the descriptive drawings. In particular, the descriptive drawings should reflect the issue of material differentiation.

- Plans, sections, elevations and axonometric views 1:200.
- The drawings should show the climatic border and the means of vertical circulation such as stairs, ramps or a lift.
- Without knowing the actual building material, we can only draw the walls of a unified thickness.
- Use different line weights and line types and maybe fill tones to express material differentiations.

4 建造 | construction

这个设计研究的最后阶段的重点是从模型材料到建筑材料的转换，即如何通过建造的手段来实现建构构思。这里所说的建造并非实际去建造的活动，而是从建造的角度来研究设计的意图和实现之间的关系，即从构思形式向建造形式的转换。具体的作业包括通过一个透视和立面局部的拼贴练习来研究从抽象材质到建造材料转换的可能性，在此基础上再制作一个1:20的局部模型来探讨建造与建构表达的问题，最终是局部的立面和剖面图的制作。

关于建筑材料我们引入三个基本的概念，即构件、拼接和层次。这里我们要特别区分构件与要素之间的关系。一个要素如板在模型材料中可以是一块硬纸板，而在建造材料中它必须有若干的构件组合而成。反之，砖作为建筑材料，可以被认为是块体，但是砖不但可以建造出块的实体，也可以建造出作为板的墙和作为杆的柱。而拼接和层次正是构件组织的两种基本关系，拼接指的是构件并置的关系，层次是纵向的层叠关系。

从抽象材质到建造材料的转换是通过一个拼贴练习来完成的，拼贴的素材来自于那些最接近于前述材料研究的已建成建筑的图片，再进而通过对相关的建筑剖面图来学习建筑材料的组织方式。学生不是直接将这些知识照搬到自己的设计中，而是将此作为设计的参照，进行必要的减化和转换。

这项工作主要在1:20的模型上来完成。模型选取一个典型的局部，研究的重点放在建构表达方面，而不是构造细部，即研究建造的不同处理方式如何支持建构意图的充分表达，主要体现在构件在横向和纵向的组织关系，即拼接和层次两种关系。因此，我们并不强调模型制作中材料的真实性。就是说，没有必要用真正的混凝土来制作混凝土墙体。这里，我们利用模型材料的象征性来表达建筑材料。

最后，有关的研究成果以局部剖面和立面的方式纪录下来，图的比例为1:50。这组图实际上综合了前述的两项工作的内容。剖面图纪录建筑材料的组织，立面图表现了材料和建造的表达。

Villa Savoye, Le Corbusier, 1931

Barcelona Pavillion, Ludwig Mies van der Rohe, 1929

Jacobs House, Frank Lloyd Wright, 1937

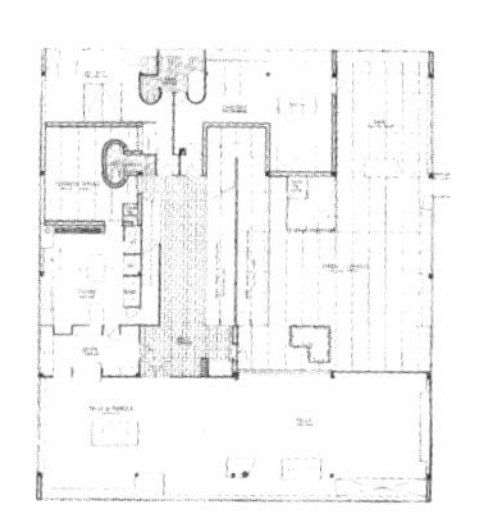

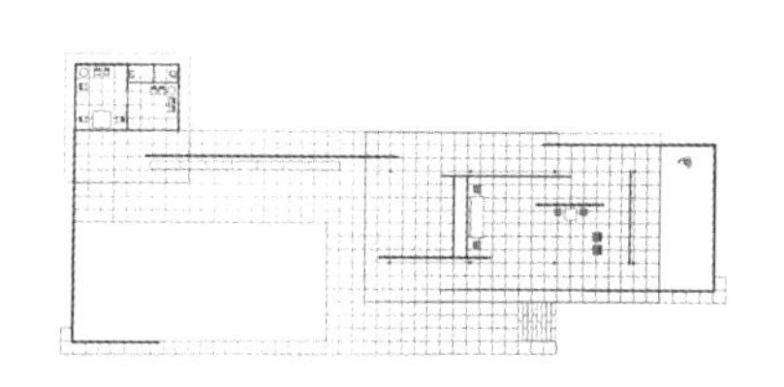

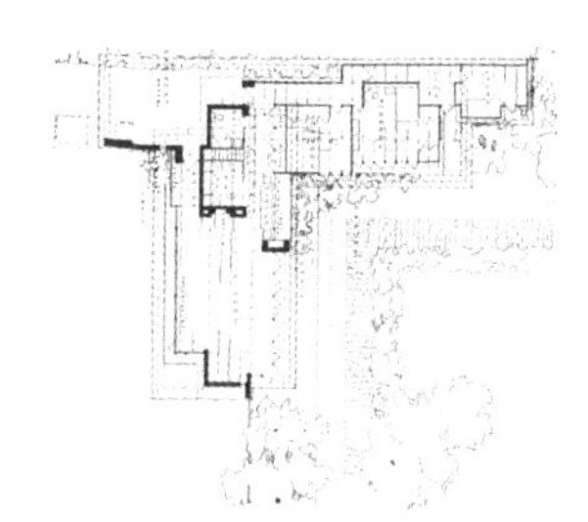

In this phase we try to deal with the transformation from working with model material to working with building material. The primary challenge in this study is to find a proposal for construction and at the same time to retain or even strengthen the tectonic concept developed until now.

We work with three basic concepts regarding building material: component, joint and layer. Component refers to the hierarchical nature of construction elements. For instance, a plywood panel as a component has dimension limitation and might be composed of several sub-components. To make a wall, these plywood panels have to be further combined together. We use two keywords, joint and layer, to distinguish the organisation of components in two directions, lateral and in depth.

The transition from model material to building material is not trivial. We approach this issue from two directions, expression and construction, and in three media, collage, model, and drawing. In a collage we try to apply existing expressions. In a model we try to construct with components. And in a drawing we try to link construction with expression.

For the collage, a partial elevation or a perspective view is outlined based on the last phase. Cut-outs from views of existing buildings are copied or scanned from books or magazines and then used to render the outline. As the surfaces are taken from existing buildings, they not only represent material, but also contain information about construction aspects, like subdivision in components, substructures, joints and layers.

The model is a partial model in the scale of 1:20 of a representative part of the model. As the model material now indicates building material, and to emphasise construction, we recommend the use of only one model material. The model itself is constructed with components and should show their order, lateral joints and the layering in depth.

The drawing in the scale of 1:50 of a partial section with a corresponding partial elevation links the two previous steps. The section clarifies graphically the relationship of components, and the elevation checks the expression of material and construction.

构思与实现 | concept and realisation

4.1 照相拼贴研究

我们以照片剪贴的方式来开始建筑材料表达的研究。一片板片可以以不同的建筑材料以不同的建造方式来建造。板片的表达不仅与材料有关、也与建造的方式有关。

- 在先前的建筑图中选取一个局部立面，放大到1:50，以及一个内部空间透视图。
- 在建筑杂志和书籍中寻找与自己的设计构思相近的以建成建筑的图片，把它们用在自己的立面和透视图中。
- 同时收集有关那些以建成建筑的剖面资料。
- 可以用两种方法来做照片剪贴，用图片、剪刀和胶水，或者用电脑软件。如用电脑软件，切勿用软件中的建筑材料图库来做拼贴。

4.2 局部模型，1:20

我们通过一个局部模型来研究空间限定构件如何转换成建造构件。研究的重点不是构件的节点构造，而是与构件的连接的层次有关的位置、比例和形式问题。

- 在1:100模型上选取一个最能代表材料区分以及包含立面和空间的部分。
- 参照照片剪贴。
- 在这个模型中模型材料的作用在于象征或提示所选择的建筑材料。如果可能，尽量采用单一模型材料，如纸板。
- 局部模型的比例1:20，放置在A3底板上。

4.3 局部剖、立面，1:50

练习的最后，根据照相拼贴和局部模型的研究来绘制局部剖面和立面图。

- 以1:20的局部模型为参照
- 图中清楚表达构件的相互关系、如连接和层次、尺寸和比例关系等。
- 剖面图要求表达构件关系，而不是节点大样。
- 立面表达光影和材质。
- 比例：1:50

4.1 photomontage study

We use the technique of collage to approach the issue of expression of construction and material. For example, instead of an abstract slab we show an existing wall which can be constructed in different ways from different types of material.

- Choose a representative part of a previous elevation enlarged to 1:50 and an eye-level interior perspective as a base.
- Copy appropriate examples from published buildings and apply them to your elevation and view.
- Keep records of detailed construction sections related to the images used.
- Use paper, scissor and glue to make the collage or use a computer. When using a computer, don't use any textures provided by the software.

4.2 partial model 1:20

We build a partial model to study how space defining elements can be constructed from components. We are not studying the technical aspects; instead we focus on the position, proportion, and form of components with their joints and layers.

- Select an area of the 1:100 model which is representative of the material differentiation and includes part of the elevation and part of a space.
- Refer to the photomontage.
- The model material used in this model is symbolic, representing or pointing to the building material chosen. Use only one model material if possible, like cardboard or wood.
- Fix the model on a stiff A3 sized base.

4.3 partial drawings 1:50

As a final step, we draw a partial section and elevation based on the construction study with the partial model, and the expression study with the collage.

- Refer to the 1:20 model and the collage.
- The relationships of components to each other, layers and joints, size, and proportion should be clearly readable in the drawing.
- The detailed section should show the principle of organisation rather than a technical implementation.
- The elevation should include shadows and indicate material textures.
- Lay it out on a A3 sized sheet.

5 习作 | work

过程与成果 | process and products

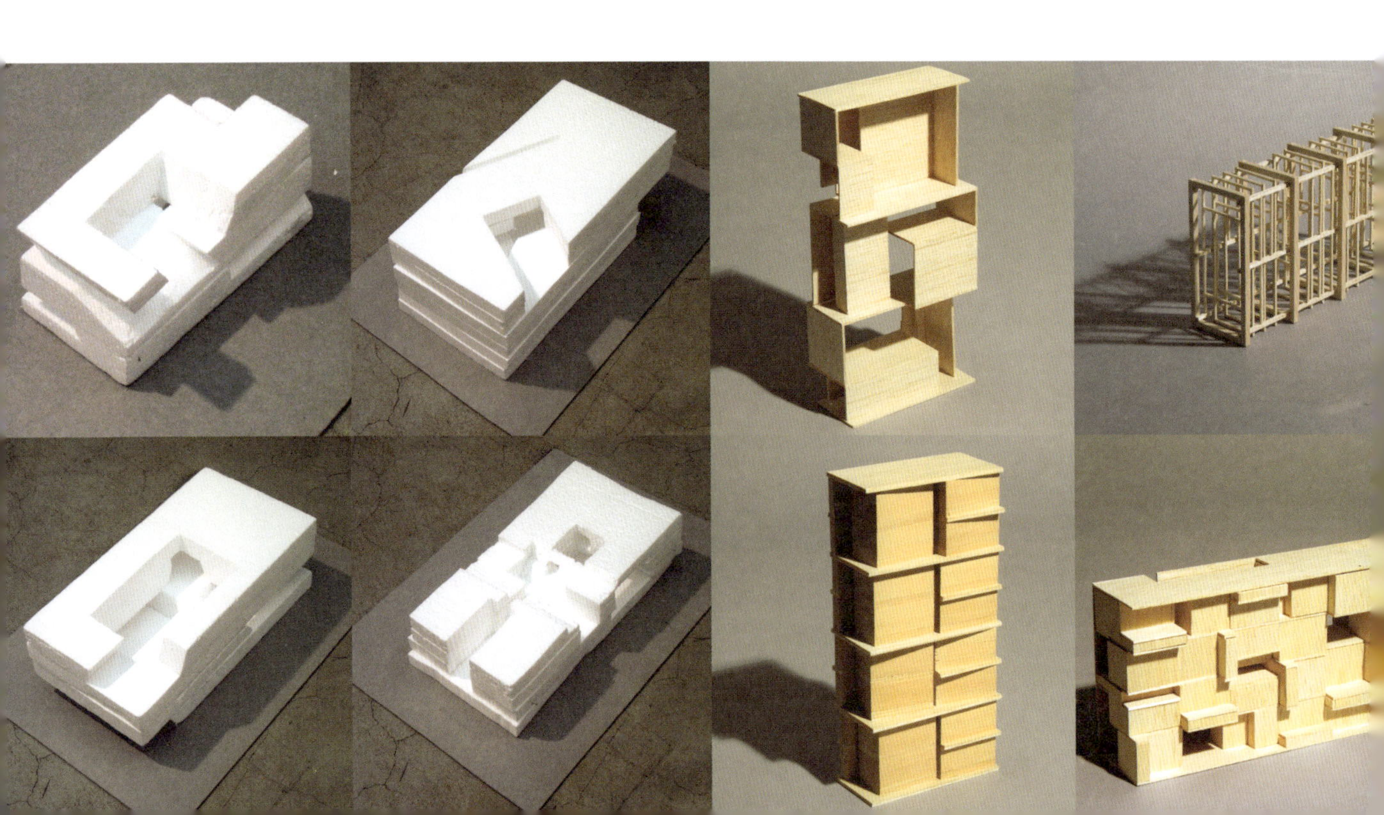

以下通过几个学生的作业来说明练习的整个过程和相关的成果。需要说明的是完成这部分练习的时间有两个版本：一个是在5周内完成的版本，另一个是在一整学期（13周）内完成的版本。这里展示的是后一个版本。两个版本的区别，其中时间长的练习，除了给于每个练习步骤予更充裕的时间来发展外，还增添了三项新的任务，使得整个过程更加完整：

1. 一个1:50的模型
2. 1:100的平、立和剖面图
3. 建筑内部空间和外部形体的表现图

On the following pages, the process and the products of the exercises are illustrated with selected student projects. When the programme was expanded from five weeks to the whole term, a last phase was added to the process, as more time became available. The final presentation always consists of the process and results of all four previous phases. In this new phase three representations emphasising the result or product were added:

1. Model 1:50
2. Plans, sections and elevations 1:100
3. Key views of interior space and exterior form

体块 · 平房 | block • flat

Chan Chi-hong, 0708T1Y2

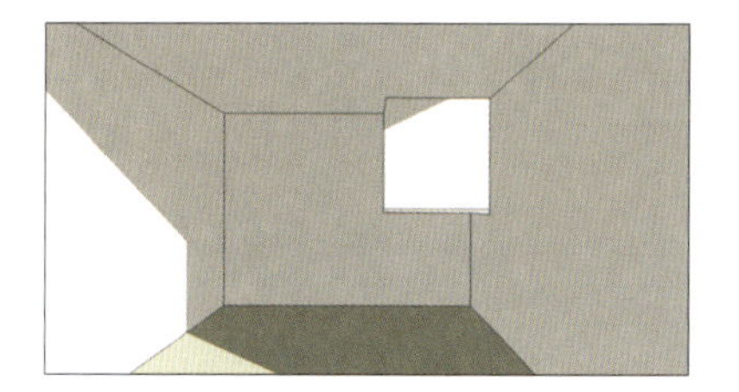
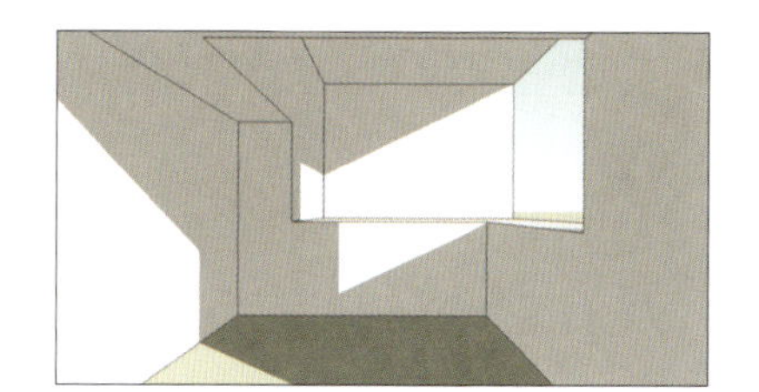
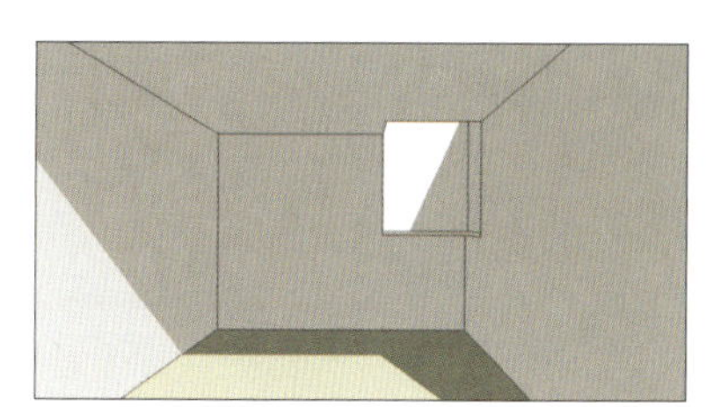
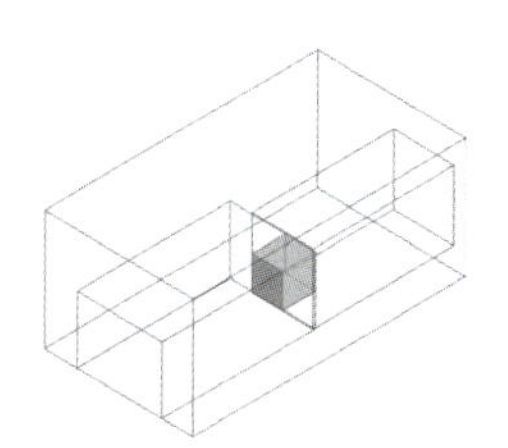
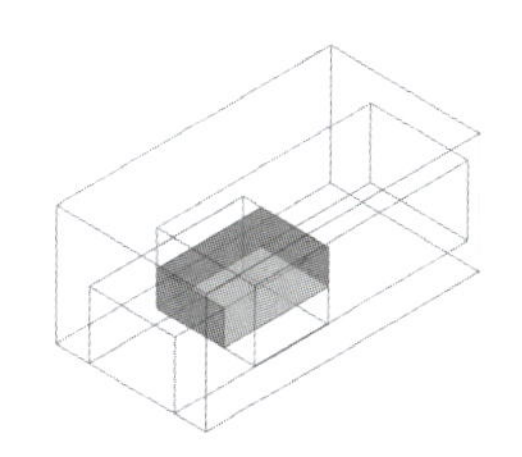
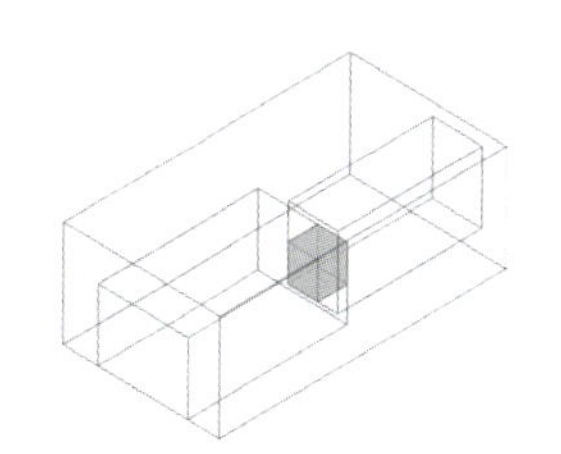
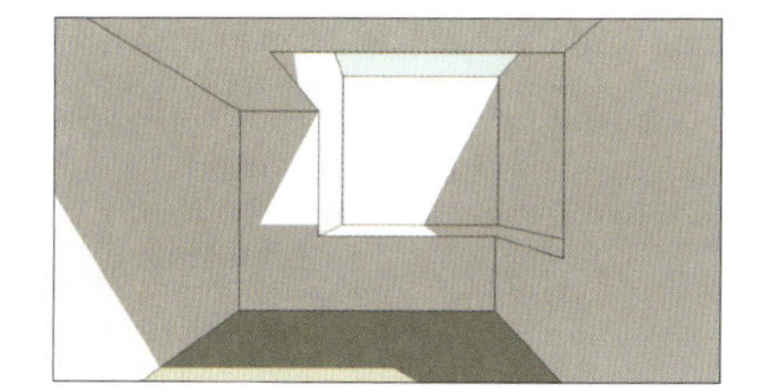
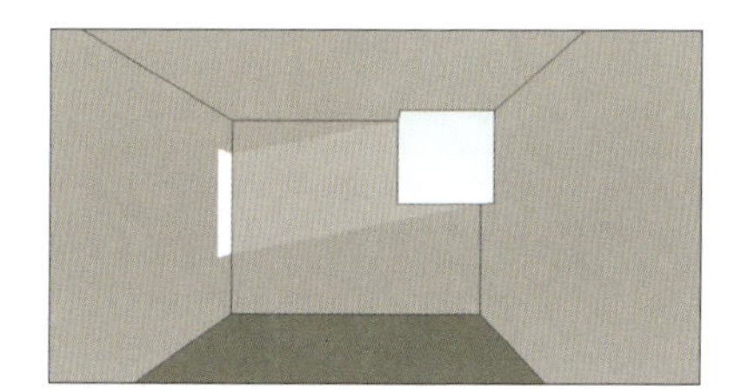
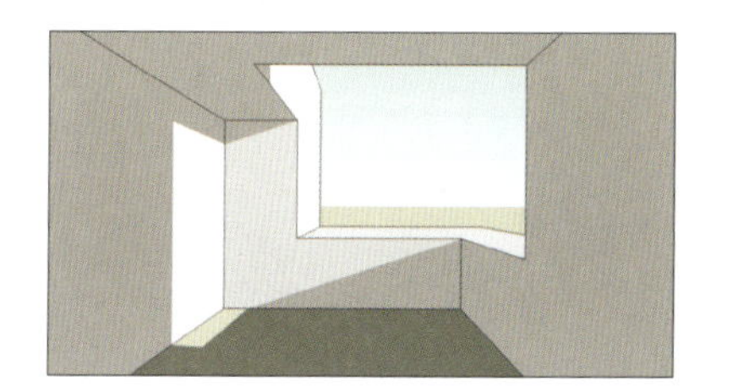
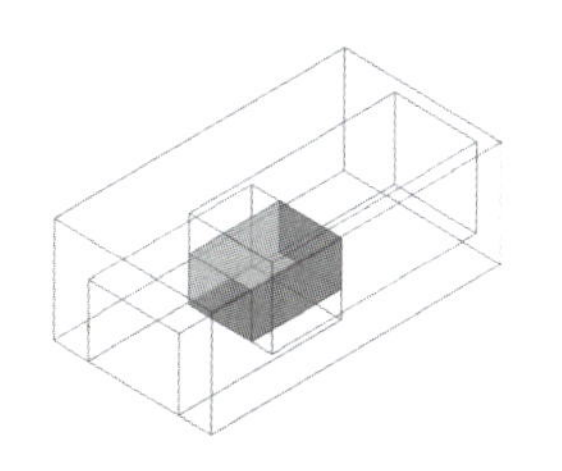
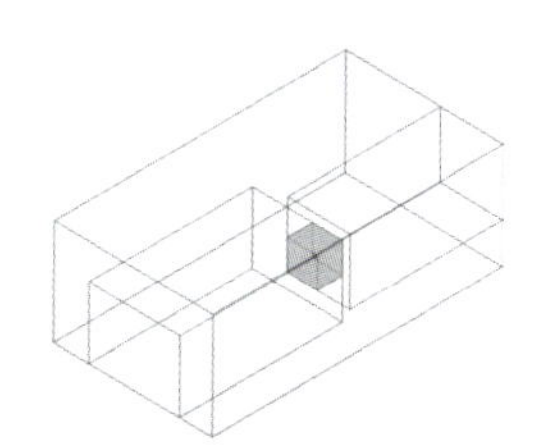
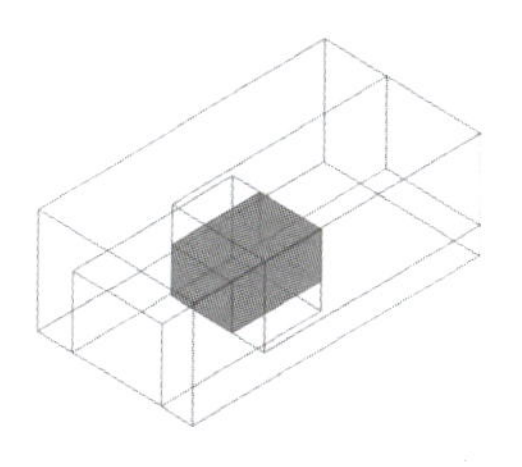
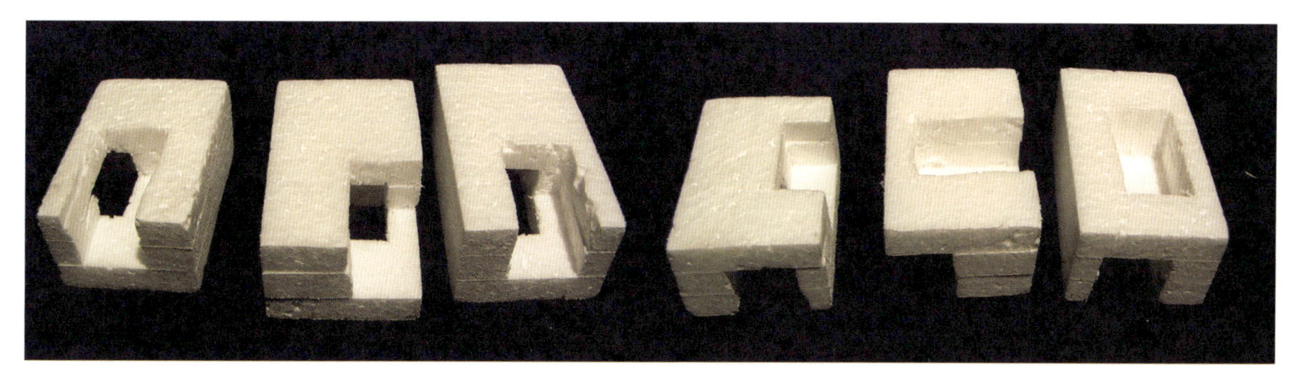

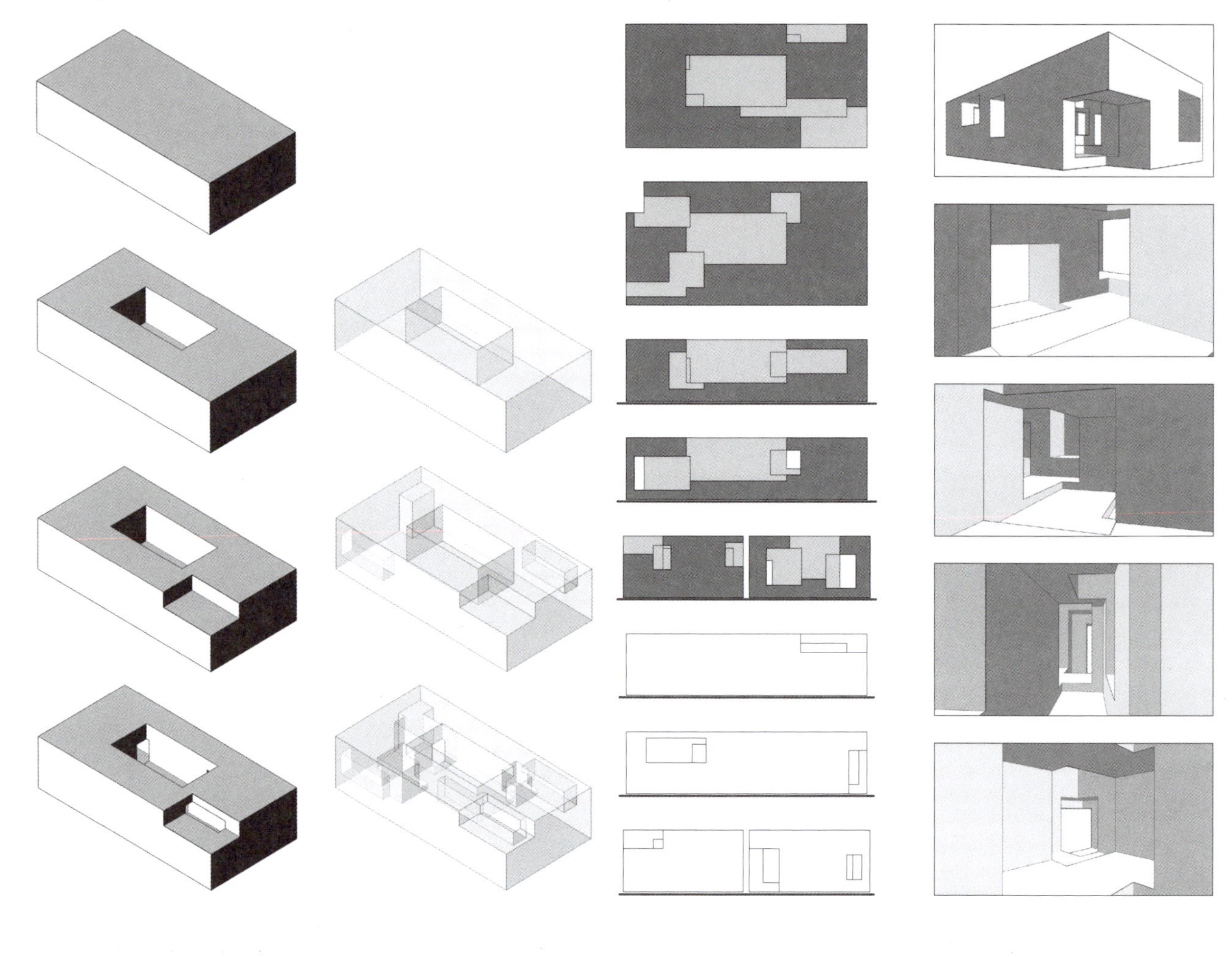

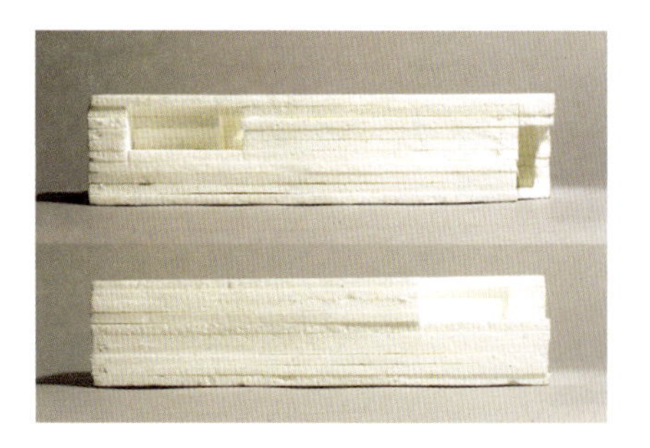

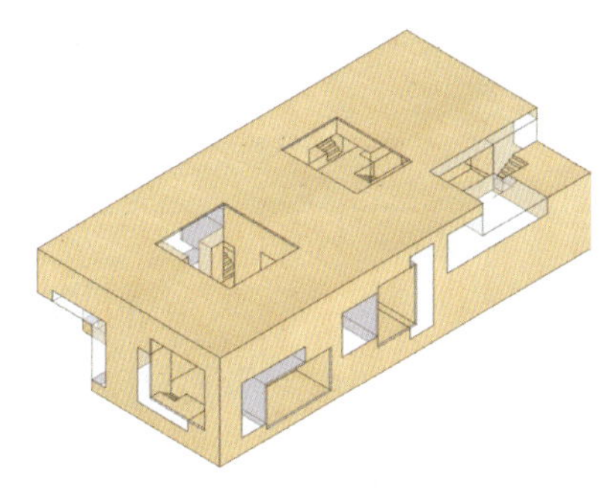

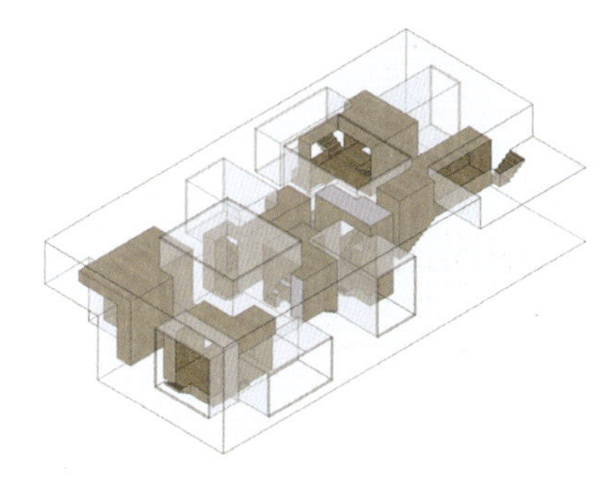

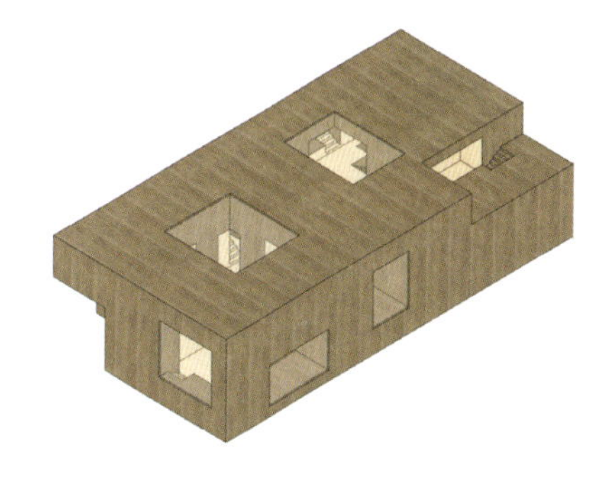

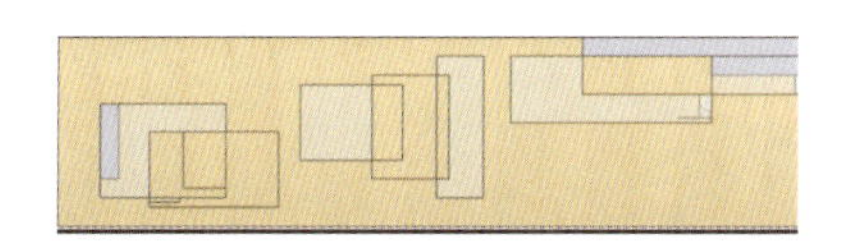

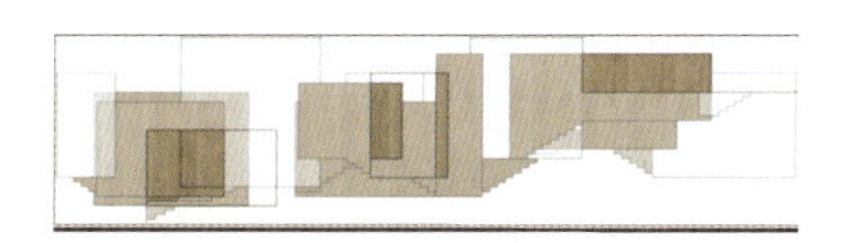

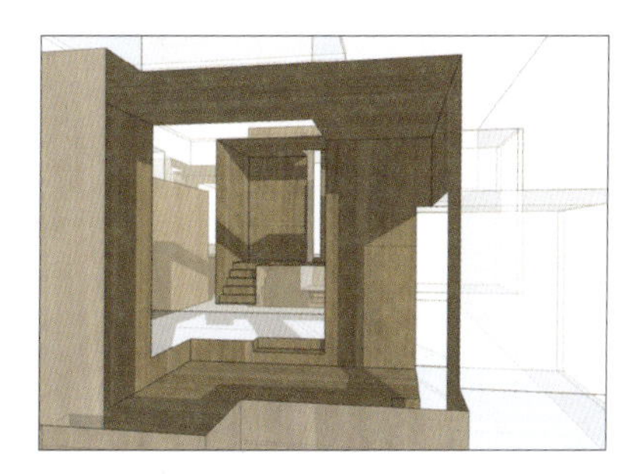

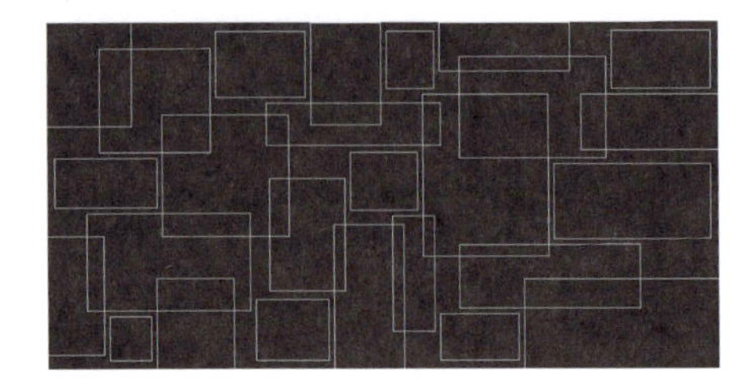

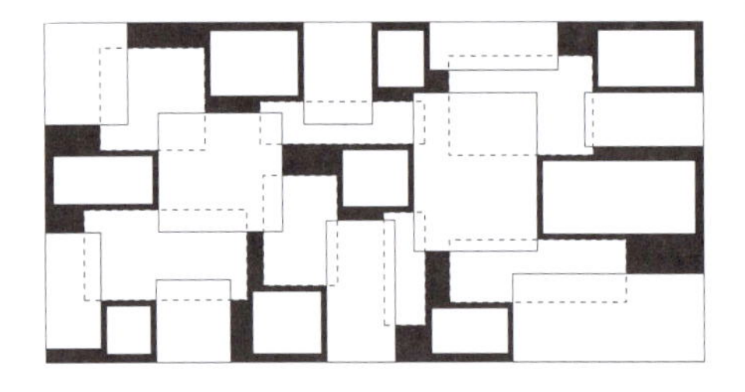

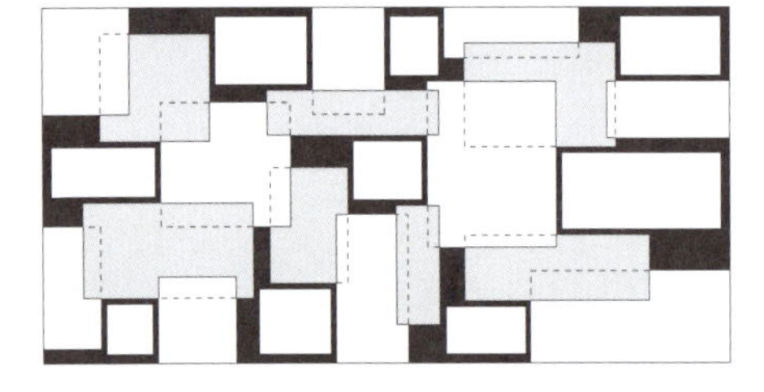

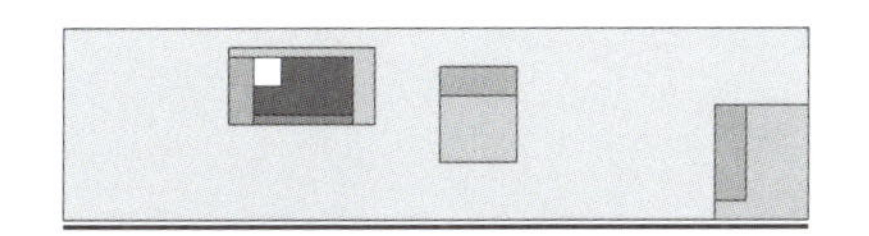

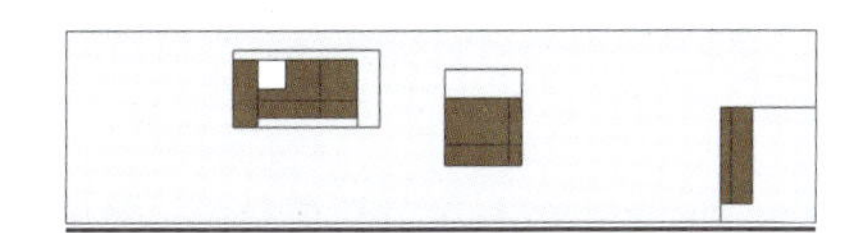

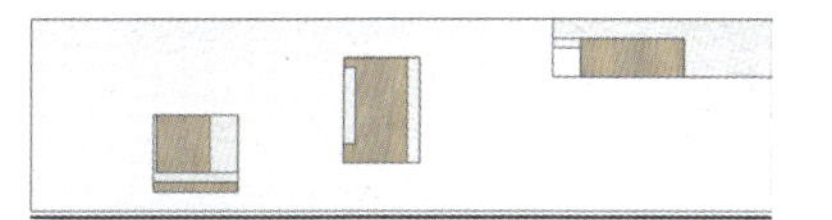

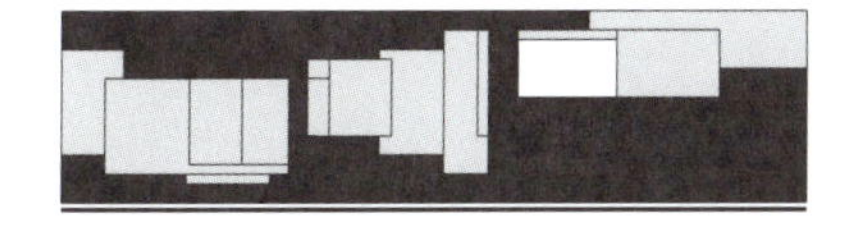

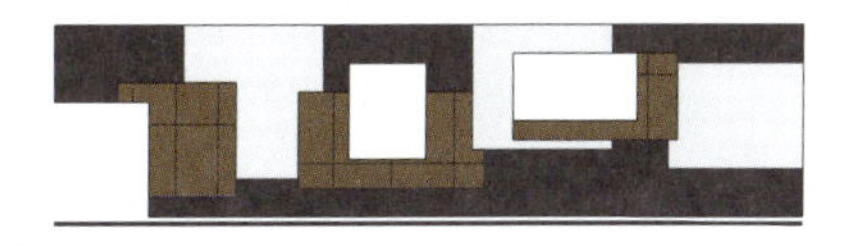

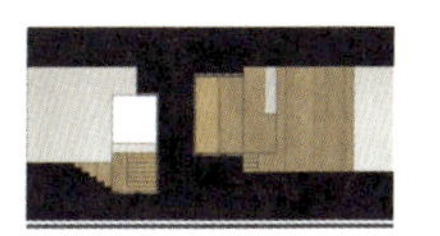

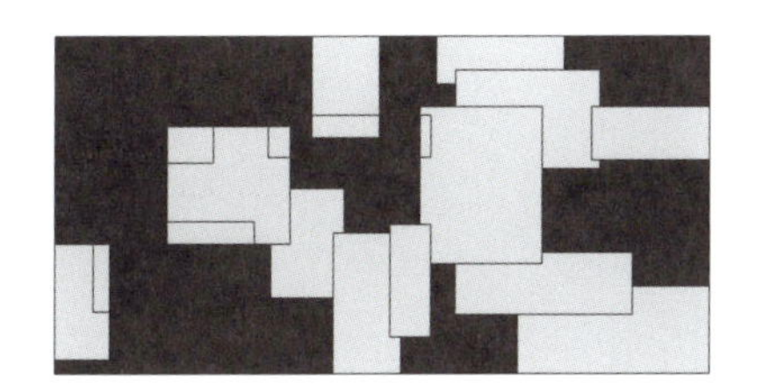

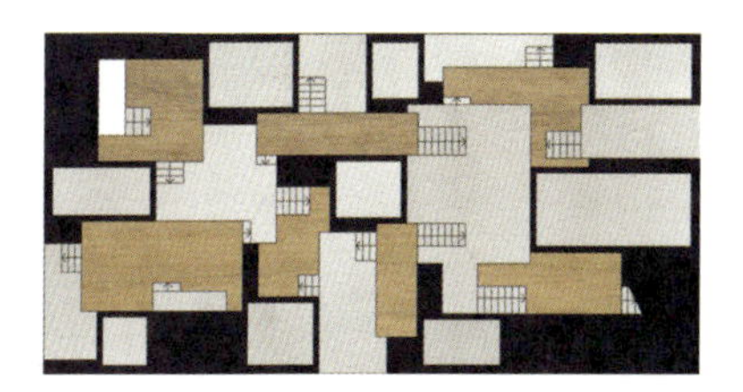

abstraction

articulation

realisation

plan A 1:200

plan B 1:200

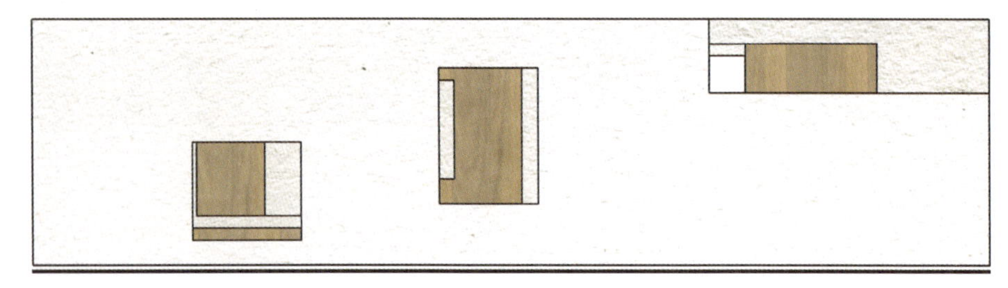

elevation 1:200

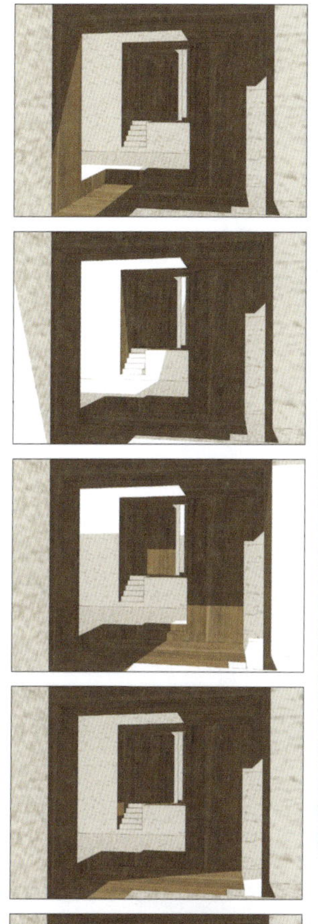

sun study june/december – morning/noon/ afternoon

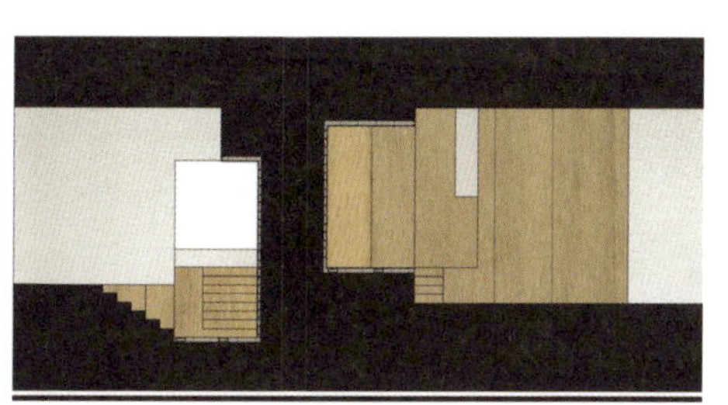

section 1:200

板片 · 塔楼 | slab • tall

Kelvin Mo Kar-him, 0708T1Y3

概念 | concept

抽象 | abstraction

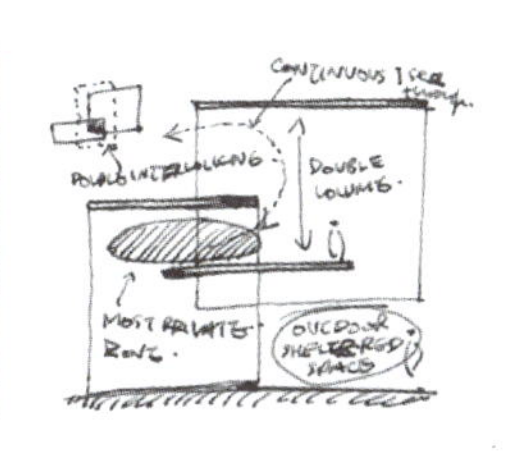

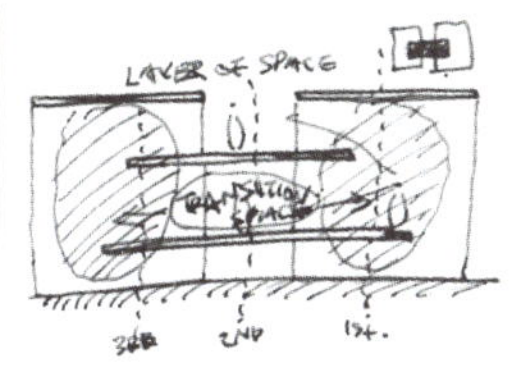

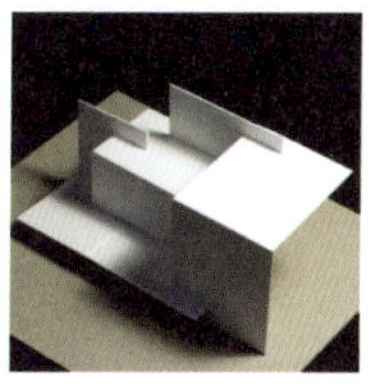

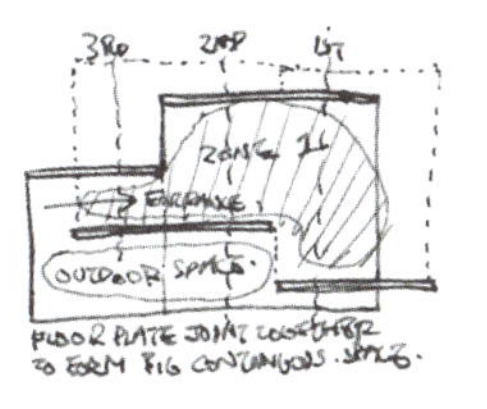

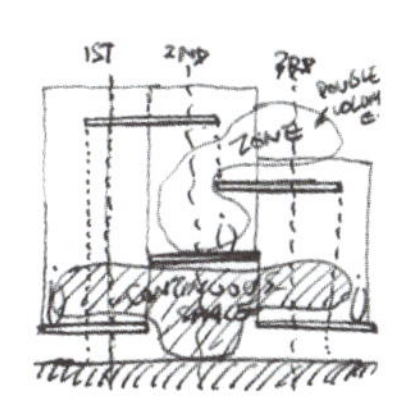

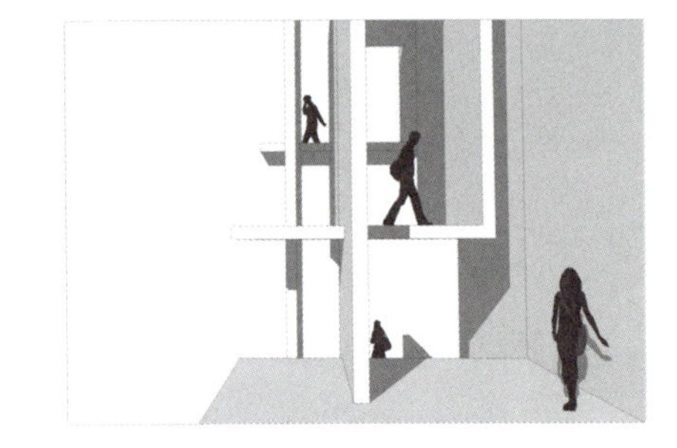

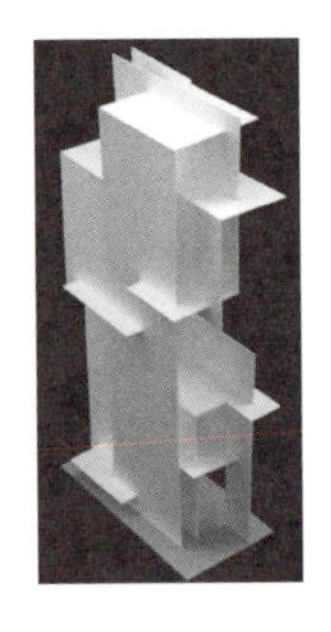

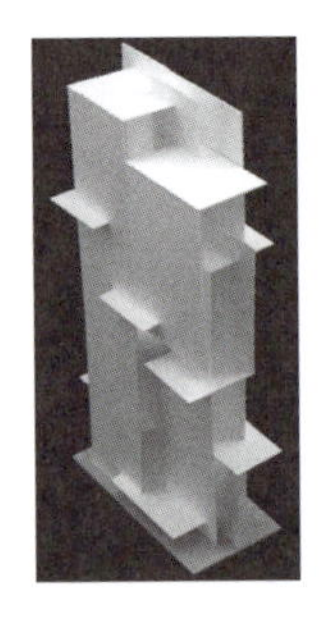

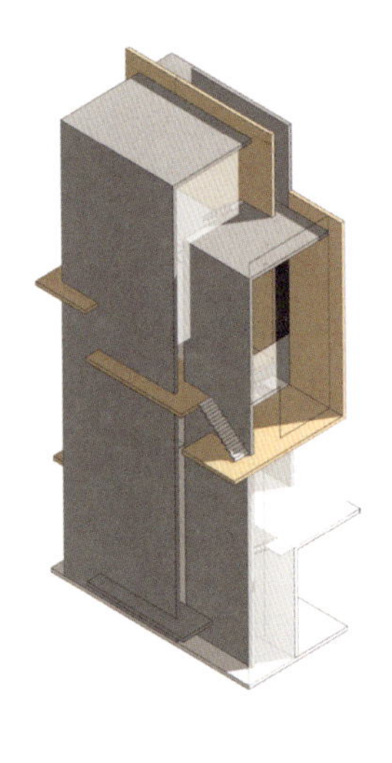

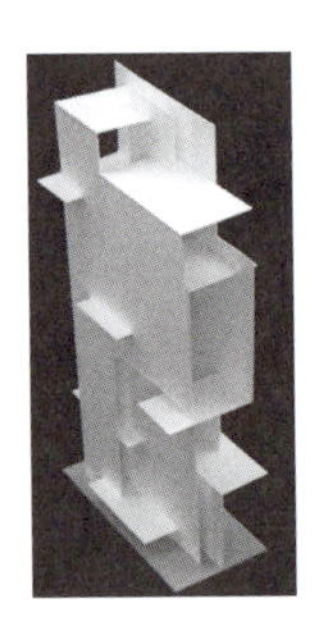

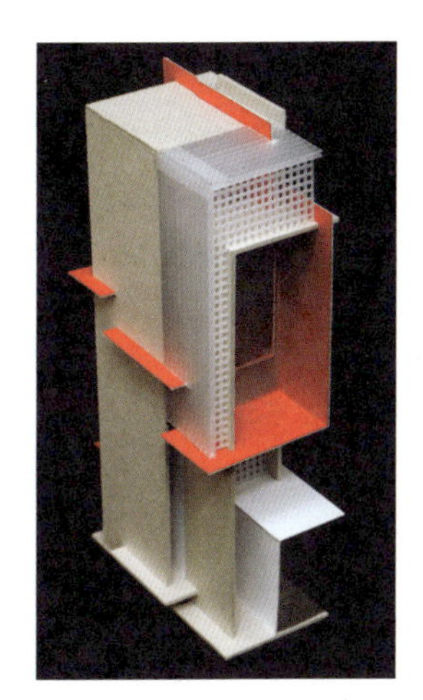

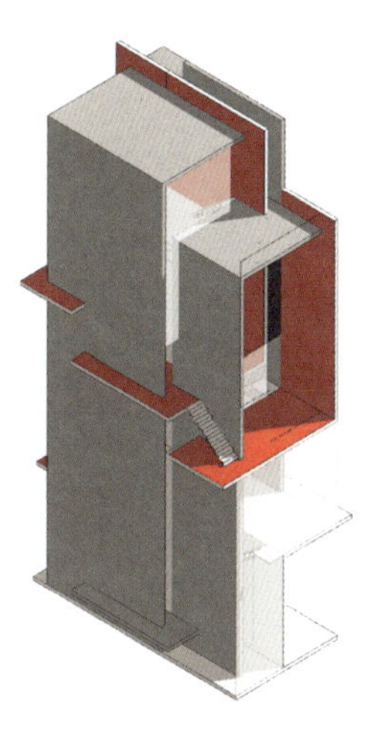

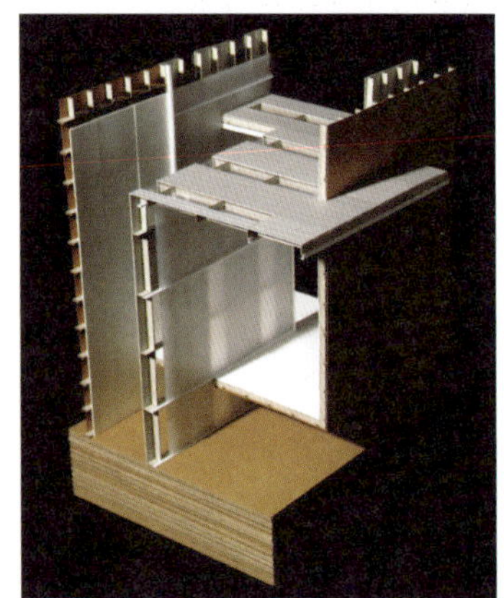

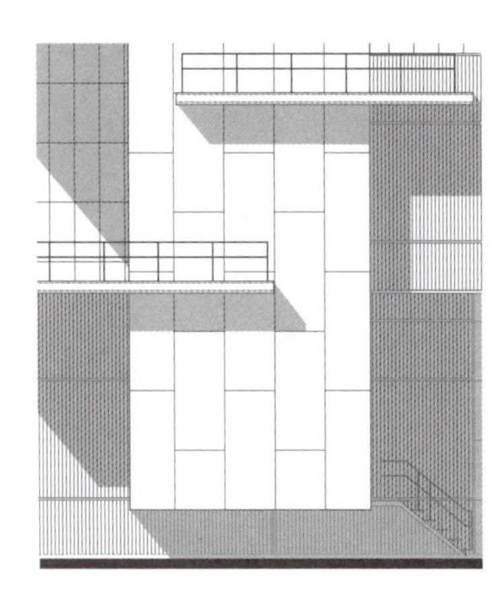

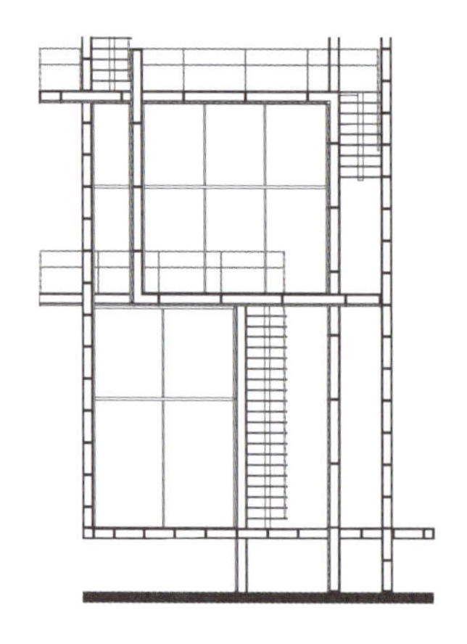

成果 | product

elevation

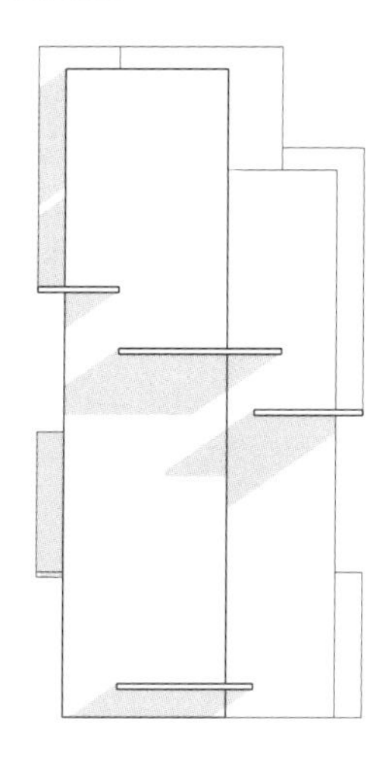

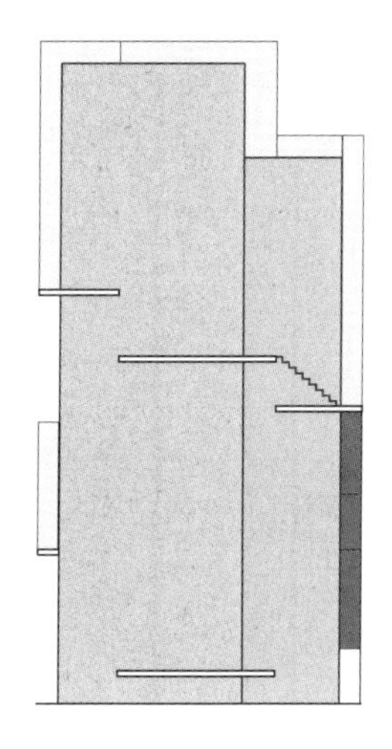

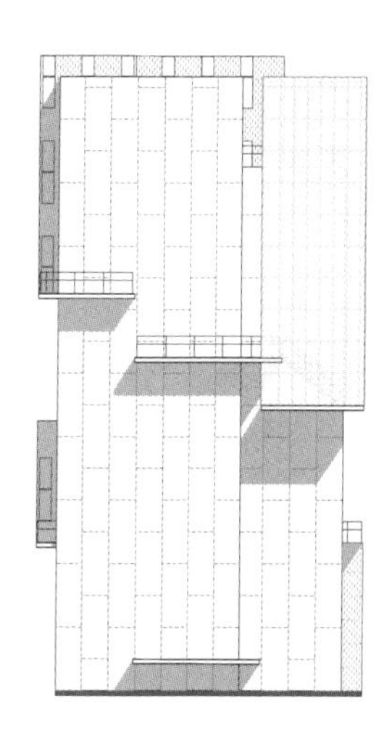

section

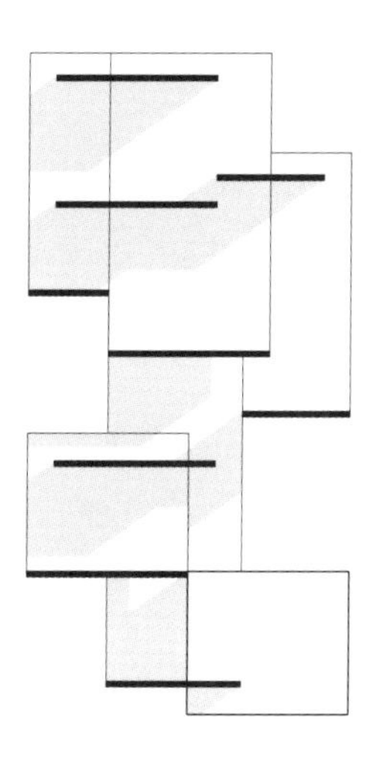

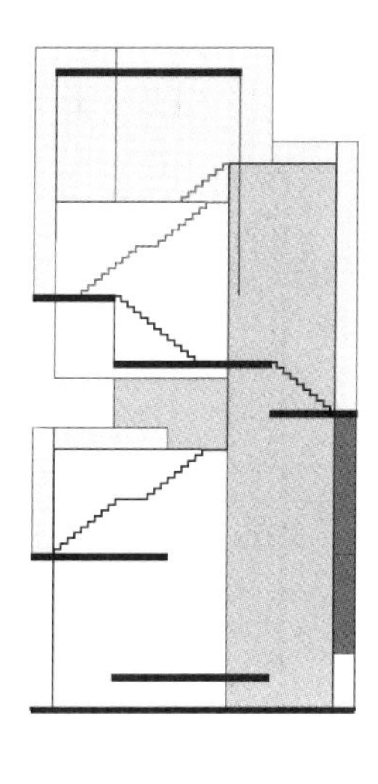

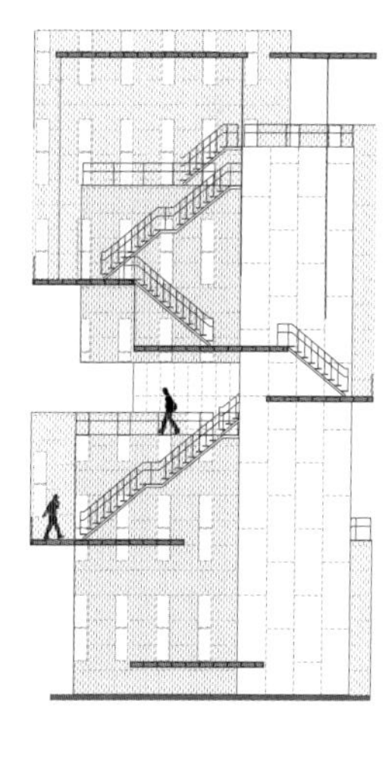

plan

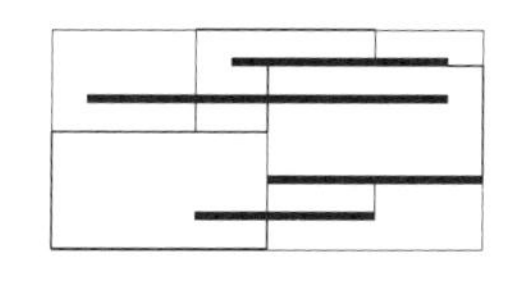

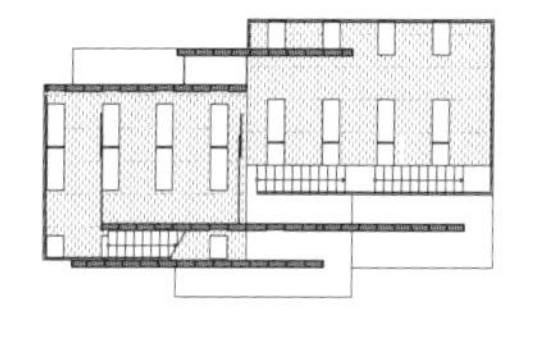

abstraction

materiality

construction

section AA'

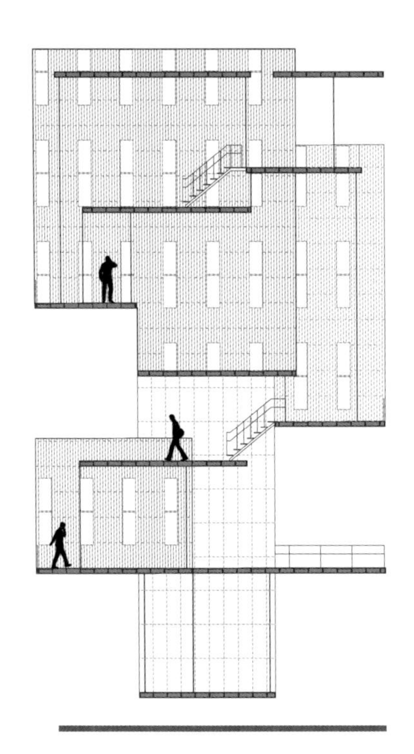

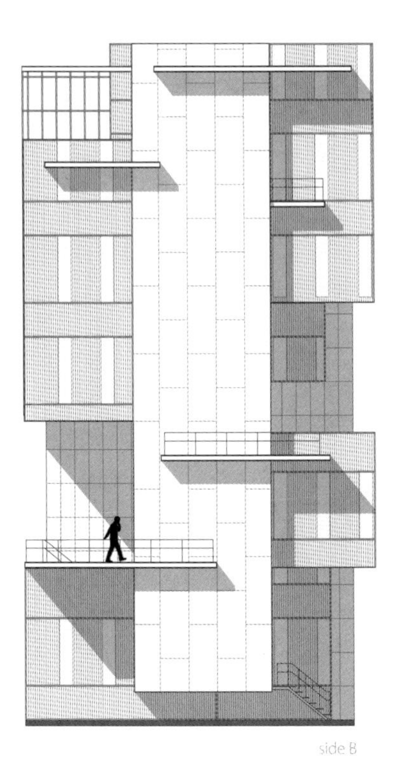

side B

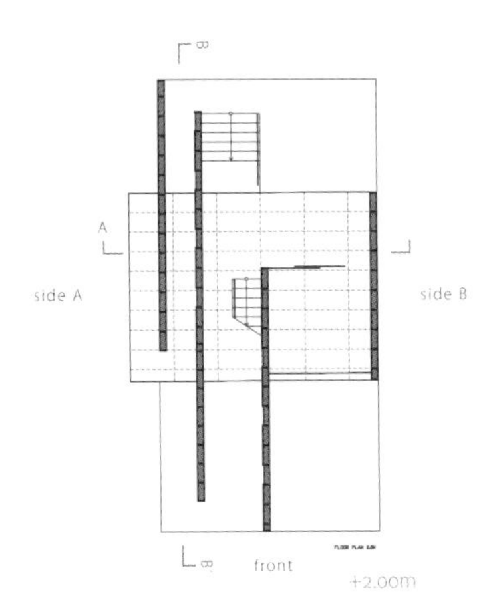

+2.00m

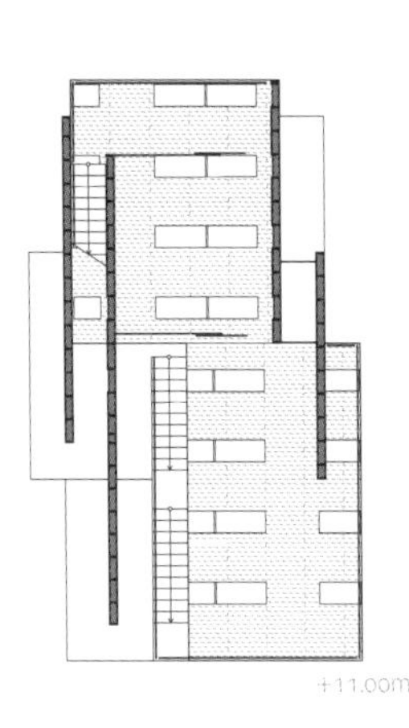

+11.00m

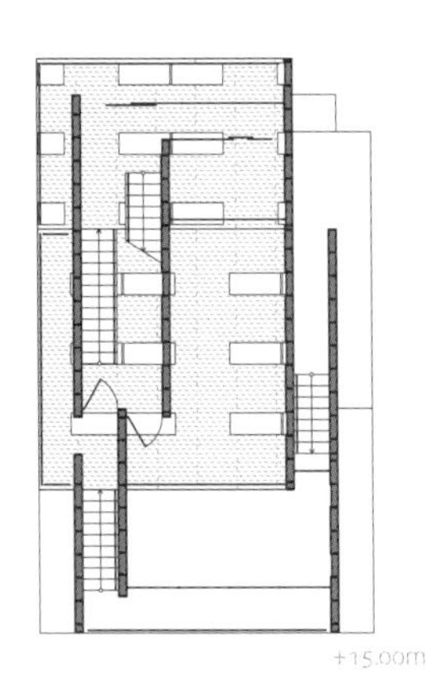

+15.00m

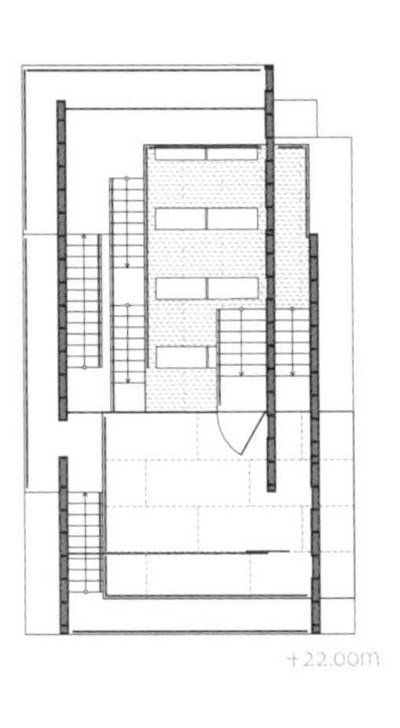

+22.00m

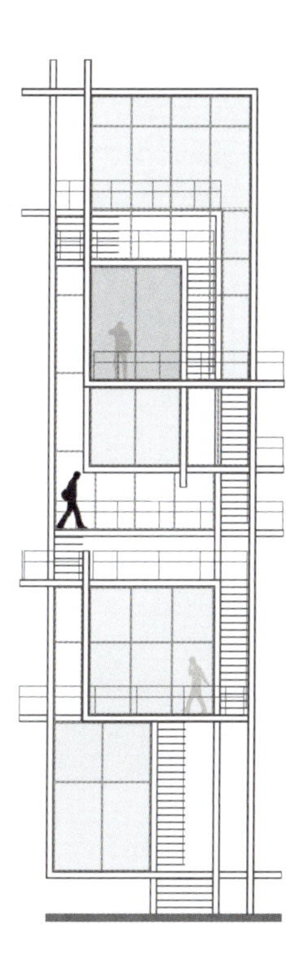

杆件 · 板房 | stick • thin

Ian Lam Yan-yu, 0708T2Y3

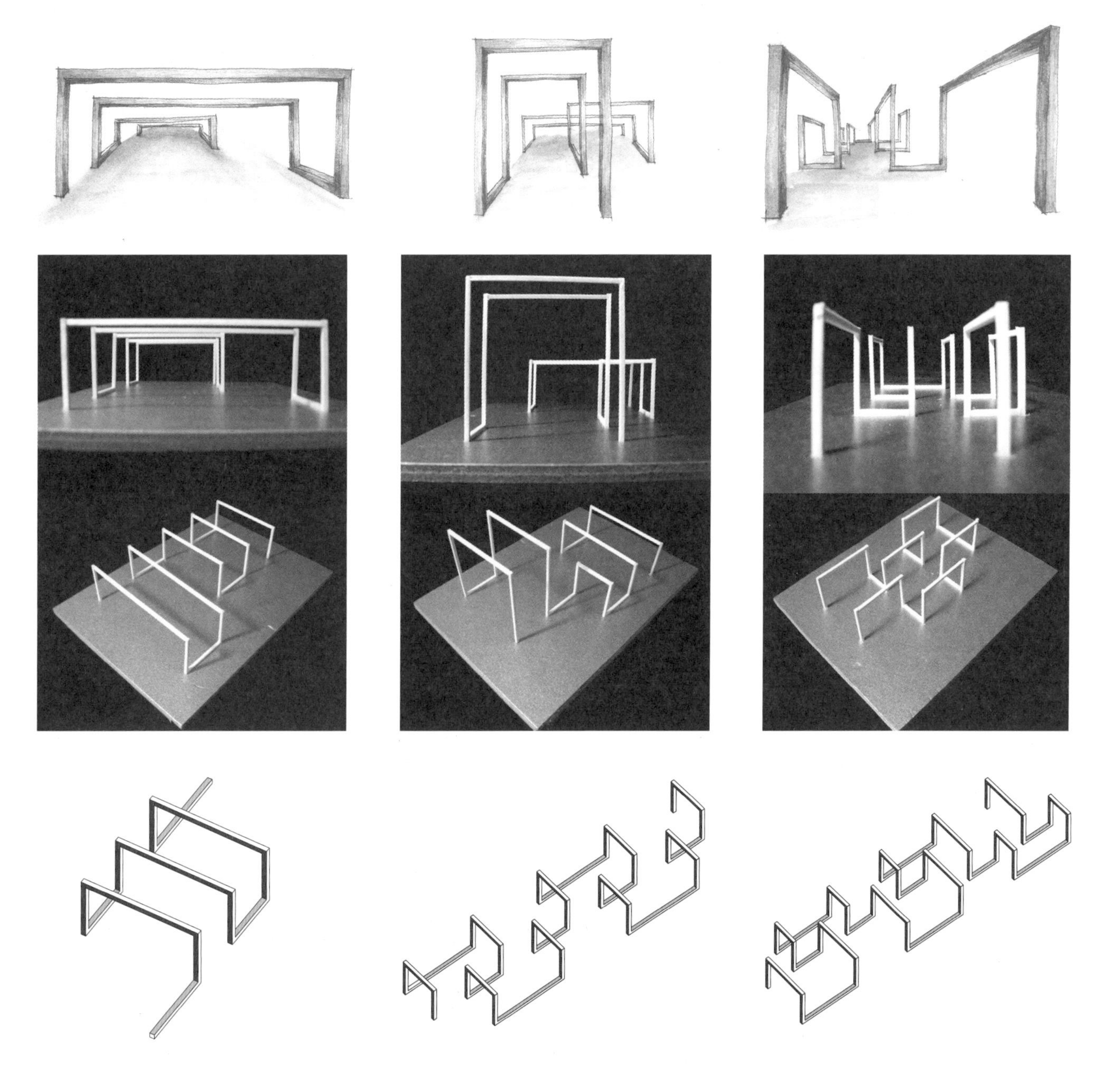

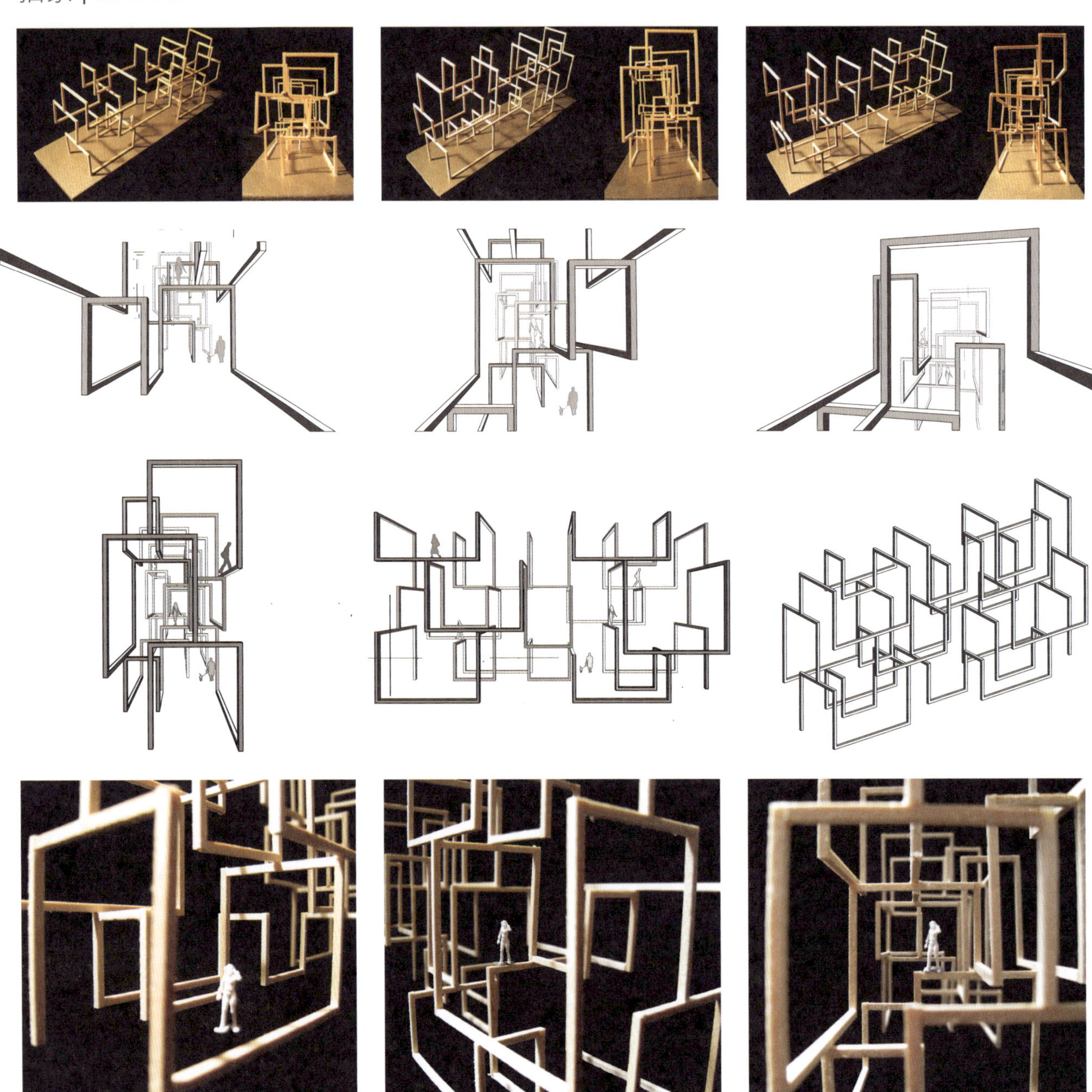

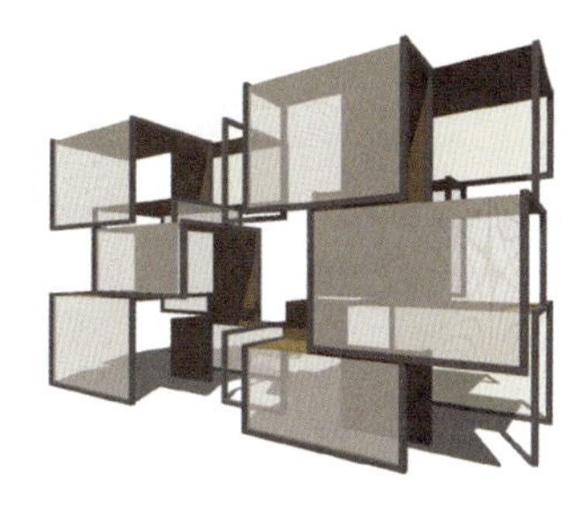
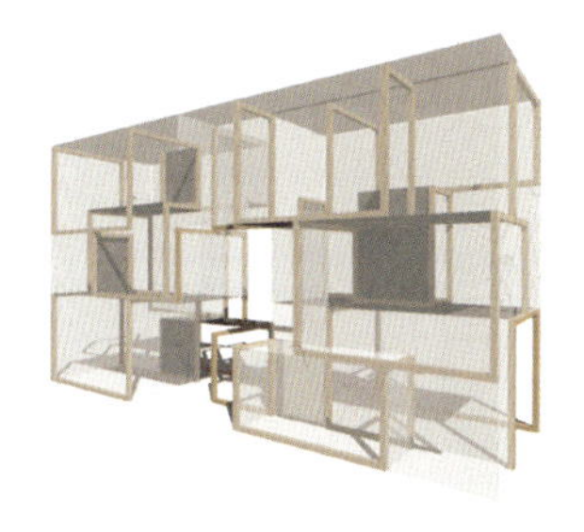

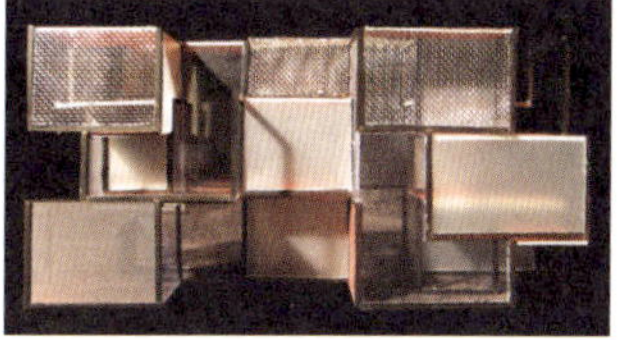

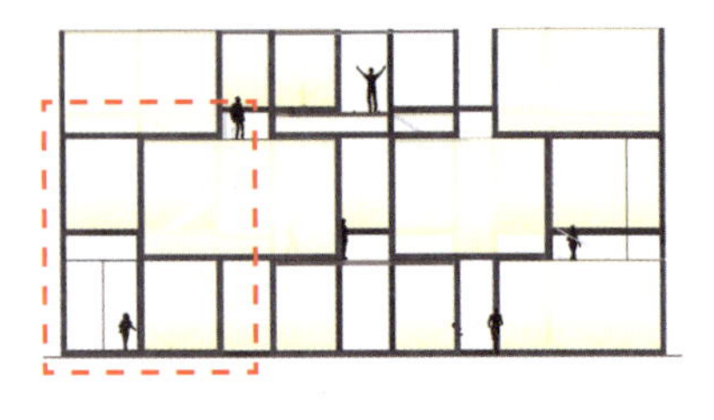

1 wooden slab

2 c-channel

3 steel tube

4 louver

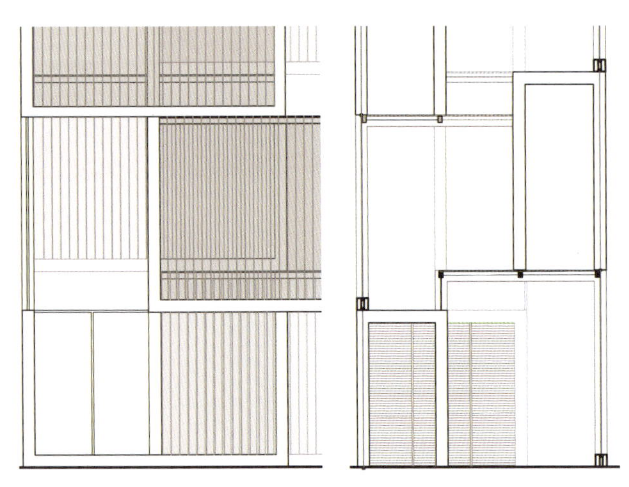

abstraction | articulation | realisation

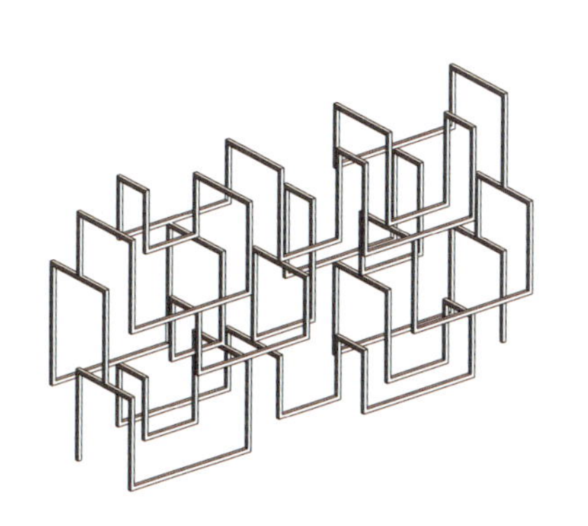
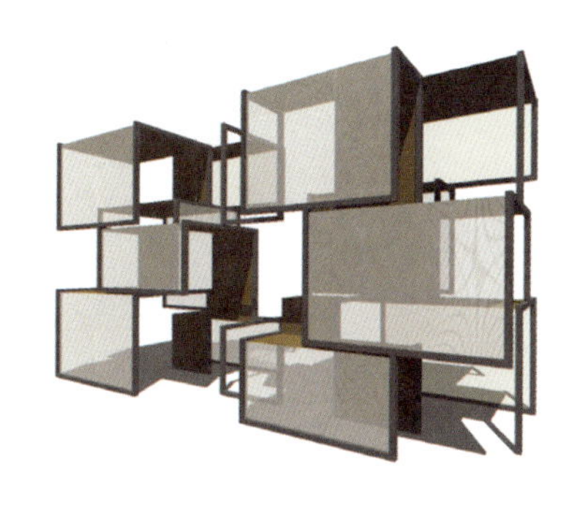

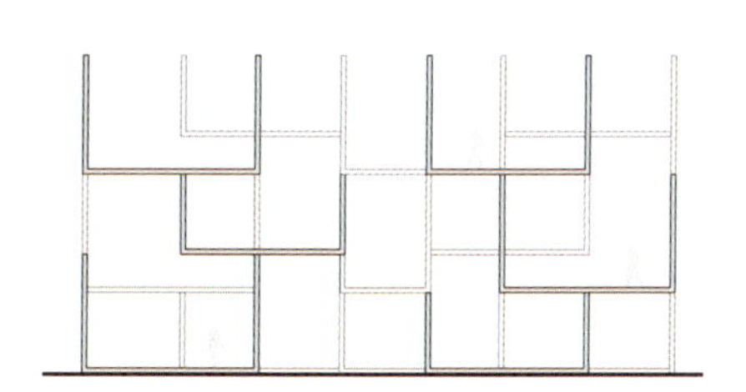

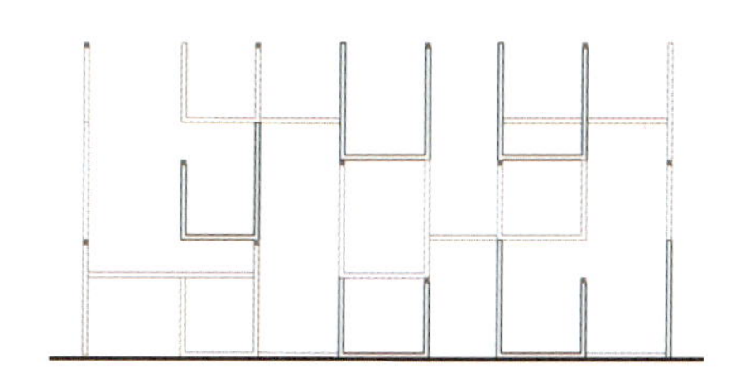

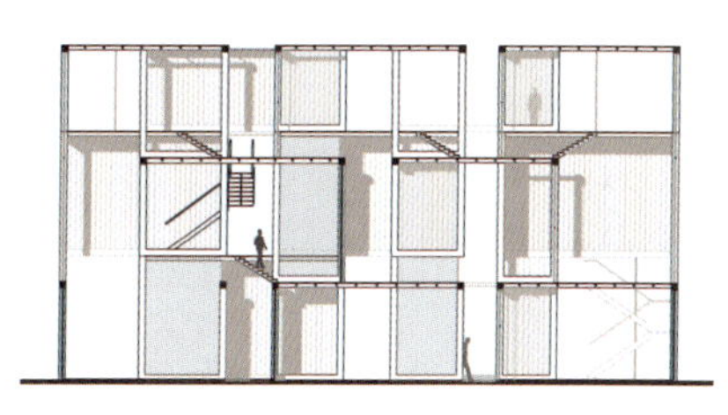
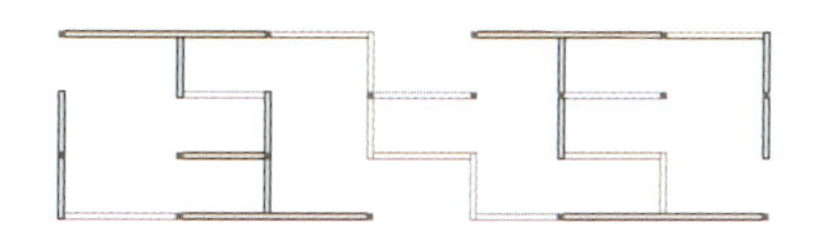
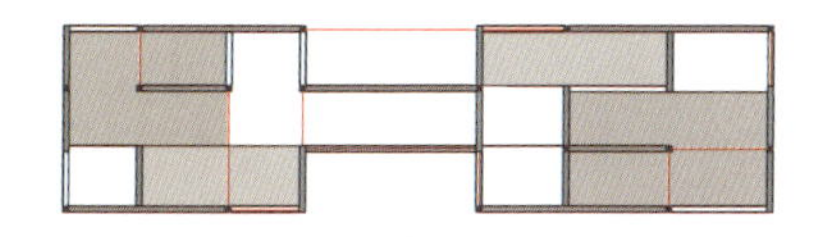
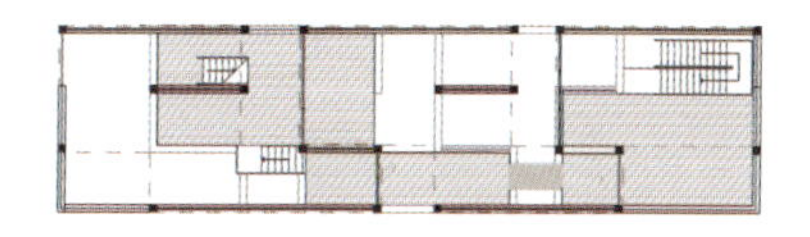

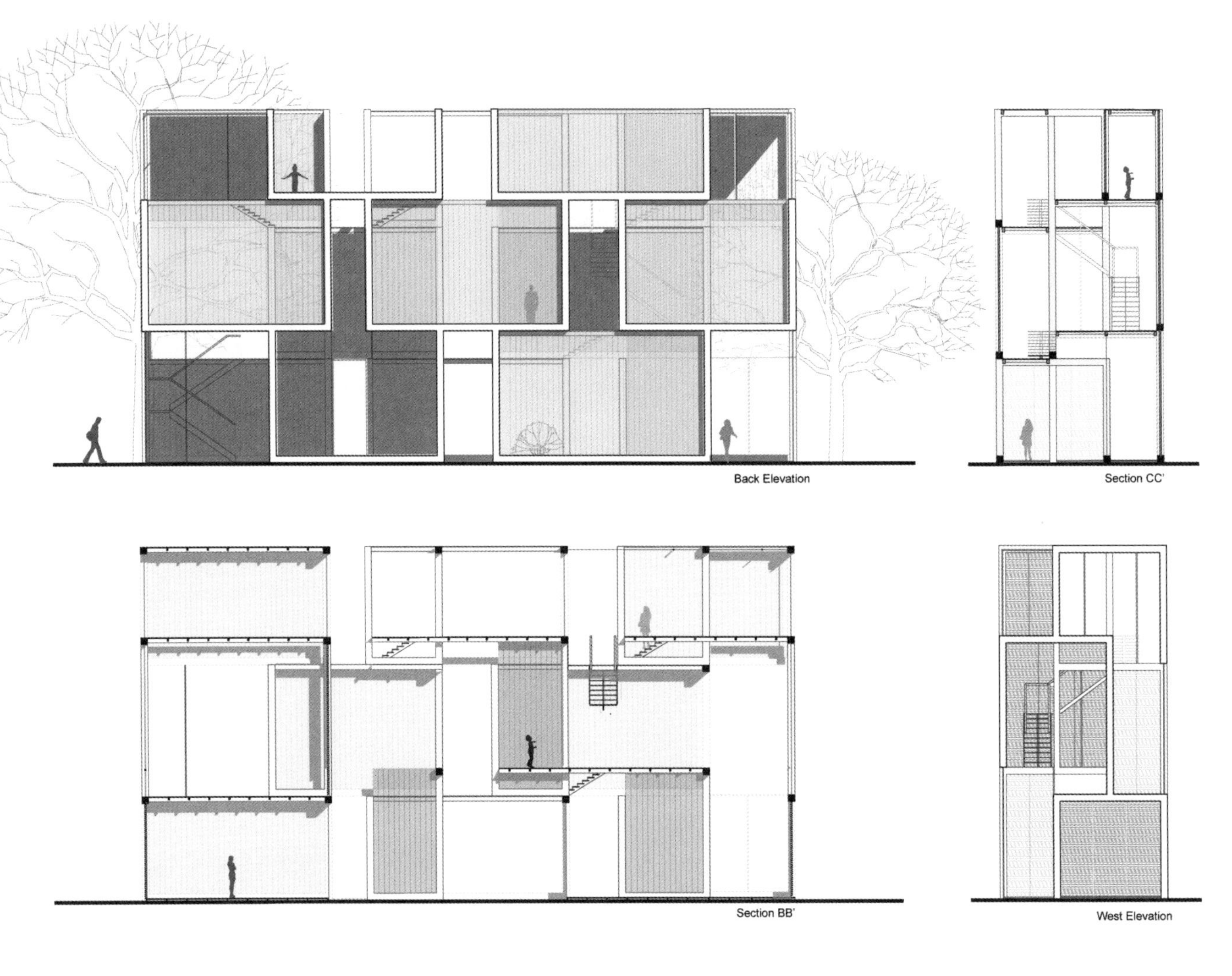

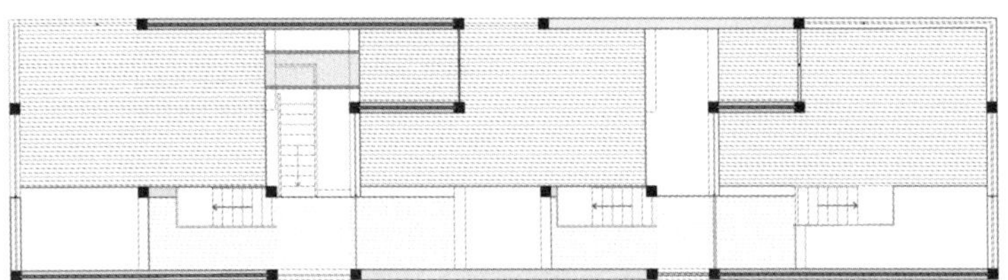

2/F Plan

4 研究 | study

体块 block
板片 slab
杆件 stick

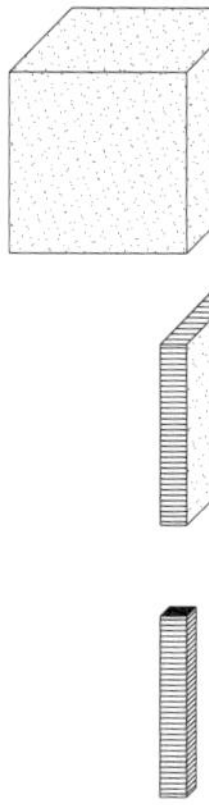

体积 | 表面 | 边缘
mass | surface | edge

体积 | 表面 | 边缘
mass | surface | edge

体积 | 表面 | 边缘
mass | surface | edge

一个关于空间的研究 | a formal investigation of space

在前一章中所描述的练习，从特定的意义上来说非常接近于科学实验室中的实验。我们先提出了一个有关设计工作方法的假说，即生成空间的材料和所生成的空间之间存在内在的关系。进一步说就是体块空间、板片空间和杆件空间三种基本的类型。体块、板片和杆件只是抽象的形式概念，落实到具体的模型材料，就有体块的操作、板片的操作和杆件的操作。那么，以模型作业为基础的设计过程究竟对建筑的空间形式有什么影响？如何借助模型工作方法来发现新的空间形式？

为了能够证实先前提出的假说，我们设计了一个实验的操作过程，即设计研究的四个阶段。进一步，我们必须将所要研究的问题进行必要的处理，排除掉一些不必要的因素，使得要研究的问题能够突出出来，比如将建筑的功能内容抽象为人在空间中的活动，将场地因素抽象为一个建筑体与地面的三种摆放关系。这些对建筑问题的抽象化处理的惟一一个目的就是使得我们能够在一个接近于实验室的环境下来观察和研究体块、板片和杆件三种要素的操作所产生的空间特征。

我们在现实的观察中一般很难找到非常纯净的体块空间、板片空间和杆件空间的建筑。这是因为，一方面建筑师需要具有非常明确的空间概念和追求才能做出极端的形式，另一方面现实的建筑总有这样或那样的条件制约，使得空间的纯净表达变得困难。但是在实验室的条件下，追求形式和空间表达的纯净和极端就变得可能和非常必要。当然，我们必须要把实验室中的研究和现实中的建筑设计两者区分开来。我们相信当一个设计者能够在实验室的抽象环境下对三种空间限定要素的空间潜能有所认识，他才最有可能在现实的建筑设计中实现空间形式的纯净表达。

在这个实验中，我们把体块空间定义为勾勒性空间，把板片空间定义为模棱两可的空间，以及把杆件空间定义为调节性空间。这三种空间特征应该是三种要素最纯粹的空间形式。我们在设计的研究中发现，除了这三个最极端、最明确的空间特征外，与所采用的模型材料的物理特性以及操作方法相关，所产生的空间有很多的变化形式，比如体块的操作也可能产生模棱两可的空间，有如杆件的操作可能与体块的操作相类似，等。只有通过不断地实验研究，我们才能了解空间限定要素的操作和生成的空间之间的丰富内涵。

这一章的目的不是罗列体块、板片和杆件三种空间限定要素生成空间的所有可能性，就一个开放的和无止境的研究而言，这几乎是不可能的。但是，我们希望通过每个要素10个案例的介绍来展示如何展开研究的可能性。

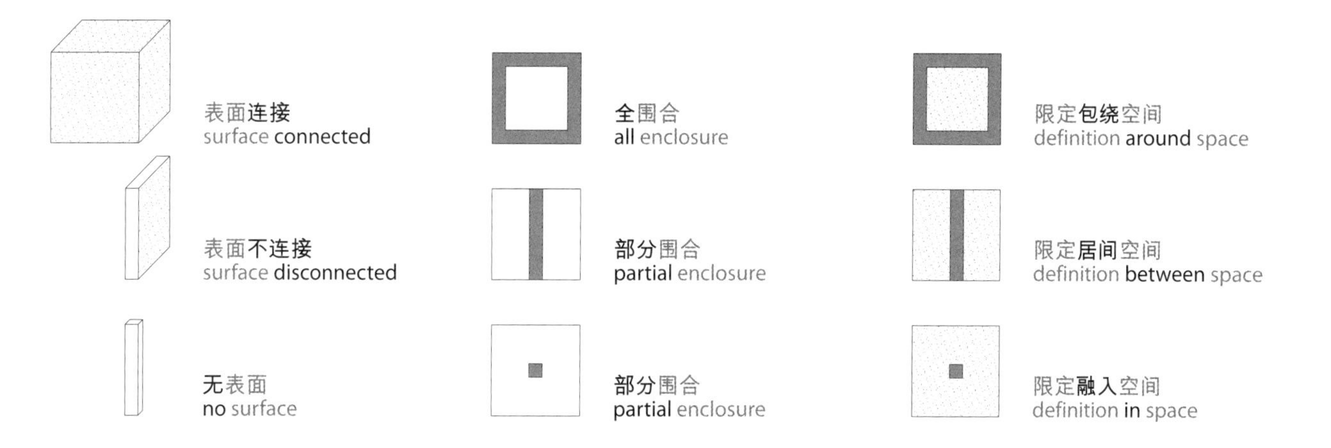

In the last chapter, we described the exercises and illustrated the results with student work. Now we would like to look more closely at the nature of the exercises. Rather than considering them sections of a design process which moves from given conditions to a new proposal, or from a problem to a solution, we understand that these exercises are more like laboratory experiments. They are based on the hypothesis that there is a relationship between the properties of space defining elements and the character of the perceived space they form, specifically, that the use of the three element types – block, slab, and stick – leads to three distinct space types – outlined, ambiguous, and modulated space.

To study this, we set up design experiments. They are structured into four phases of exercises. As experiments, certain issues are abstracted or isolated compared to a design project, while others like site and use are reduced. This allows us to focus on key issues, which we hopefully can explore more deeply.

The conditions in a laboratory and in a real design situation are different. In reality, pure cases are not a priority, are therefore rare and then usually represent design experiments. In experiments pure cases can be set up and studied.

The element type of the space definition determines the type of space. In a design, elements of unclear or ambiguous type like slab-like blocks, stick-like slabs or block-like sticks, etc. can be used. This also depends on the scale of the element, as building material adds further dimensions. Also, various element types and variations might be used in a design. In our experiments we use only one type of element to create pure instances of the type. This is important to find out the relationship between space defining element and perceived space. But each of the three element types offers a range of possibilities. For the principle with which the original given element is manipulated to be able to define and differentiate space, we use the term operation. We want to study how different operations – single, in sequence, or in combination and specific to each element type – result in variations of each space type.

The ten examples for each element type show, without exhausting the possibilities, a wide range of variations within each space type. The materialisation phase further widens this range through the interpretations of the elements' properties. The role of structure and the necessity of additions are interesting questions we deal with, but they are not emphasised in the descriptions.

1 体块 | block

体块作为一个形式要素，它的基本特征是由表面包裹的实体。如果把它剖切开来，其内部应该是实心的。体块的表达主要取决于体积的表面。体块对于空间生成的作用有几个方面，一是体块以其体积来占据空间，二是体块可以利用其外表面来界定外部的空间，三是体块的实心内部也可以产生如“盒子”般的空间，这个空间的基本特征是它被周围的表面所包裹，形成“包裹性”的空间。

板片的操作大致可以归纳出几种，但决不限于这几种：一种是在一个大的体块上通过挖去、推挤、位移等方法来形成空间，这是一种减法的操作；另一种是通过若干小的体块的堆积、组合、排列等方法来形成空间，这是一种加法的操作。前者的空间是实体上被缺损的部分，后者的空间是实体之间的虚空。但是，实际操作的可能性取决于特定体块模型材料的特性。

Create space: A block has a mass – which can also be interpreted as space – and a continuous outer surface. The surfaces and their edges zone the surrounding space. There are two basic types of operations to create space within the block: first, by cutting smaller pieces and placing them to create space in-between, and second, by removing parts which creates space as an inversion of the mass. The space defining element is on the outside of the space. Space is perceived through the continuous surface surrounding or outlining it, and the zones formed by convex edges. The change of the viewing angle gives another view of the same space or reveals another part of this space. The movement through space doesn't change the perception of the space much except when passing convex edges.

Because of the duality of mass and space, and the contrast between surface, mass, and space, the material differentiation offers many interpretations and strong articulations of the concept developed in the abstraction phase.

体块空间的最基本操作之一是以体块来占据空间，同时产生体块与体块之间的空间。体块的大小和形状随体块的功能内容而定，因而大小不一。这种情况下欲建立一定的空间秩序往往比较困难，不过我们还是可以看出作者尝试线性组织和中心性组织两种方法的努力。对体块的不同材质的区分皆在于表达空间组织的特点，而建造的秩序着重表达水平向的楼面板和体块之间的对比。

A big block is cut into smaller blocks of various sizes, according to the space requirements for use – the programme. These smaller blocks contain space and define in-between space for another type of use – circulation. They are sandwiched between floor slabs, which support the continuity of the surfaces outlining the in-between space. At the periphery, this supports the expression of construction as stacking. The space is organised through sequences of linear and central configurations, which are articulated by material differentiation.

Louis Yim Ying-kit, 0203T2Y2

将生成空间所需的体块简化为一种或几种单元，从而赋予体块本身一种内在的秩序。该设计采用了两种单元体的不同组合。单元体以及其内部的空间具有“图”的性质，而单元体界定的外部空间则具有“底”的性质。在材料区分的阶段又引入了第三种半透明的单元（有两种不同的尺寸），且该半透明的单元与红色的单元发生重叠的关系，更加丰富了空间和形式关系。

The operation is developed over several stages. A block is cut regularly and when the pieces are shifted, they define space between them. There are cuts in all three main directions and horizontal shifts in two directions. The modularity creates a recognisable order with blocks perceived as figures and the in-between space as ground. At the stage of material differentiation, part of the in-between space is reinterpreted as translucent blocks. The material differentiation allows multiple readings of the expression.

Object - figure

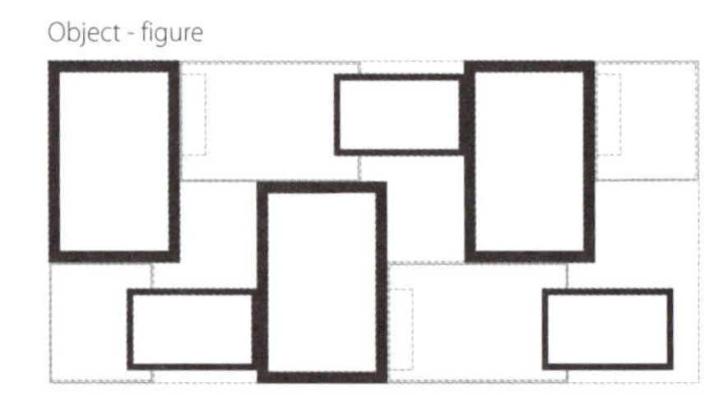

Upper Floor Plan

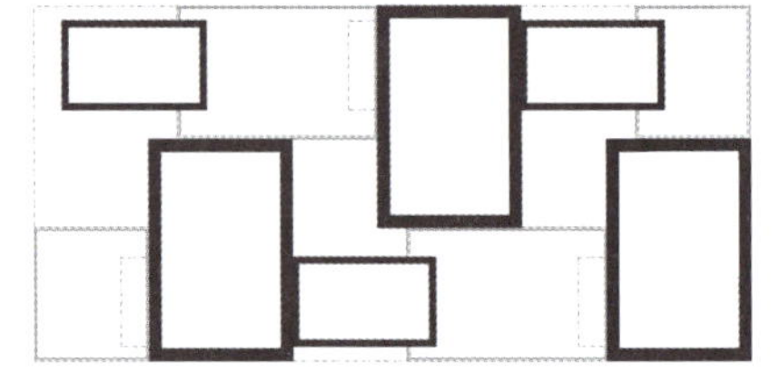

Ground Floor Plan

Zephyr Lam Wing-him, 0304T2M1

Space - ground

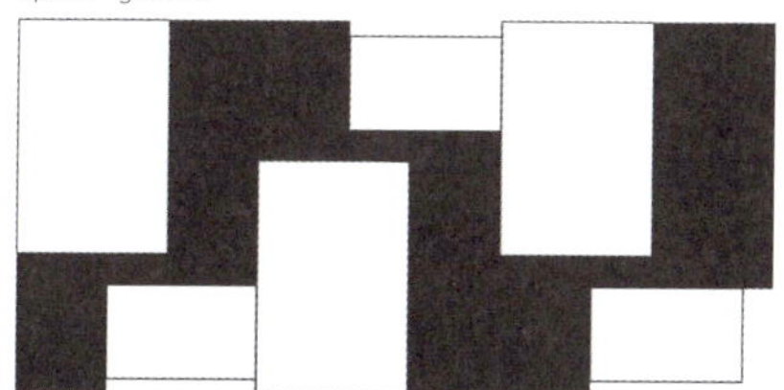

Upper Floor

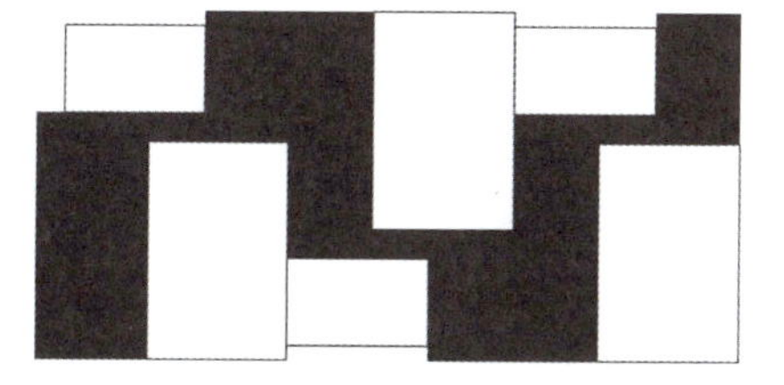

Ground Floor

这个设计的研究的焦点在于体块之间的间隙空间。单元体块在水平方向上有规律的旋转产生不同的间隙，形成单元体块的单调和间隙空间的丰富两方面的强烈对比关系。在最后的设计中单元体块进一步区分为大小两种类型，暗示功能区分的可能性。位置保持不变的水平楼板则起到强调体块旋转的对比作用。

A primary operation cuts a big block in three directions regularly into smaller blocks. A secondary operation shifts the small blocks horizontally apart. And a tertiary rotates the blocks by different degrees in each layer. The resulting spaces are studied in a systematic way in phase one, showing how uniform blocks can form rich space. In the second phase, possible use and the tall form trigger differentiations. The introduction of horizontal slabs, partially following the outline of the form, modifies the expression by contrasting with the blocks and strengthening the perception of their rotation.

Nelson Tam Sin-lung, 0102T1M1

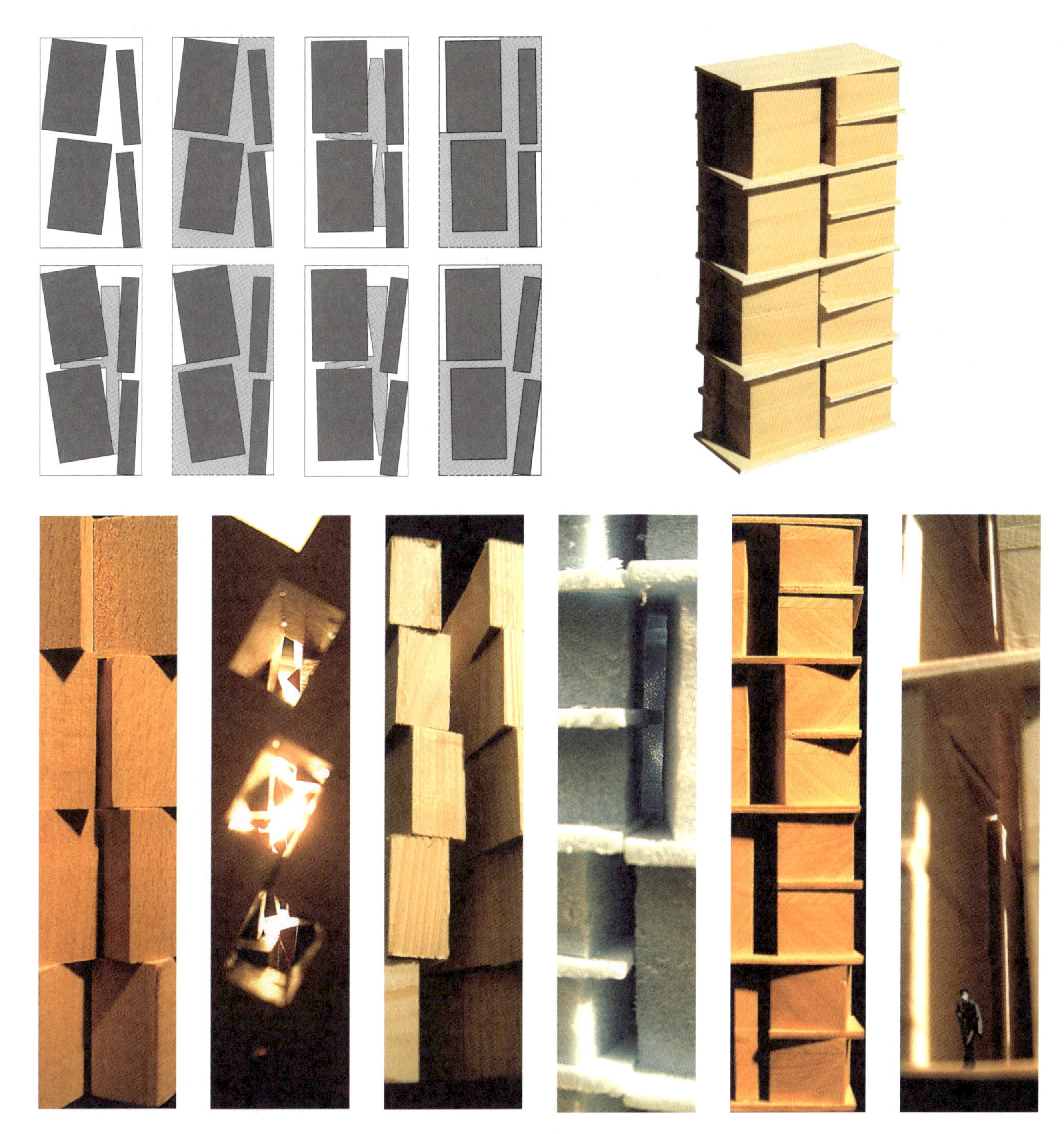

相同的单元体块通过水平向的位移来生成空间。而套筒单元的引入使得这一简单的操作变得不平凡。套筒成为单元体块的延伸，它的长度可以随需要而作自由的调整。就建筑的整体而言，套筒成为单元体块之间的连接，以及起到保持该板房的基本体量的完整的作用。

Also here, a primary operation cuts a big block into smaller blocks in three directions. A secondary operation shifts these small blocks in one direction, creating spaces between them. In the third phase the material differentiation is used to interpret the small blocks as sleeves for the outer and the inner surfaces. These two sleeves can also be shifted against each other. The material differentiation, especially using translucency, results in a rich expression through layering, and at the same time strengthens the reading of the overall form and the operation of shifting.

Eureka Chu Lai-nga, 0506T2Y3

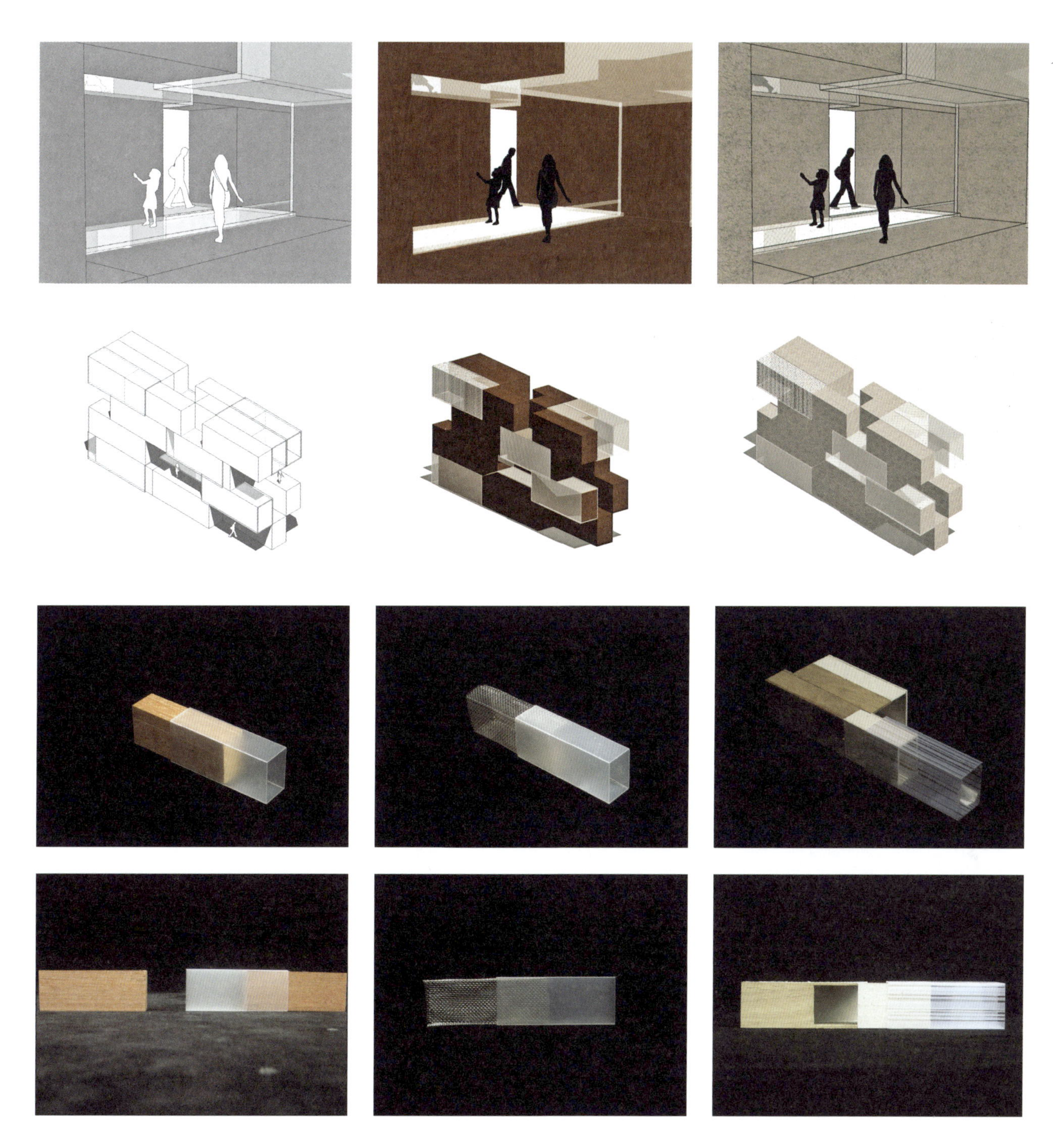

挖去是体块空间的一个最基本的操作，即从一个实体中挖去一部分来生成空间。这个设计特别反映了挖去操作所产生的空间的内部特性：空间在体块的内部，在外面除了几个开口外，几乎什么也看不到。内部被挖去的空间具有明确的形状，可以用另一个体块来表达，即实体与空间的转换。

In a primary operation the tall block is cut into several regular horizontal layers. In a secondary operation each layer is cut differently but always in a way that the four cutting planes meet. In a tertiary operation, the central small blocks are removed to create space. This space is continuous but varies from layer to layer. It is also completely inside the form and touches the periphery mostly in relatively small slits. One model shows the inversion of space and mass, demonstrating that the two are basically the same. In the materiality phase, one block on every second layer is reinterpreted as translucent.

Kelvin Chan King-wai, 0304T2M1

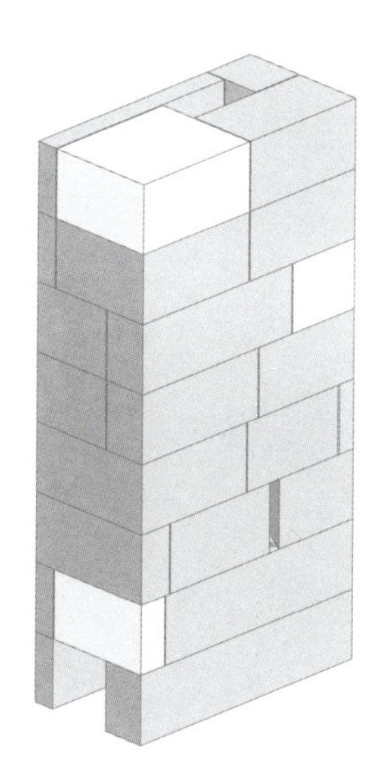

这个设计采取了挖去的操作方法，其特点是从体块的边缘向里作挖去的动作。挖去的两个体块以相切的关系相连接，形成相切的空间关系。其操作的过程通过一系列的示意图而表露无疑。材料区分主要表达实体和空间的关系，即原始体块的表面是一种材料，挖去的空间表面是另一种材料，而不去刻意在各相切的空间之间作进一步的区分。

With carving we describe an operation which cuts into the mass of the block and removes part of it which becomes space. Here, this carving operation is applied from different directions and in varying sizes but always from an edge into the two adjoining surfaces without reaching another edge. Each carving is just deep enough to reach a surface of other carvings. This creates openings between carvings and finally links all the spaces. The axonometric diagrams show that both the space and the remaining block are continuous and can be reversed. The material differentiation emphasises the contrast between the original surfaces of the block and the surfaces of the carvings.

Angela Lee Lai-wai, 0405T2Y2

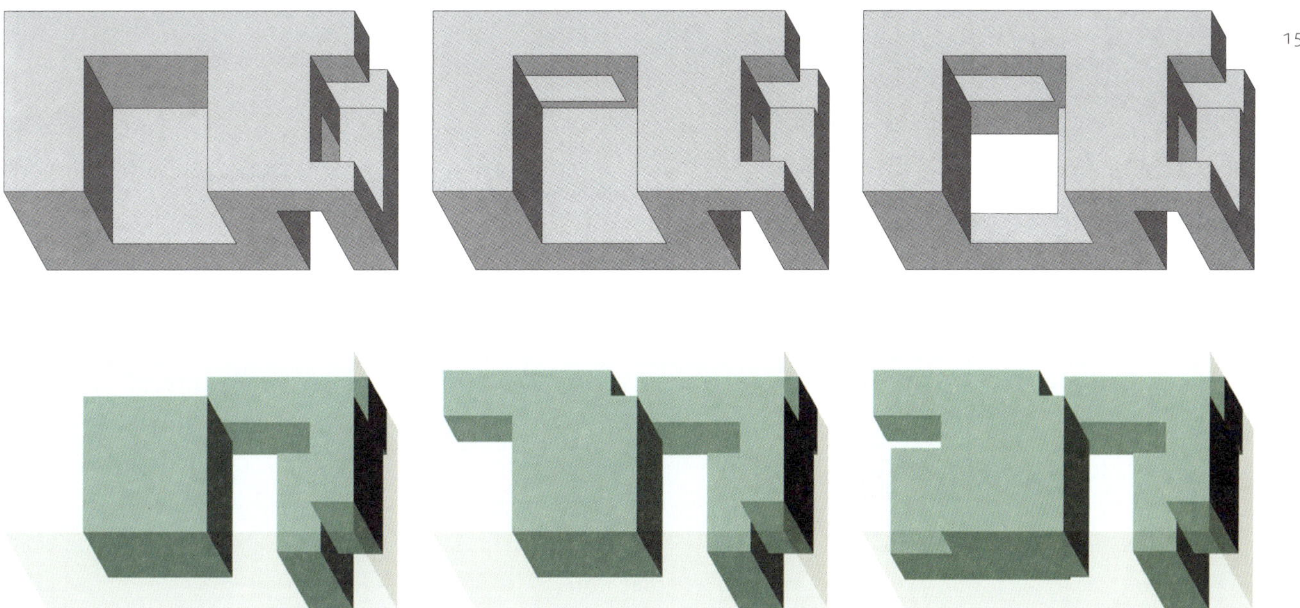

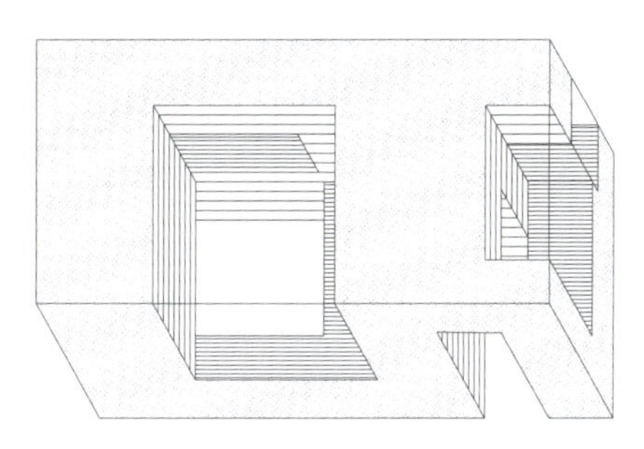

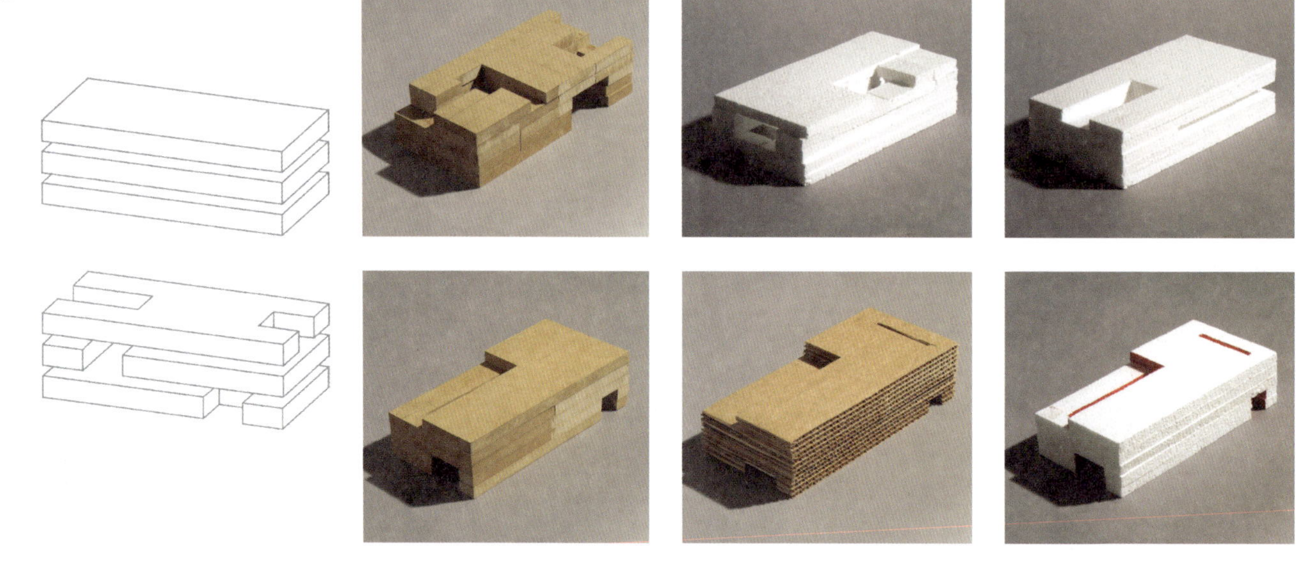

体块模型的操作受制于材料的限制往往不是在一整块的体块上作业，而是从厚板开始，即由数块厚板构成一个大的体块。这个设计利用这一条件作为空间生成的特点，通过分层的挖去作业，最后再将厚板叠加起来，无论是外部的形体和内部的空间均充分表达了该操作的特点。

A primary operation slices a big block into equal horizontal layers which are much smaller than a full storey. A secondary operation carves out part of each layer to form space which can only be inhabited when extending over several layers. In the materialisation phase, the solid block on each layer is further differentiated into solid and translucent. The layering together with the differentiation into three types of blocks or spaces – open, translucent and solid – determine the expression of the overall form and the interior space.

Hilary Ng Ka-po, 0304T2Y2

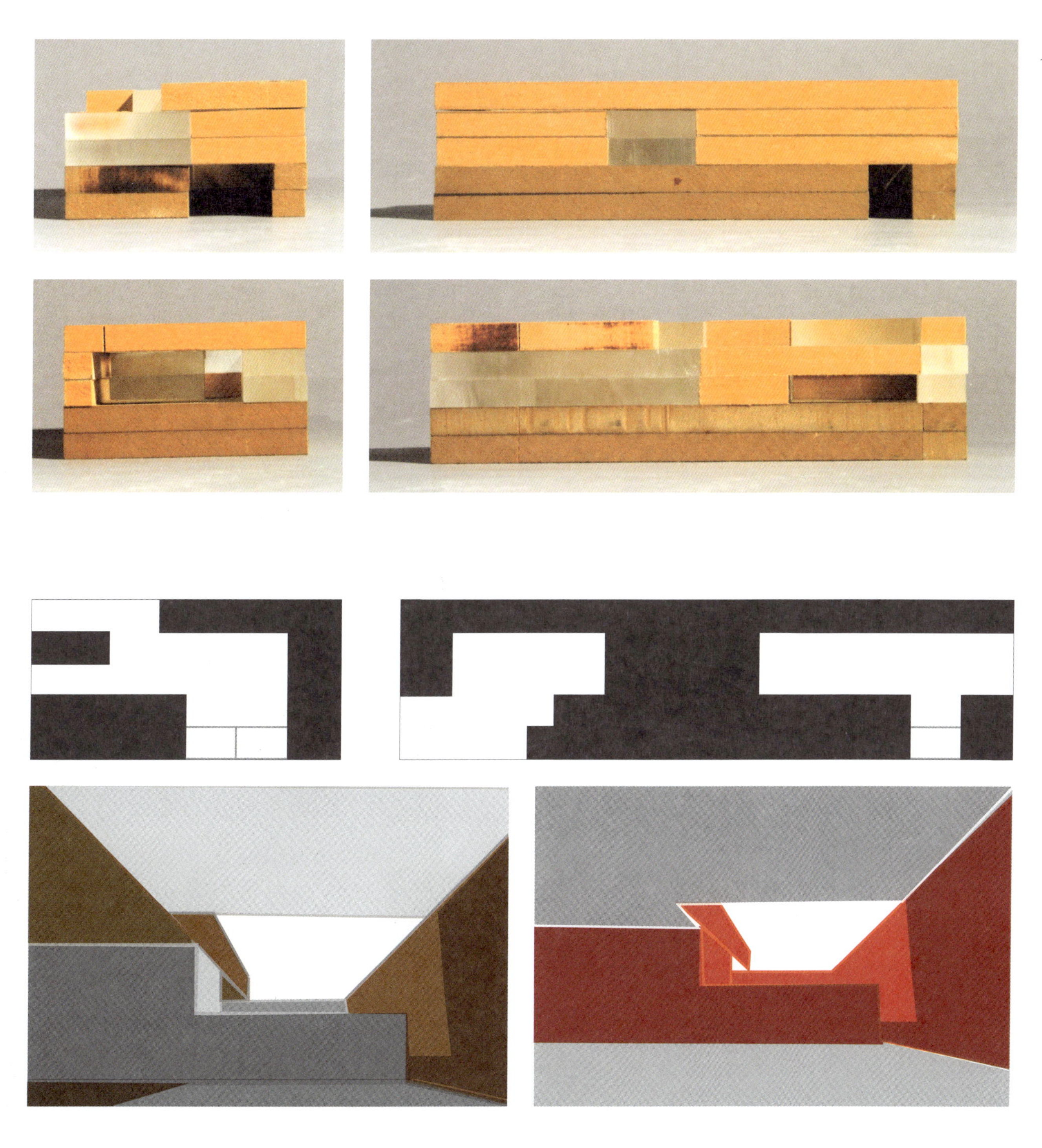

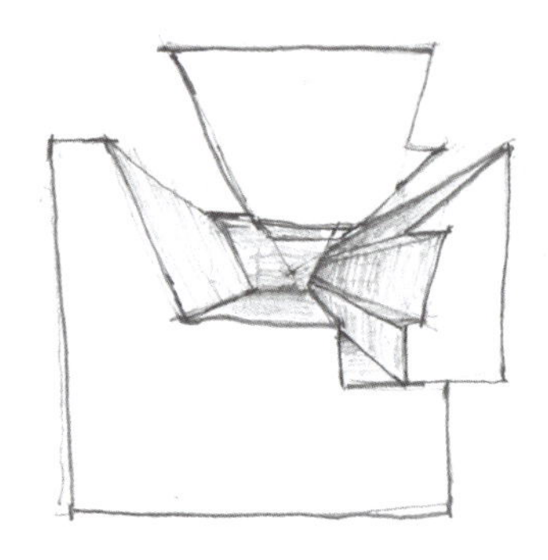

位移作为一种操作策略意味着通过从一个体块中切割和位移部分体块来生成间隙空间。通常从形式表达的角度来说，位移的运动还暗示复位的反向操作。这一切割和位移操作的独特之处在于切割的断面形状。位移的结果不但形成空间，而且还提供了光线进入空间的可能。材料区分着重表达原初的体块和被切割、位移的体块之间的对比。

Apart from multiple operations, there are also combined operations. One type of combined operation can be called cut and shift. The big block is cut in a similar way as when carving, but instead of removing the cut part, it is shifted. Here the cut contains angled surfaces which will create additional space when shifted. Some of this additional space can't be inhabited. Instead it is used to bring light into the interior space or to allow a view from the interior to the outside. In the materialisation phase the cut and shifted smaller blocks are made from a different material, and the cut surfaces of the original block are differentiated from the original surfaces.

Eve Leung Ka-u, 0405T1Y3

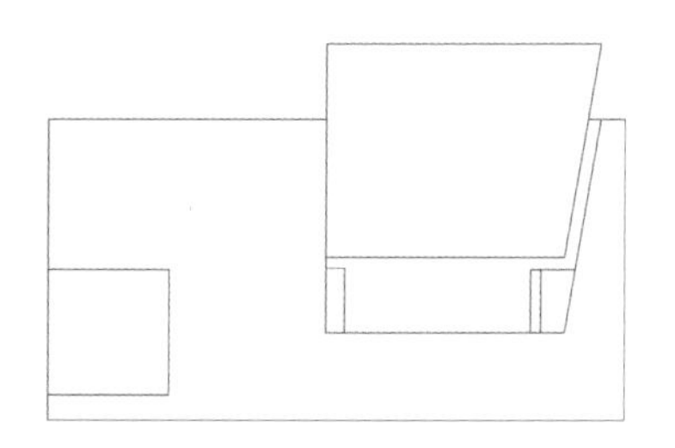

这个设计的基本操作方法是在体块中通过挖去一部分来生成空间，研究的关注点在于空间和空间之间的咬合关系。可以想象成这是穿越整个建筑的一系列的连续的公共空间，而咬合的部分则是两个空间之间的交通空间。这种空间的咬合关系在实体和空间的反转模型研究中得到充分的反映，即空间的咬合关系以实体的咬合关系来表示。在材料和建造的阶段如何来表达不同空间的区分和连接则是另外的一个重要课题。

The carving operation cuts and removes mass from the big block which becomes space. The carving is done with overlapping cuts. This creates clearly defined secondary spaces which are shared by two primary spaces each. When a reversed model – the space as mass – is built, a single mass consisting of interlocked smaller blocks becomes visible, hinting at the continuity of the space through overlapping. In the materialisation phase, the outer surfaces of the original block are differentiated from the cut inner surfaces. These inner surfaces are extended beyond the outer surfaces to strengthen the expression of carving and the contrast of space and mass.

Chan Pui-ming, 0405T2Y3

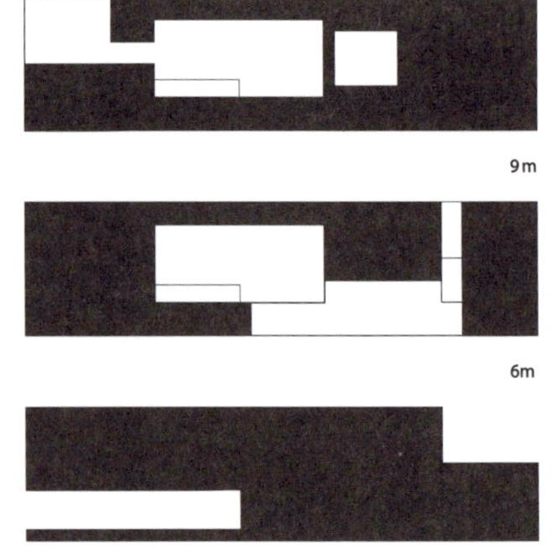

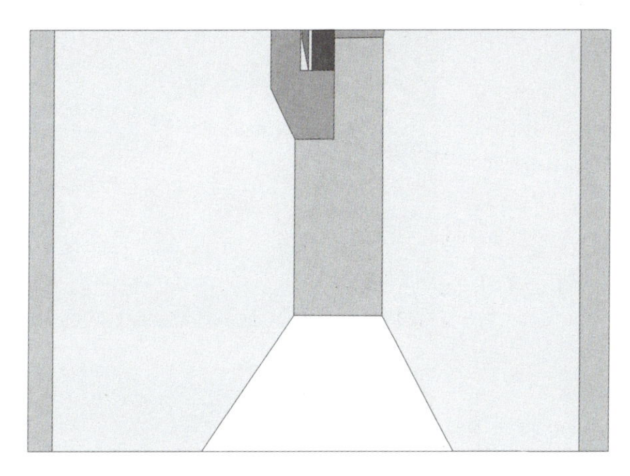

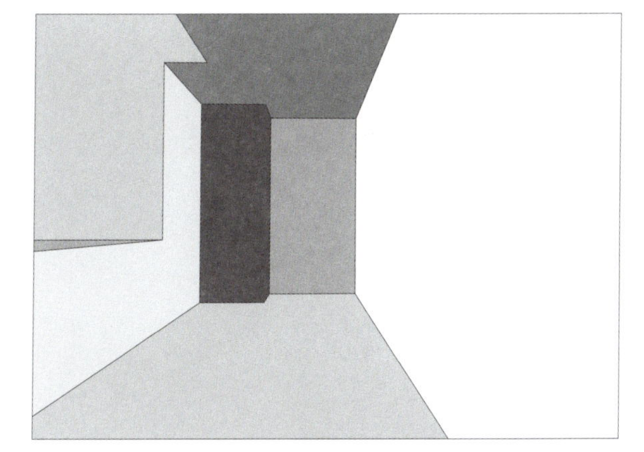

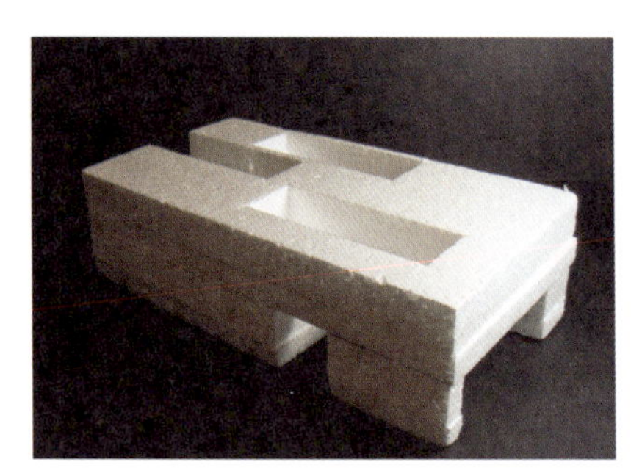

最初的研究是一个简单的挖去操作。将挖去的部分复原形成体块内部空间的实体模型表达。对该实体模型的进一步观察，发现被挖去的部分具有单元体块的特征，由此而产生一个设计研究的新课题，即单元体块的叠加和体块挖去两种操作的结合。最后的结果充分表达了两种操作的特征：从内部看是体块的叠加，从外部看是体块的挖去。

This example demonstrates that the development can be made parallel with two complementing models – one as a mass from which space is carved out, the other as smaller blocks of two sizes cut from a bigger one, and combined to define space between them. In the materialisation phase, the differentiation is used to strengthen this aspect. The modular spaces have opaque surfaces, whereas the original block has a translucent surface. In the elevations, this creates a complex expression with a lateral composition of the carvings, and the composition in depth through the accumulation of layers.

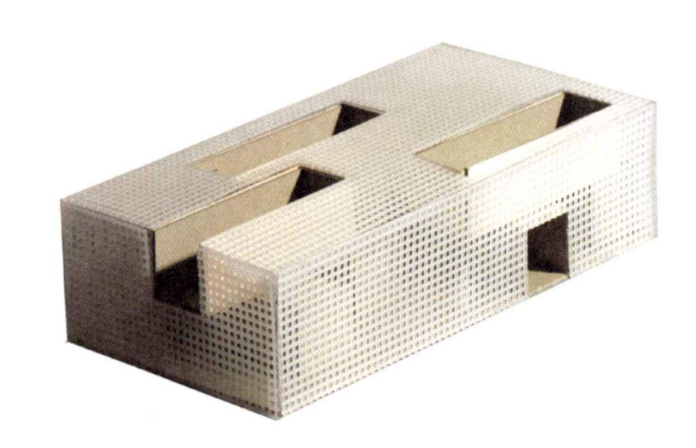

Sofia Meyer Suarez, 0708T1Y2

2 板片 | slab

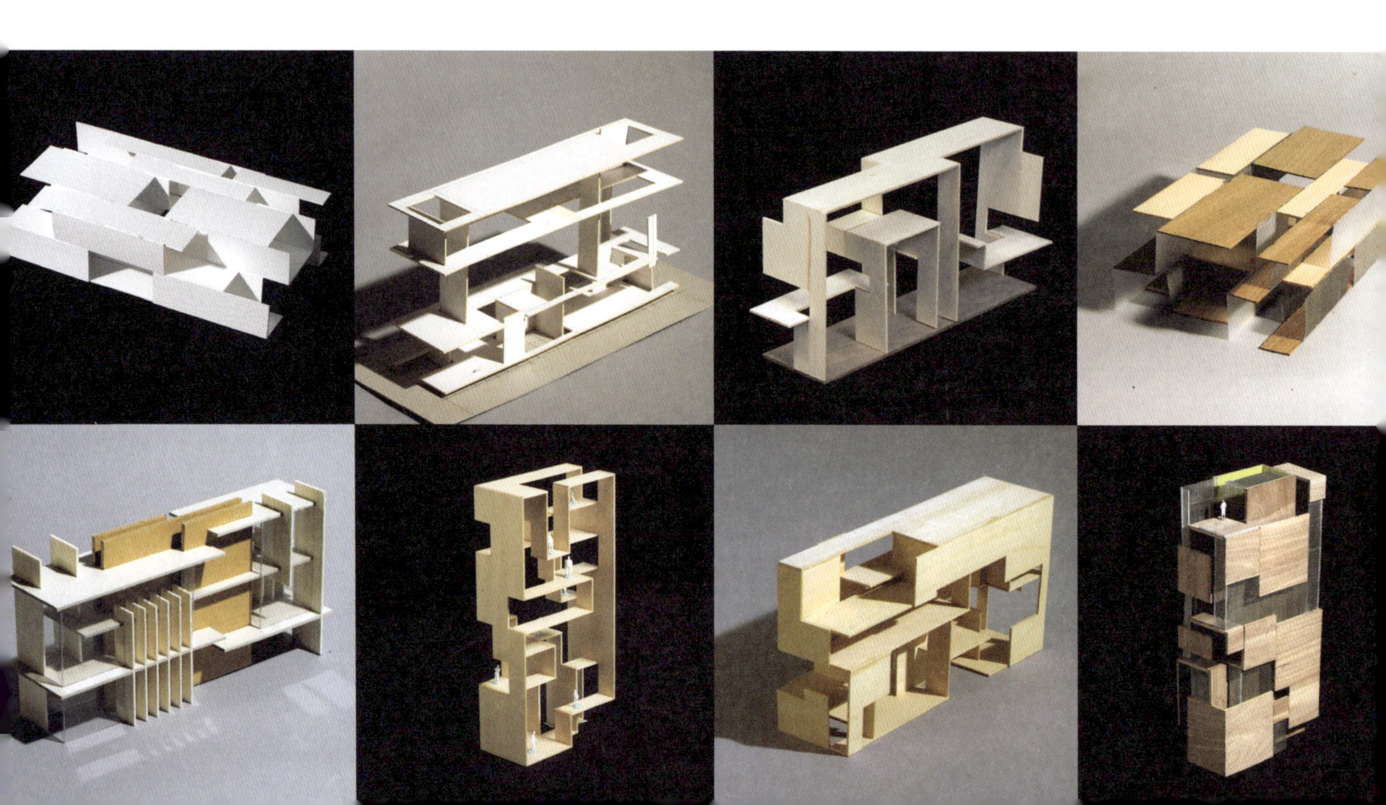

板片作为一个形式要素，它的基本特征是三个方向的量度中有一个相对于其他的两个要特别小，成为一片薄板。两个相对较大量度的面形成表面，而相对较小量度的面形成边缘。板片要素的表达并不在于它的表面，而在于它的边缘。板片对于空间生成的作用主要是以表面来界定空间。它的主要特征是空间的不完全围合。所谓“模棱两可”的空间指的是板片界定的空间边界的这一不明确性。

板片的操作大致可以归纳出几种，但决不限于这几种：一种是一定数量的板片通过特定的连接而形成一个结构和空间体；另一种是一定数量的板片构件的组合形成一个结构和空间体，往往具有单元组合的特性；还有一种是在一张板片上通过特定的操作来形成一个结构和空间体，如弯折、切割、推拉等。但是，实际操作的可能性取决于特定板片模型材料的特性。板片要素的操作体现了强烈的结构性。

Zone space: A slab has two surfaces and a continuous edge. The surfaces separate two spaces and the edges zone the surrounding space in two directions. There are four basic types of operations to differentiate space: first, by cutting smaller pieces and combining them to create space in-between; second, to score and fold the slab to increase the degree of definition; third, by cutting, scoring and folding the slab to increase definition and differentiate zones; and fourth by cutting holes into the slab to differentiate zones. The space defining element is between spaces. Space is perceived in three ways: through the overlapping zones formed by the edges, as clearly separated by a surface and as partial enclosures formed by several surfaces. The change of the viewing angle can change the perception of the space when looking perpendicularly, parallel or diagonally at surfaces. The movement through space changes the perception of the space continuously, moving from zone to zone.

In the materialisation phase, especially the two-sidedness of the slab offers opportunities of interpretation and articulation.

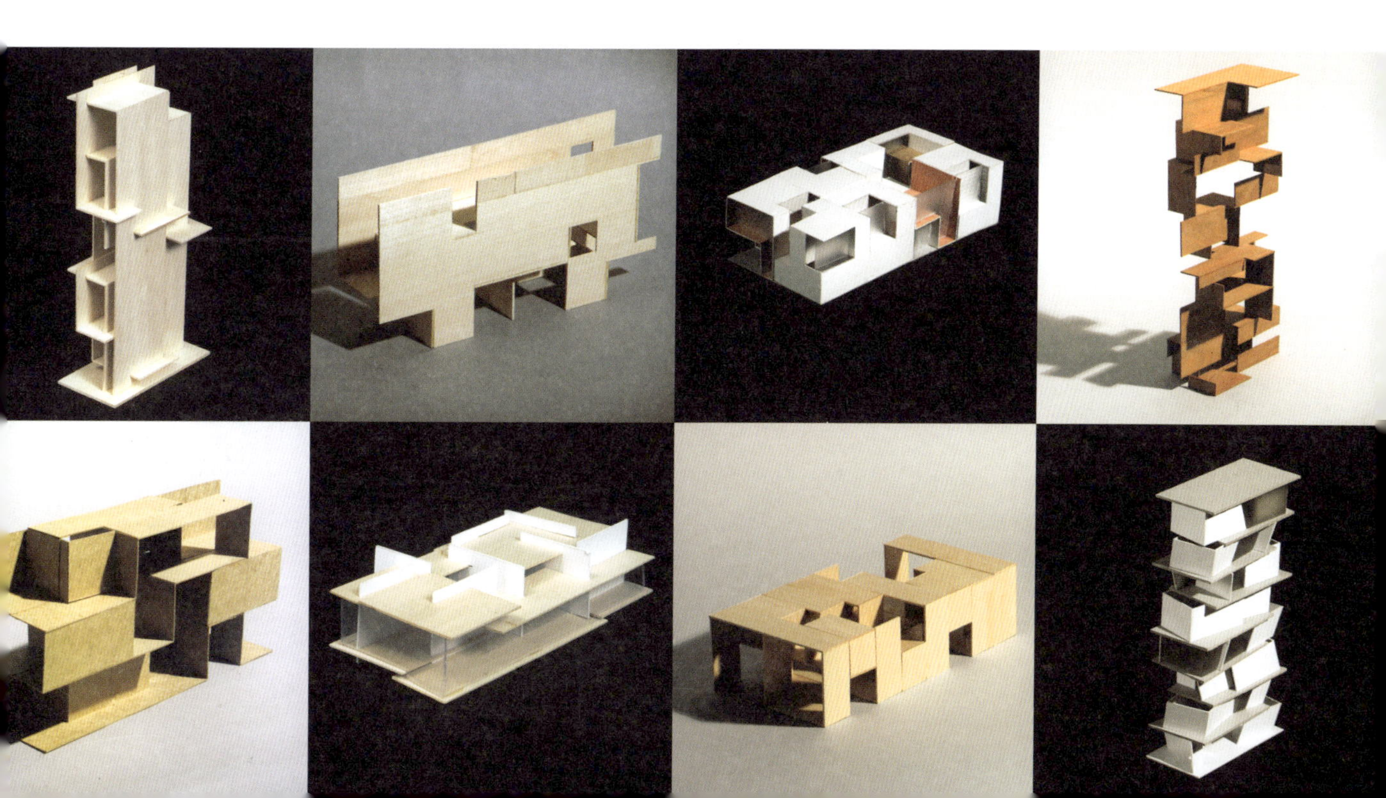

这是最直接的板片空间生成方法。水平板片对空间作垂直的分割，垂直板片对空间作水平的界定。空间在垂直方向上的连通靠水平板片的缺失来实现，垂直板片同时具有结构支撑的作用。材料研究强调水平和垂直板片的区分，以及垂直板片中结构要素和围合要素的分别。

A primary operation cuts a big slab into three types of smaller slabs. A secondary operation interlocks the first two types – two big standing slabs and a series of identical horizontal slabs at equal intervals. A tertiary operation carves – cuts and removes – pieces from some horizontal slabs. And a quaternary operation inserts slabs of the third type perpendicularly into the openings. The slabs modify the space in various ways. They separate spaces. Slab edges define spatial zones. Slab surfaces in one direction connect spaces which are zoned by other slabs. In the materialisation phase, the three types of slabs corresponding to the three main directions of space are differentiated.

Tiffany Tse Hoi-man, 0304T1M1

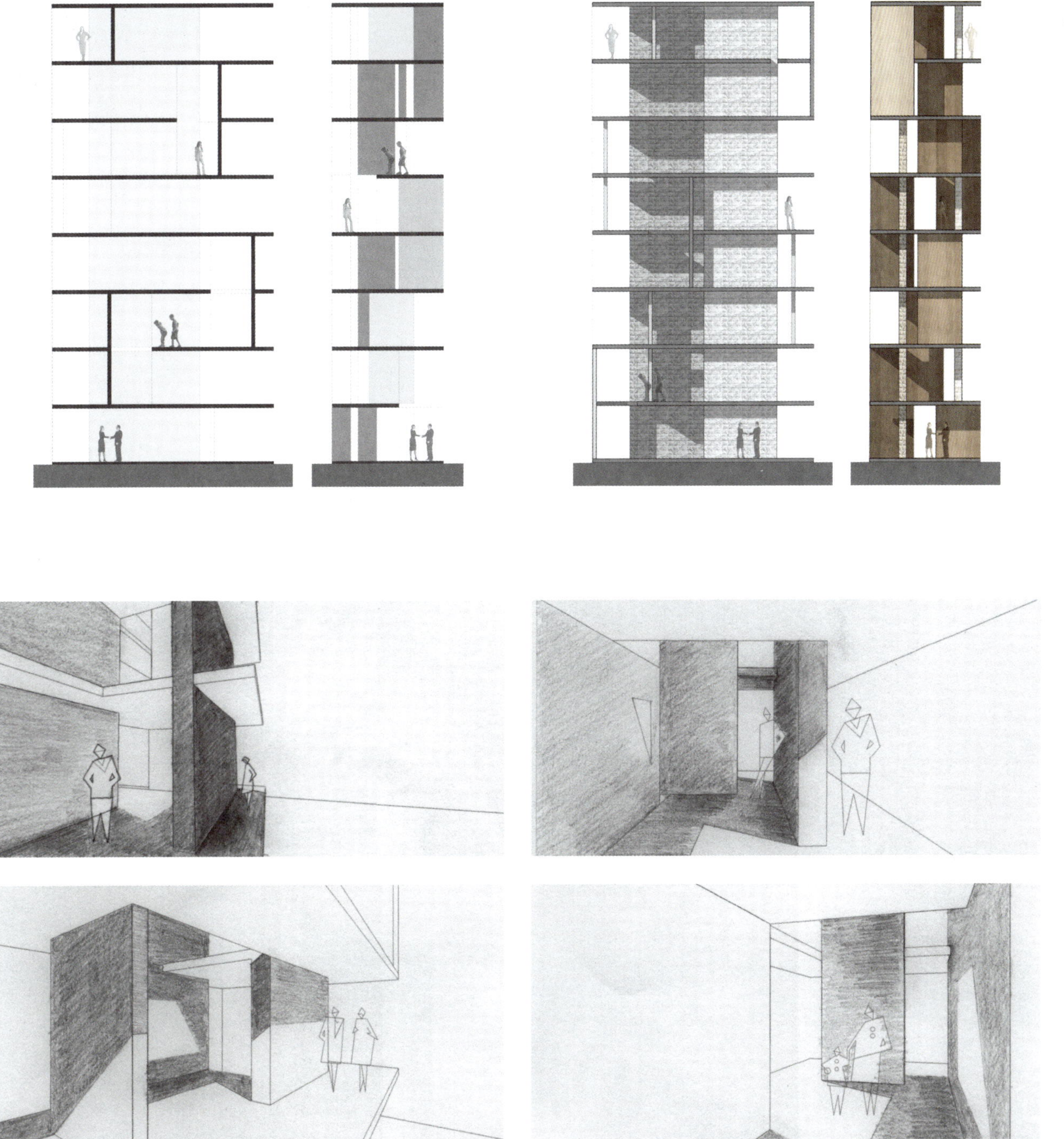

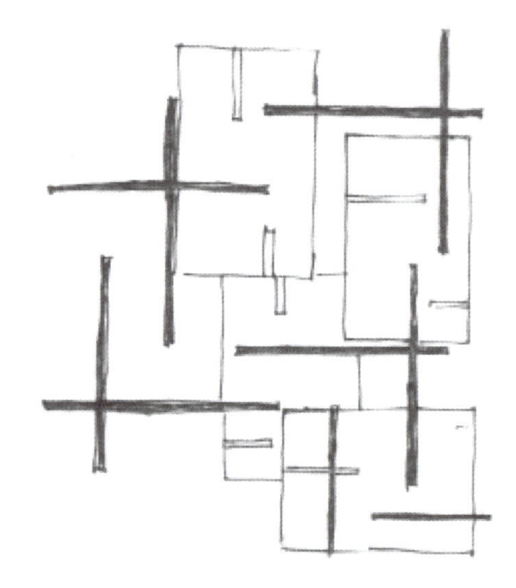

板与板之间如何连接似乎是板片操作的一个最基本的问题，既要解决结构稳定性的问题，同时要追求板片空间的表达。咬合和错位是两种表达板片结构的最常见的连接方式。咬合产生很强烈的板片连接表达，但是这种板片的咬合关系却很难在空间体验时看到。板片错位的连接方式似乎更有利于在保持板片结构的表达的同时实现流通空间的表达。总的来说，在研究中要注意把着眼点从连接问题转到空间问题上来。

In a first attempt, a primary operation cuts identical smaller slabs from a big slab. A secondary operation cuts slits into the pieces. And a tertiary operation interlocks the pieces at the slits or attaches them perpendicularly avoiding aligned edges. In a second attempt, the smaller slabs have varying sizes, slits are only made when necessary and the balance between the two tertiary operations shifts in favour of attaching. The non-alignment of the slabs emphasises their character as elements, and creates the ambiguity of separated and connected spaces and spatial zones.

Becky Cheung Chi-ling, 0203T2Y3

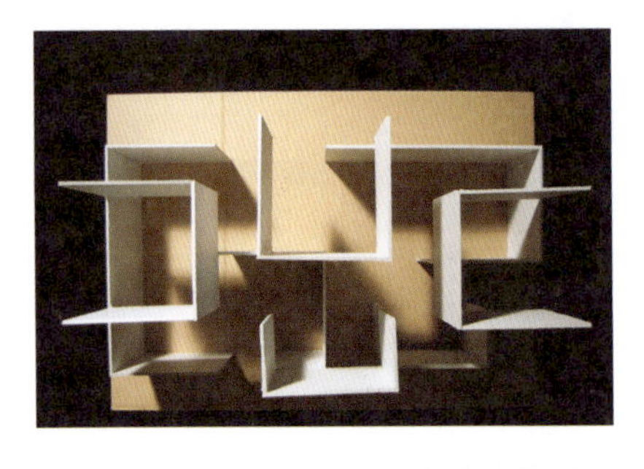

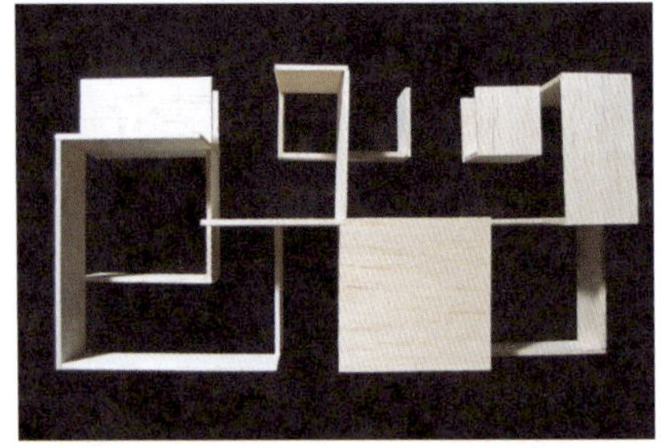

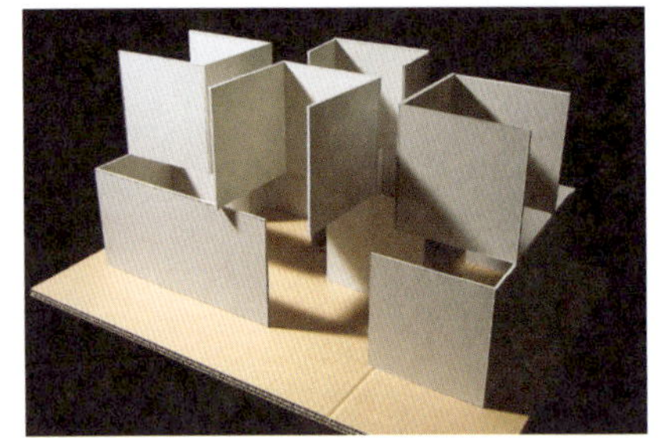

这个设计采用了“U”字形的板片要素，根据边长的变化形成四个单元类型。“U”字形的板片要素本身已经具有空间围合的特性。为了保持板片要素的可读性，设计中特别研究不同的单元连接方法。

A primary operation cuts strips from a big slab. A secondary operation scores and folds the strips into u-shaped units with varying proportions. This folding defines spaces and differentiates spatial zones. A tertiary operation combines the units which modifies the defined spaces and zones. As the folding introduces some continuity of the surfaces resembling the continuity of surfaces in blocks, the combinations have to avoid further continuity to preserve the character of the slab elements.

Becky Cheung Chi-ling, 0203T2Y3

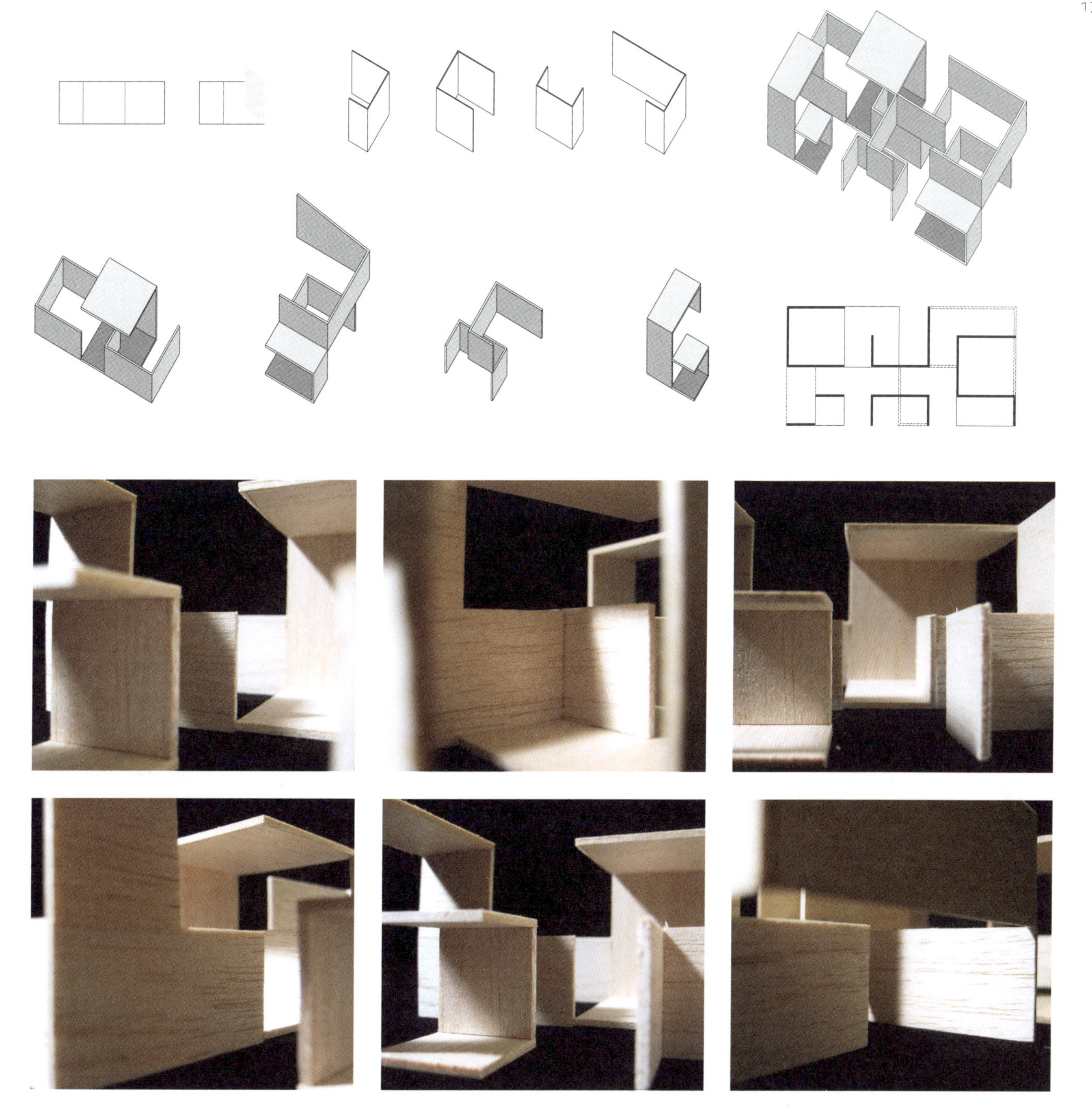

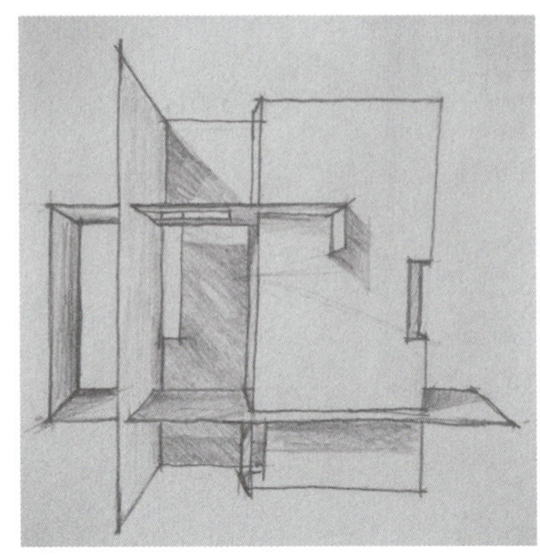

这个设计的起点是研究两片弯折纸板呈直角相交所产生的空间，这不仅意味着两片纸板相交形成的结构体，而且还意味着不同方向的空间的交互作用。在纸板上的开口除了起到空间联系的作用外，主要是想表达两片纸板相交的关系。最后的设计保留了结构和空间的双重特性，但是垂直的板片似乎主要起结构作用。

In the concept phase a primary operation cuts two strips from a big slab. A secondary operation scores and folds these strips into spirals which define and zone space. A tertiary operation interlocks these two spirals which further modifies the defined and zoned spaces. The interlocking avoids alignment. A quaternary operation cuts some holes into some surfaces which further increases the spatial differentiation. In the abstraction phase, a different combination is used, the fourth operation is omitted and additional slabs perpendicular to the spiral surfaces are added for structural reasons. In the materialisation phase translucent slabs are added to provide more enclosure.

Billy Chu Bong-yin, 0506T1Y3

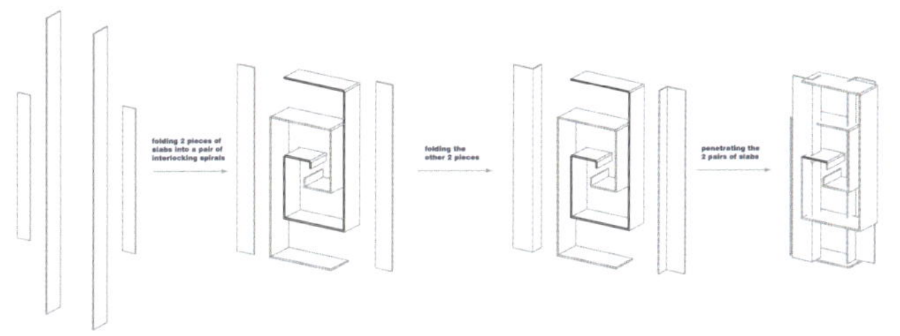
folding 2 pieces of
slabs into a pair of
interlocking spirals
folding the
other 2 pieces
penetrating the
2 pairs of slabs

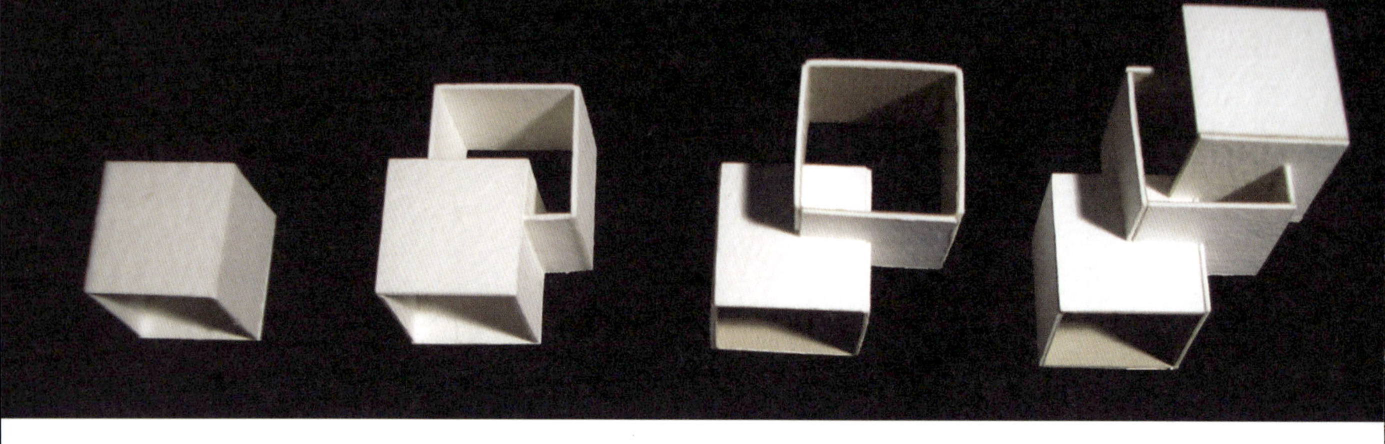

三个套筒的咬合关系研究。与其他套筒结构不同的是，三个套筒的开口方向呈直角的关系相咬合，如此可以形成有分有合的空间关系。研究的一个要点在于咬合关系的空间表达。因为，两个套筒的咬合只能从模型的角度才能看到，而在空间中只能看到几个面相交的死角。解决的方法之一就是在相交处开槽，让一片板从一个空间穿越到另一个空间。

A primary operation cuts a big slab into strips of varying proportion. A secondary operation scores and folds these strips into sleeves which define a space each, and zone the surrounding space. A tertiary operation interlocks these sleeves perpendicular to each other, avoiding alignment. This defines more spaces and creates more spatial zones. A quaternary operation cuts slits into surfaces along the intersecting lines to reduce surface continuity and avoid block-like spaces, increasing the zones further.

Yuki Ng Yee-ki, 0506T2Y3

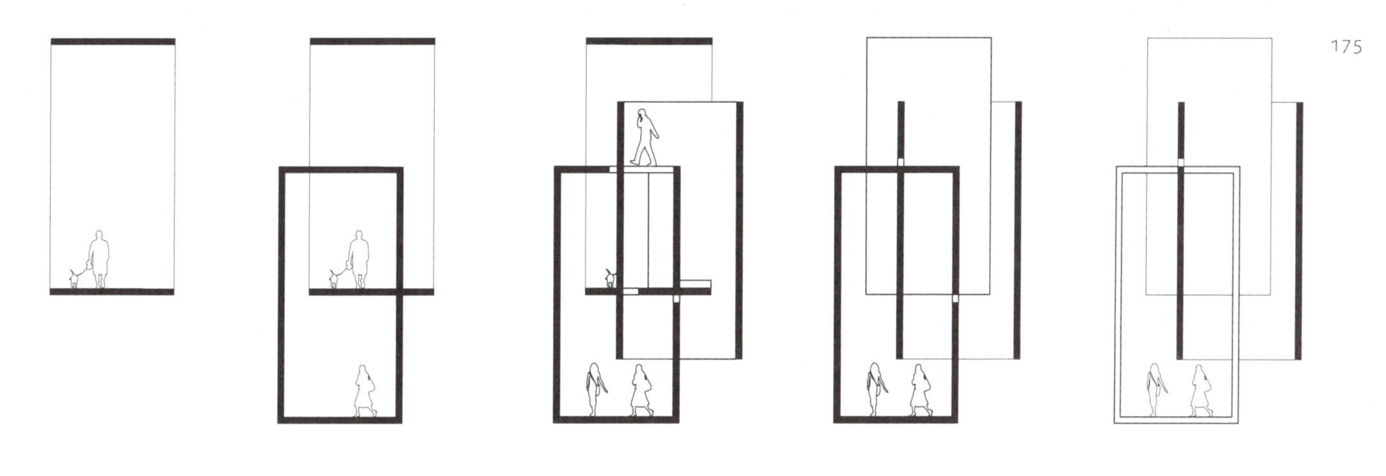

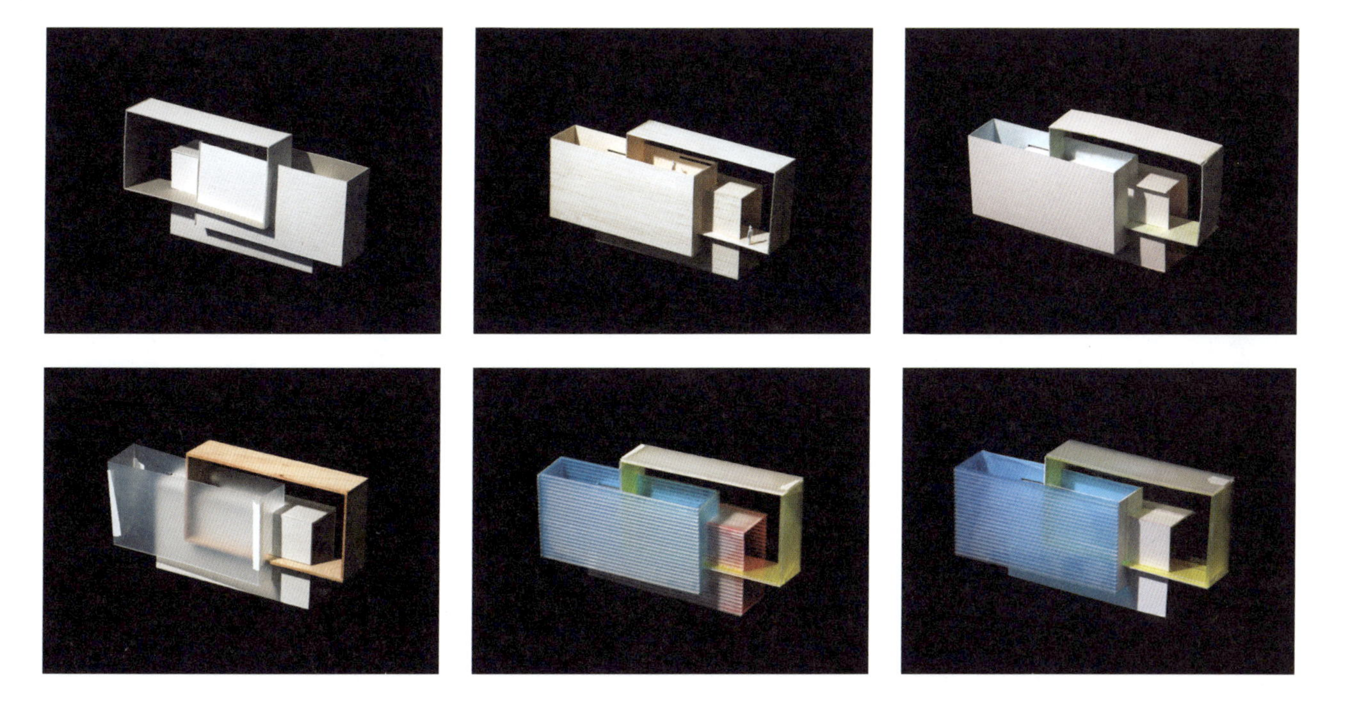

一个平面的板片通过切割和弯折而形成三度的空间结构。一个呈垂直状态的板片，切割和弯折的操作形成开启，同时弯折的板片可以作为墙体或楼板。两个并列的板片形成建筑的两个外墙面，弯折的部分向内形成楼板，向外形成露台，水平板片的连接形成稳定的结构体。材料研究提示三种表达的可能性：弯折部分与原板片的区分，弯折操作的表达，以及两个原板片的区分。

A primary operation cuts two equal slabs from a big slab. A secondary operation sets these smaller slabs next to each other. This creates a primary space. A tertiary operation cuts u-shapes into the slabs, scores the remaining edge of the formed rectangles and folds them 90° towards the opposite slab. With this, the primary space is structured into many zones to which the created openings and the corresponding folded slabs contribute. This third operation also supports use by providing floor slabs, and structure through connections. In the materialisation phase, different interpretations of the slab surfaces which result in different expressions are studied.

Jonas Tang Chin-hong, 0405T1Y2

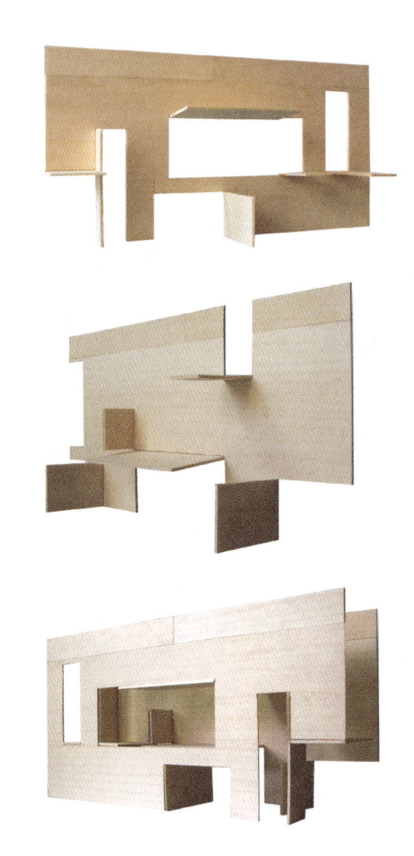

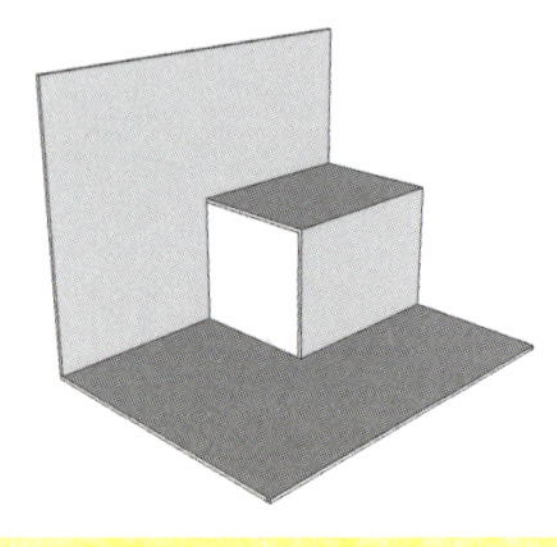

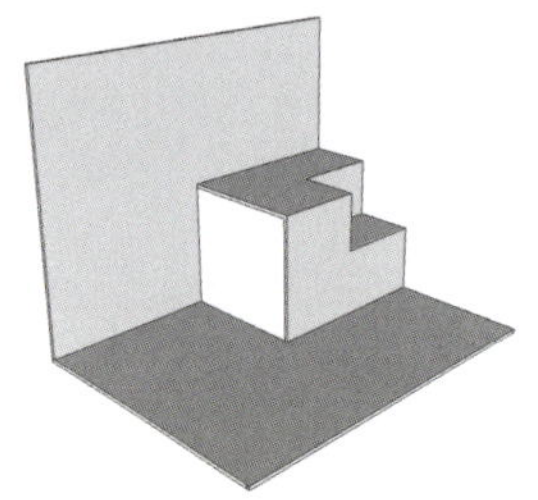

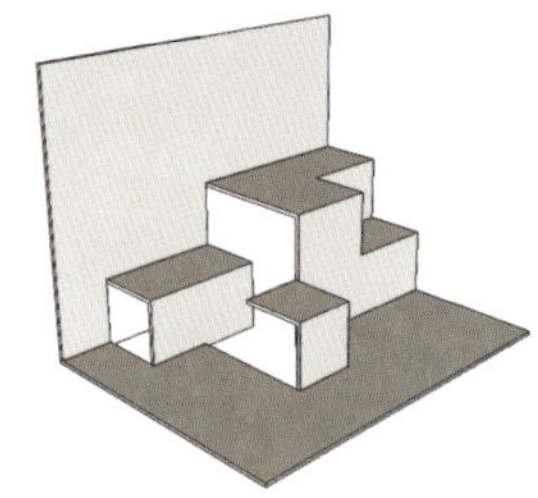

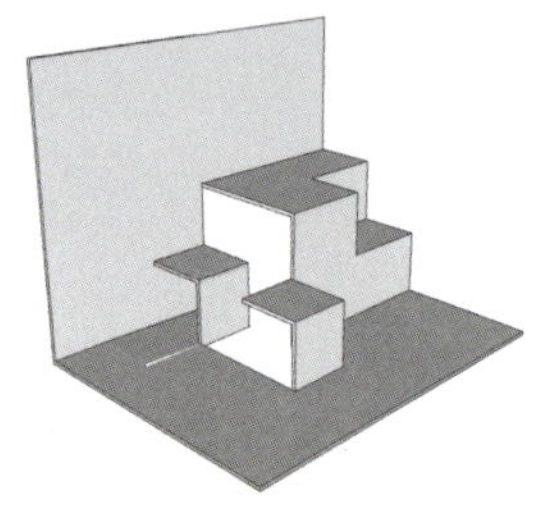

一个水平的纸片作直角的弯折而形成垂直的墙面，界定两面的空间，再在纸片上作切割和推挤的操作，就得到一个复合的空间结构，原先被墙体隔开的两面也因这一操作而相通。从透视研究中可以看到，这种操作产生的空间需要从一个特定的角度来观察。将这一概念运用到设计中去，作者尝试了几种不同的操作方法。最后的结果是三个单元体的叠加。

In the concept phase, a primary operation scores and folds a slab, defining a primary space. A secondary operation cuts one or two lines across the primary fold and reverses the fold, through which a secondary space and additional zones are created in the primary space. This operation can be repeated and also reapplied. In the abstraction phase, the primary operation is modified by forming a loop as a primary space, and finally three such loops are stacked in a way that secondary spaces are formed vertically. The perception of space changes with the viewing angle. In one direction the edges dominate, in another the surfaces do so, with more complex in-between stages.

Ieong Si-kei, 0708T1Y2

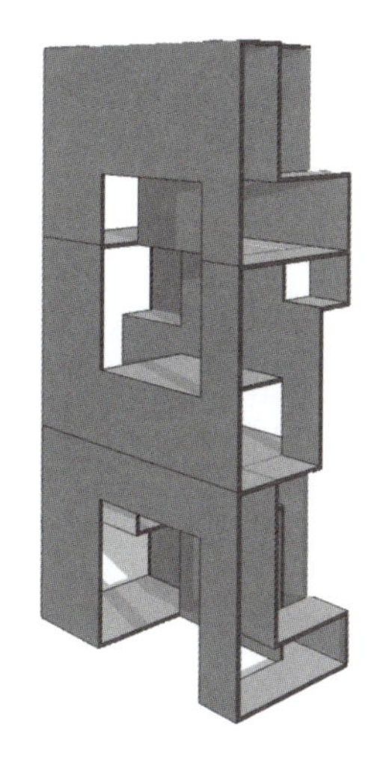

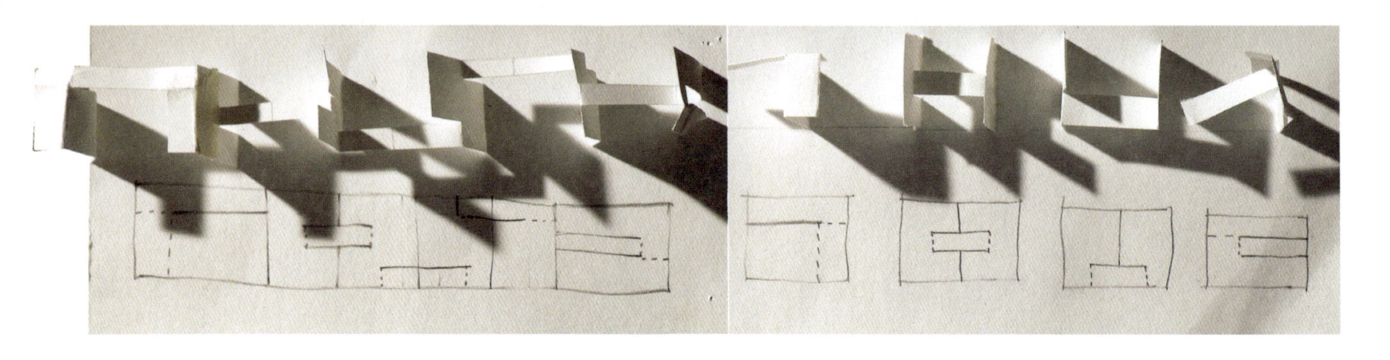

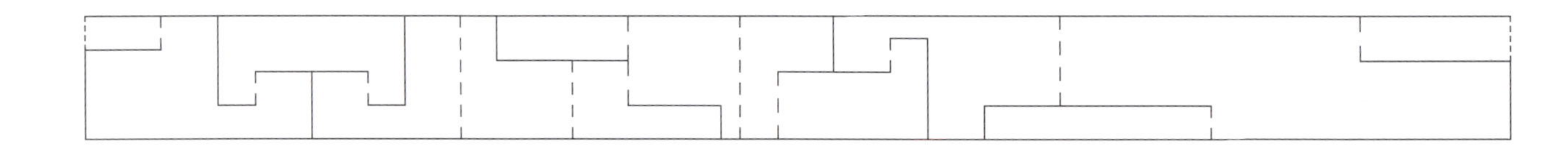

折叠操作的一个典型问题是如何从一张完整的卡纸上通过切割和弯折来完成一个空间结构的造型。设计的难点在于如何在形式的完整、稳定的结构，以及丰富的空间体验之间取得平衡。研究的过程从最基本的单一操作开始，再探索一个连贯的操作，最后形成一个完整的设计。

The original slab is a long strip. A complex operation cuts L- or T-shapes into the side of the strip, scores the strip partially or completely across the strip and folds at 90° along the scored lines. With this, spaces are defined by horizontal surfaces and vertical surfaces in one direction. The edges differentiate the spaces further into zones. The principles are systematically studied, but the final version has to be explored by trial and error. The spatial perception changes according to the viewing angle, with two contrasting perpendicular directions and oblique views.

Karen Tsui Ka-yan, 0708T1Y3

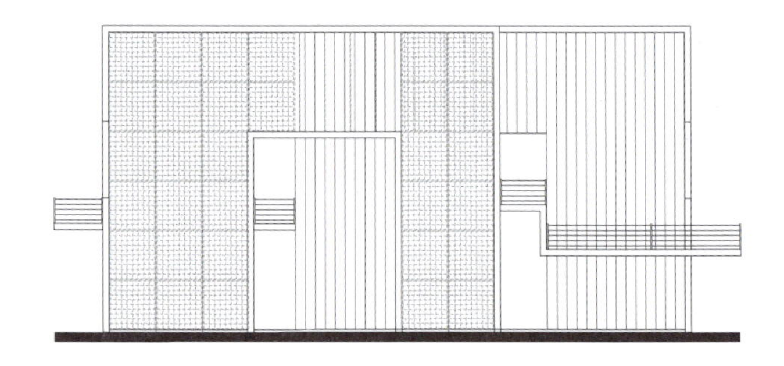

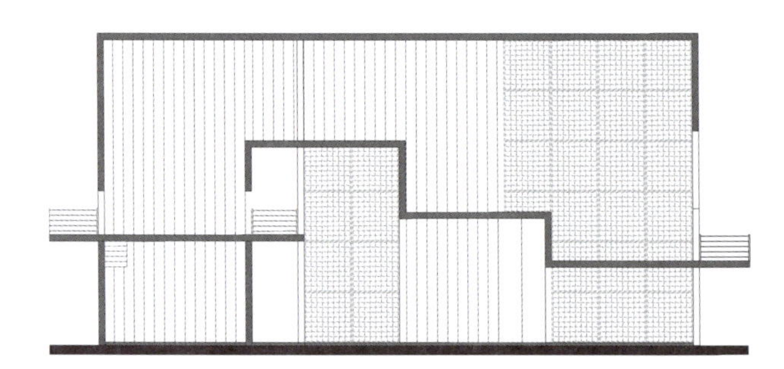

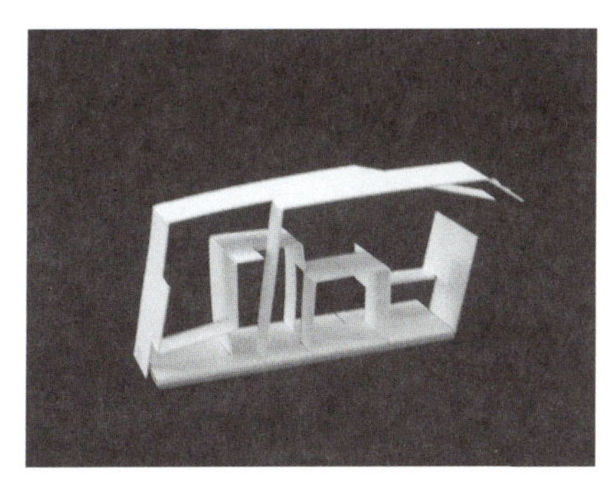
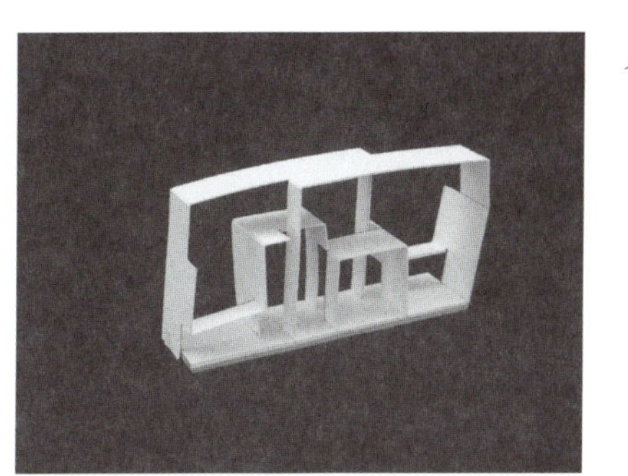

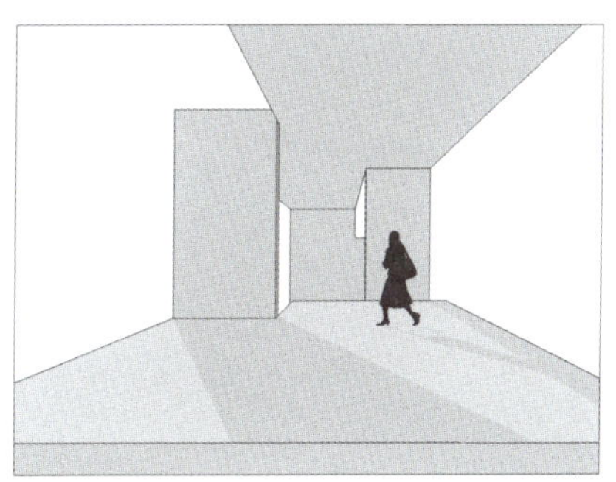
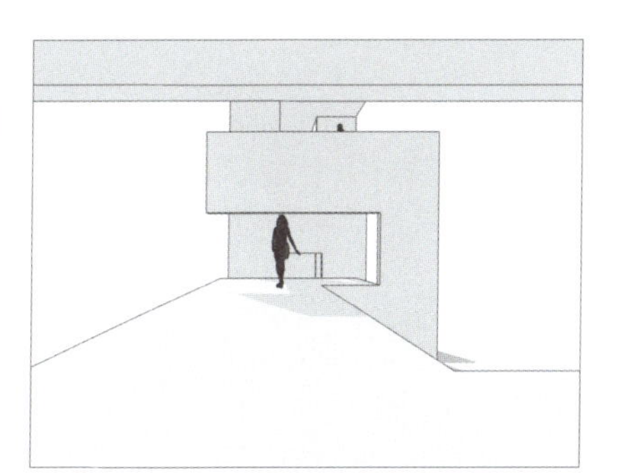

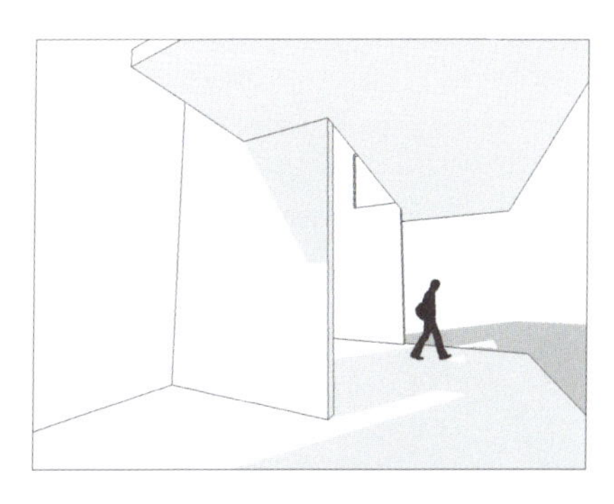

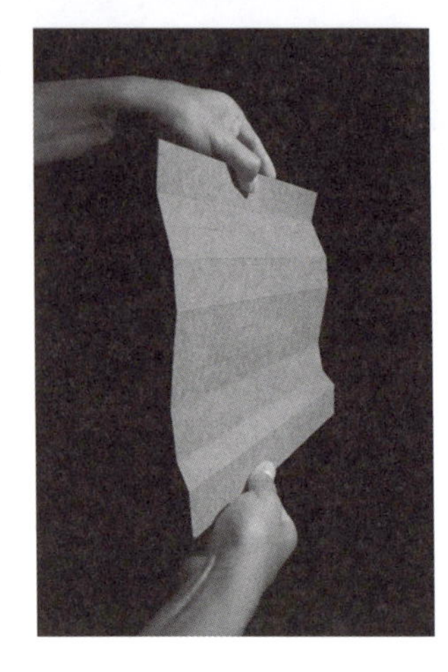

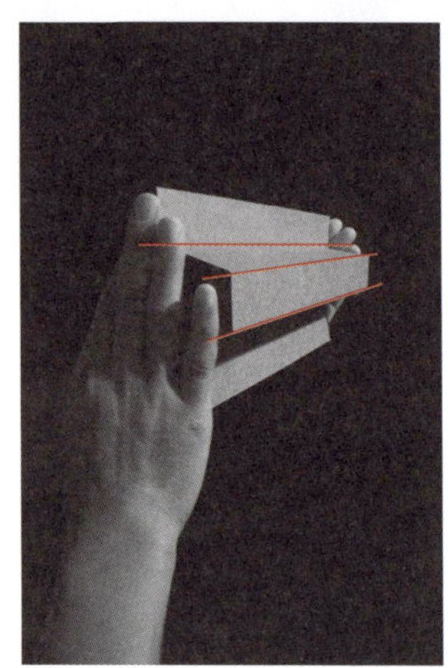

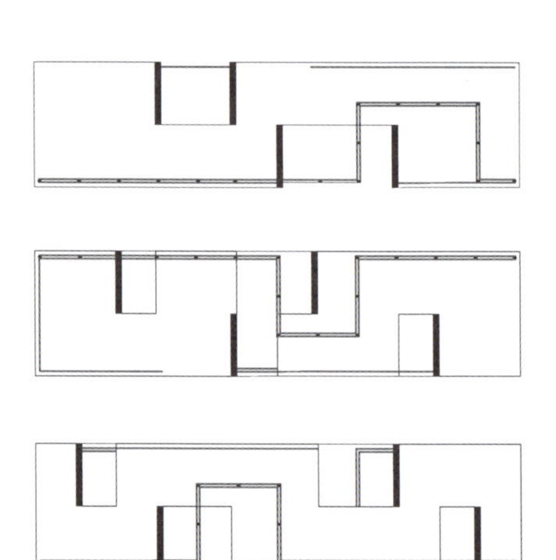

一张卡纸通过弯折而形成一个空间结构。这个设计的独特之处在于在完成一个方向的弯折之后再施加一个相对方向的挤压操作，其结果是引起纸板在另一个方向上的弯折。研究的焦点在于如何利用挤压的操作来创造更加丰富的空间关系。

The original slab has a certain proportion. A primary operation scores this slab in parallel lines and folds along these lines in alternate directions by 90°. Within the outline of the object, three primary spaces are defined. A secondary operation cuts parallel to the primary folds, scores perpendicularly and folds along these lines at 90°. This results in a shortening of the length of the primary spaces, the vertical extension of one primary space to another, a lateral layering of the primary spaces and the creation of secondary spaces. The consistency of the secondary operation is difficult to verify with the presented material.

Lau Chun-yiu, 0506T1Y3

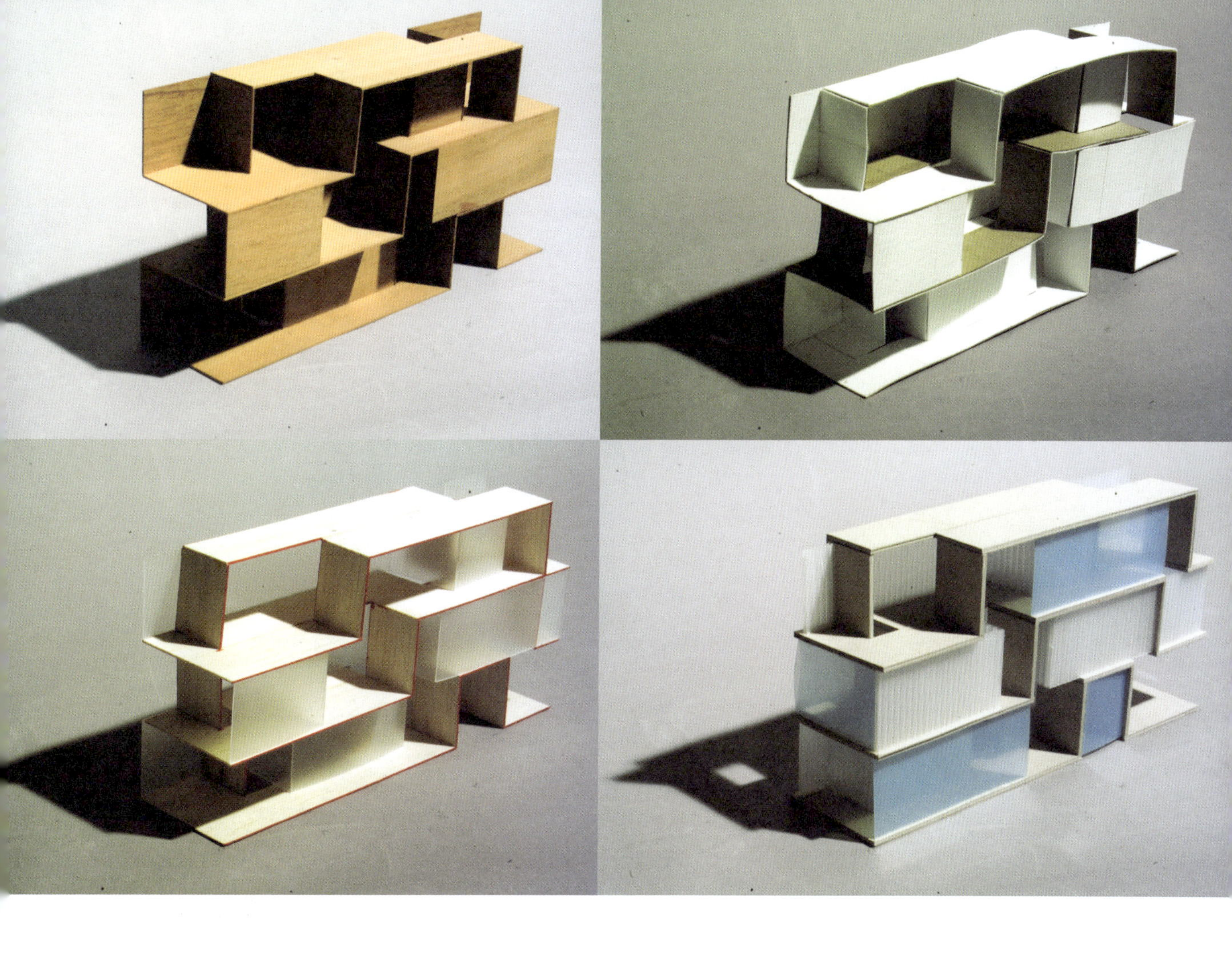

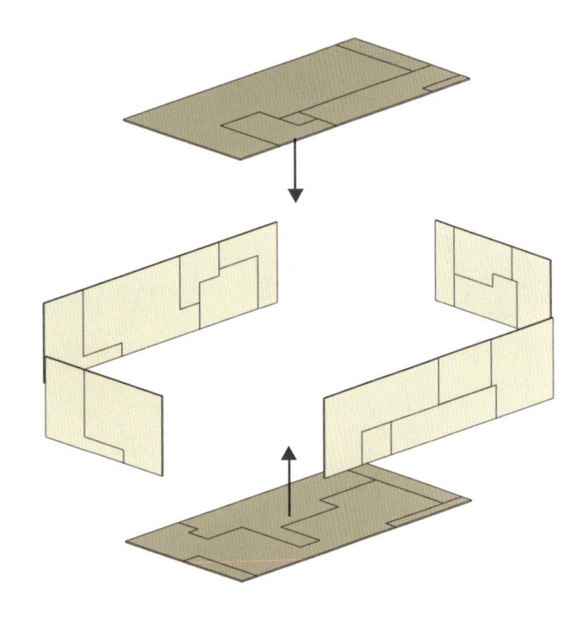

一个盒子有六个面，表现为一个实体，其中则是一个封闭的空间。通过将从盒子的六个面上切割出的形状向盒子里面推，这个设计从盒子的实体特性以及内部的包裹性空间出发，结果则是丰富的板片要素和流通空间的表达。材料区分对于表达推出去的板片和原先的板片之间的关系以及“推”的操作均至关重要。

A primary operation forms a box out of six slabs cut from a larger piece, without connecting them. This of course forms a complete block with one internal space. A secondary operation cuts lines into each surface which divide each slab into smaller slabs. A tertiary operation shifts the smaller slabs into the block to various depths. As a result, the primary space gets differentiated into smaller spaces, and the block expression completely disappears. The defined secondary spaces have at a certain depth two opposing parallel walls with complementing open and closed surfaces and corresponding edges. In the materialisation phase, the expression of this shifting or pushing of the slabs is supported by material differentiation.

Richard Li Kin, 0506T2Y2

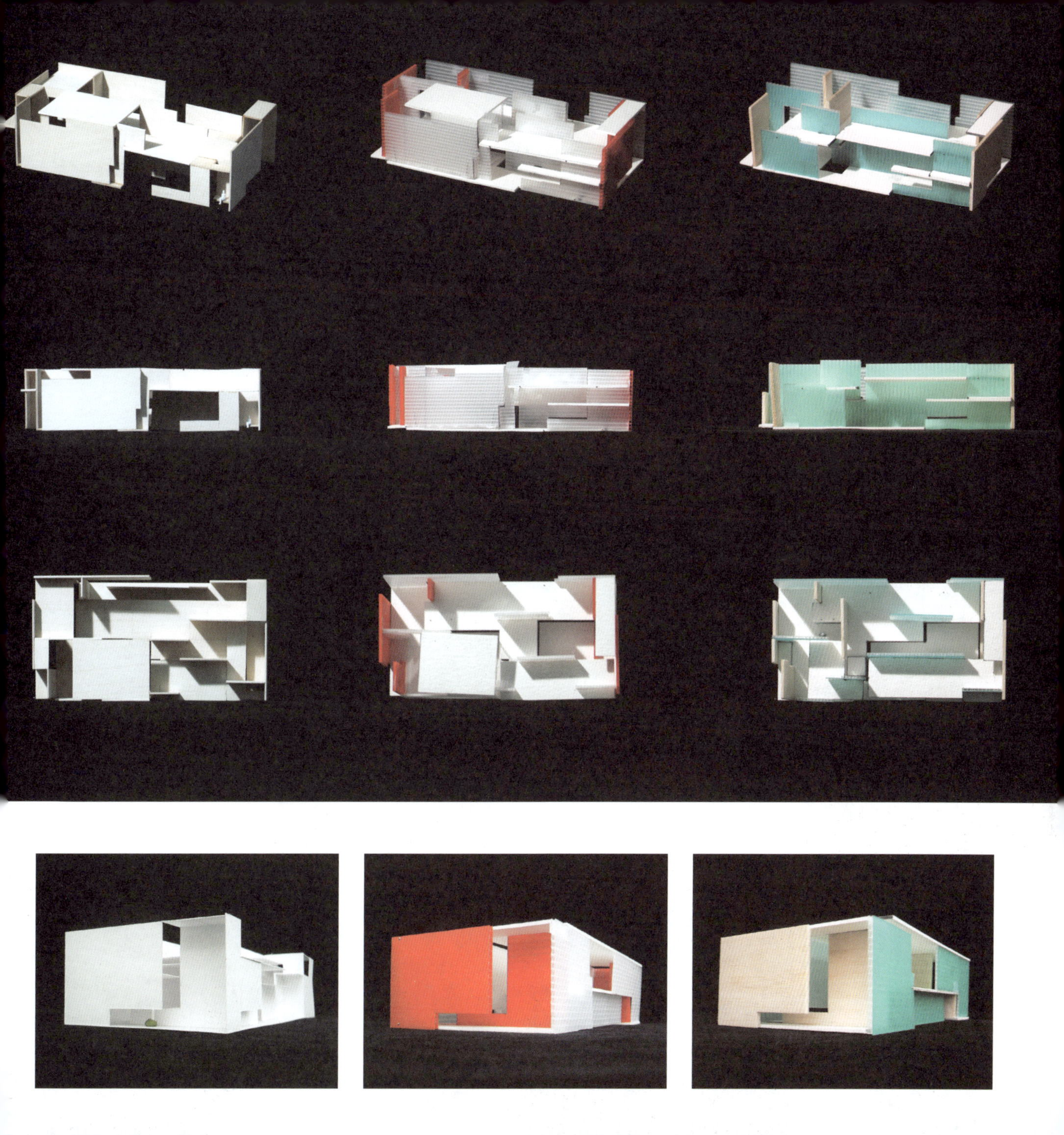

3 杆件 | stick

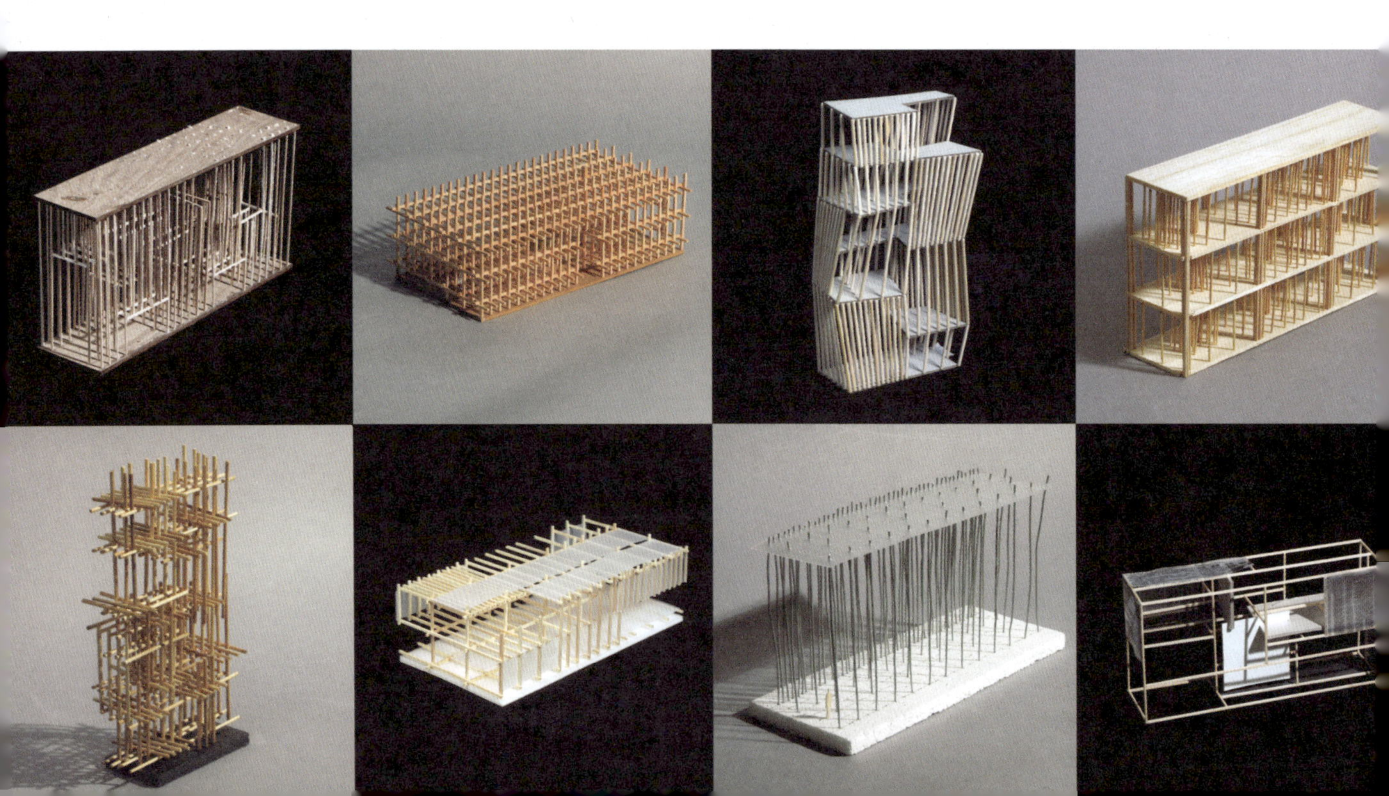

杆件作为一个形式要素，它的基本特征是在长宽高的三个量度上有一个量度占主导，形成一条线。杆件要素没有表面，只有边缘。杆件对于空间生成的作用既不在于如体块般的“占据”，也不在于如板片般的“界定”，而在于对空间的“调节”，即对空间的密度和韵律的调节。

杆件的操作大致可以归纳出以下几种，但决不限于这几种：一种是以一定数量的杆件通过特定的连接方式形成一个结构和空间体，材料的操作表达建造方式；一种是对密度的操作，犹如植树造林，是杆件生成空间的特有方式；一种是以杆件构成某个特定的构件，再通过构件的组织来形成一个结构和空间体；最后一种是在以根杆件上通过弯折的操作来形成一个结构和空间体。但是，实际操作的可能性取决于特定杆件模型材料的特性。

杆件要素的操作体现强烈的结构特性。杆件形成的空间，从使用的角度来说，需要加入板片或体块的要素来形成楼面和房间。

Modulate space: A stick has only one edge. It forms a locus defining space around it. There are three basic types of operations to modulate the space further: first, by cutting smaller pieces and placing them to create density differences; second, by cutting smaller pieces and connecting them to create multiple loci and to form frames; and third, by bending the stick and therefore changing the direction of the locus and forming some outline. The space defining element is inside the space. Space is perceived in several different ways, as surrounding a stick, as a permeable boundary, as a linear outline and in the accumulation in depth resulting in a differentiation of density. The change of the viewing angle triggers different perceptions. Some types of perception even need continuous small changes of the viewing angle. The movement through space intensifies this experience.

In the materialisation phase the interpretation and articulation is relatively limited by the sticks, but the slabs and blocks which have to be introduced for other considerations, expand the possibilities.

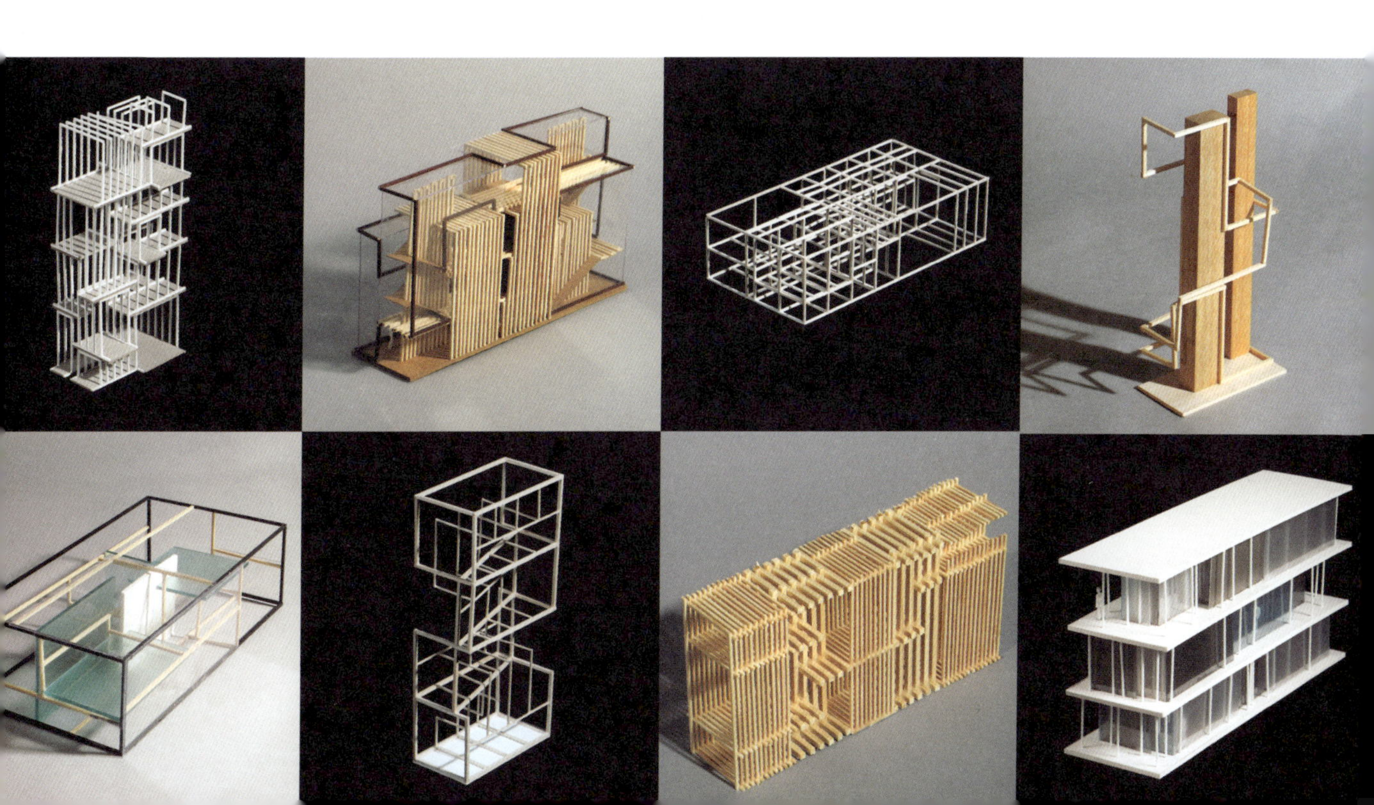

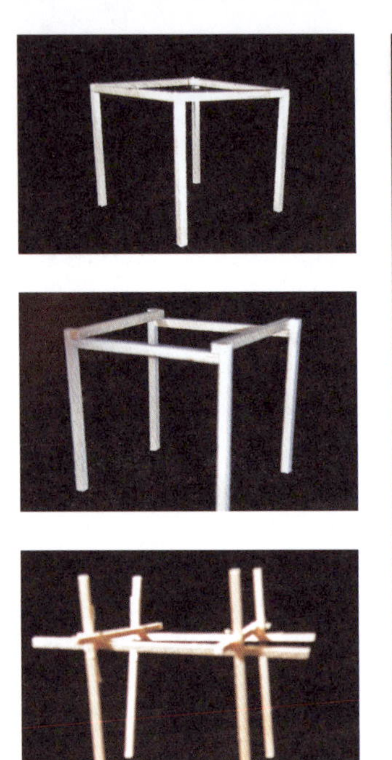

杆件的一个最基本的操作就是搭接，用若干杆件以某种特定的方式相互连接而成为一个空间结构。在这个设计中，研究的兴趣显然在于杆件的连接方式。最后的设计包含了主次梁的关系，连续的主梁强调空间的方向，次梁则给予连续的空间一种韵律的调节。

A primary operation cuts two types of sticks – one thin, the other thicker – into smaller sticks. A secondary operation joins the sticks in nodes different for each of the three main spatial directions – thick single centred, thin double centred and thick single off centre – supporting the hierarchy of direction in the flat object. As not each possible node is used, the density of the sticks varies within the overall object. In the materialisation phase, additional slabs are used to provide floors and enclosure.

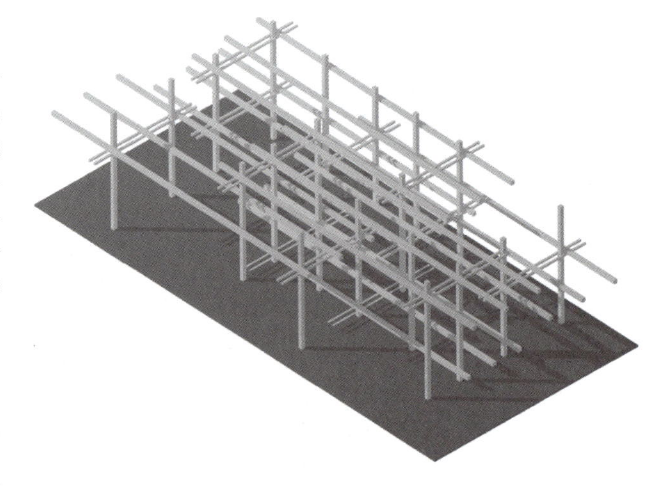

Anna Chow Tsz-kwan, 0203T2Y2

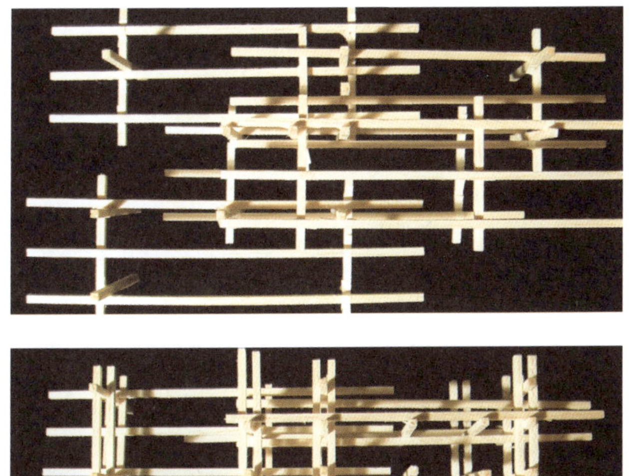

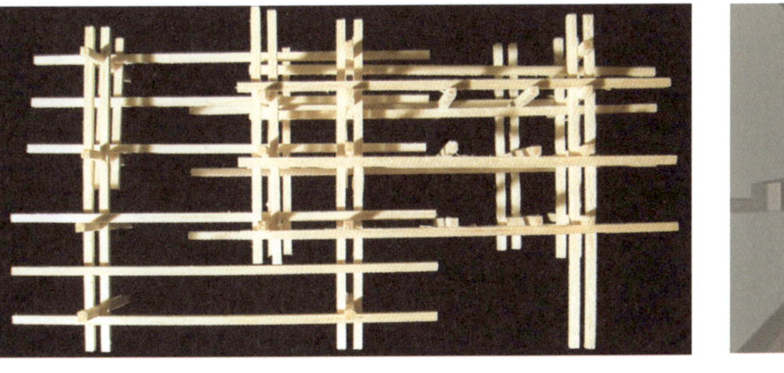

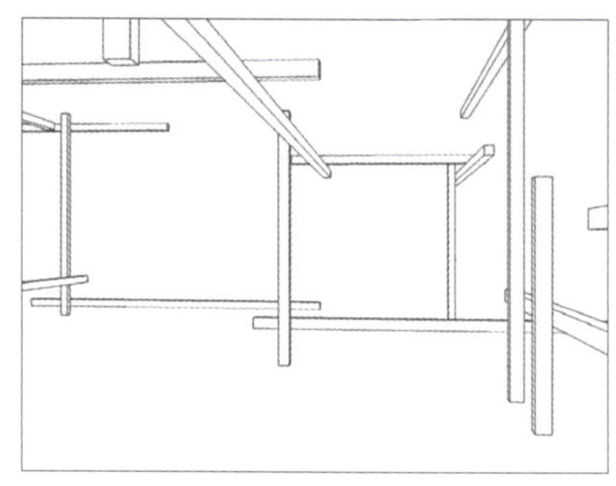

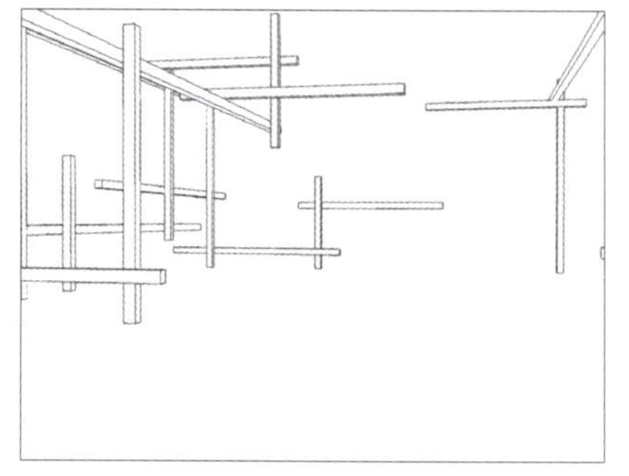

这个设计中杆件的搭接方式显然更为自由，垂直杆件的定位和水平杆件的高度均根据设计者的空间意图来灵活安排，不过结构的稳定性仍然是需要达到的目标之一。整体上，由于这种杆件使用的方式致使空间没有明确的主次方向。

A primary operation cuts one type of stick into various sizes. A secondary operation joins these sticks in nodes where two or three sticks touch at the side. As the distribution is uneven, some spatial differentiation occurs. In the materialisation phase, the sticks in the three directions get differentiated – thick vertical, thin longitudinal and with a rectangular cross-section laterally in two orientations. This gives the space a little bit more structure. Some translucent slabs and blocks are added to provide enclosure.

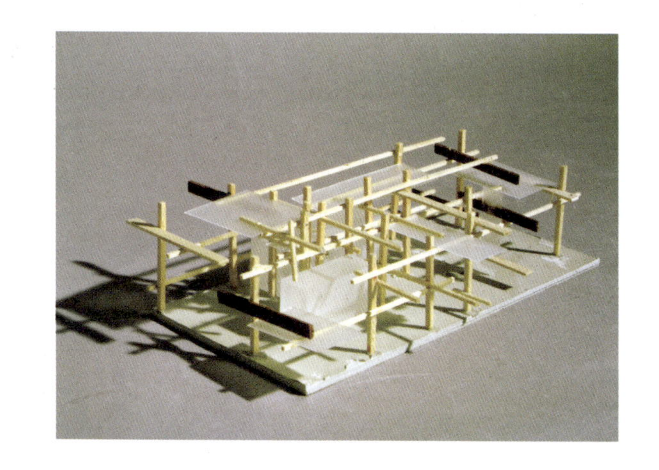

Louise Yang Lu, 0506T1Y2

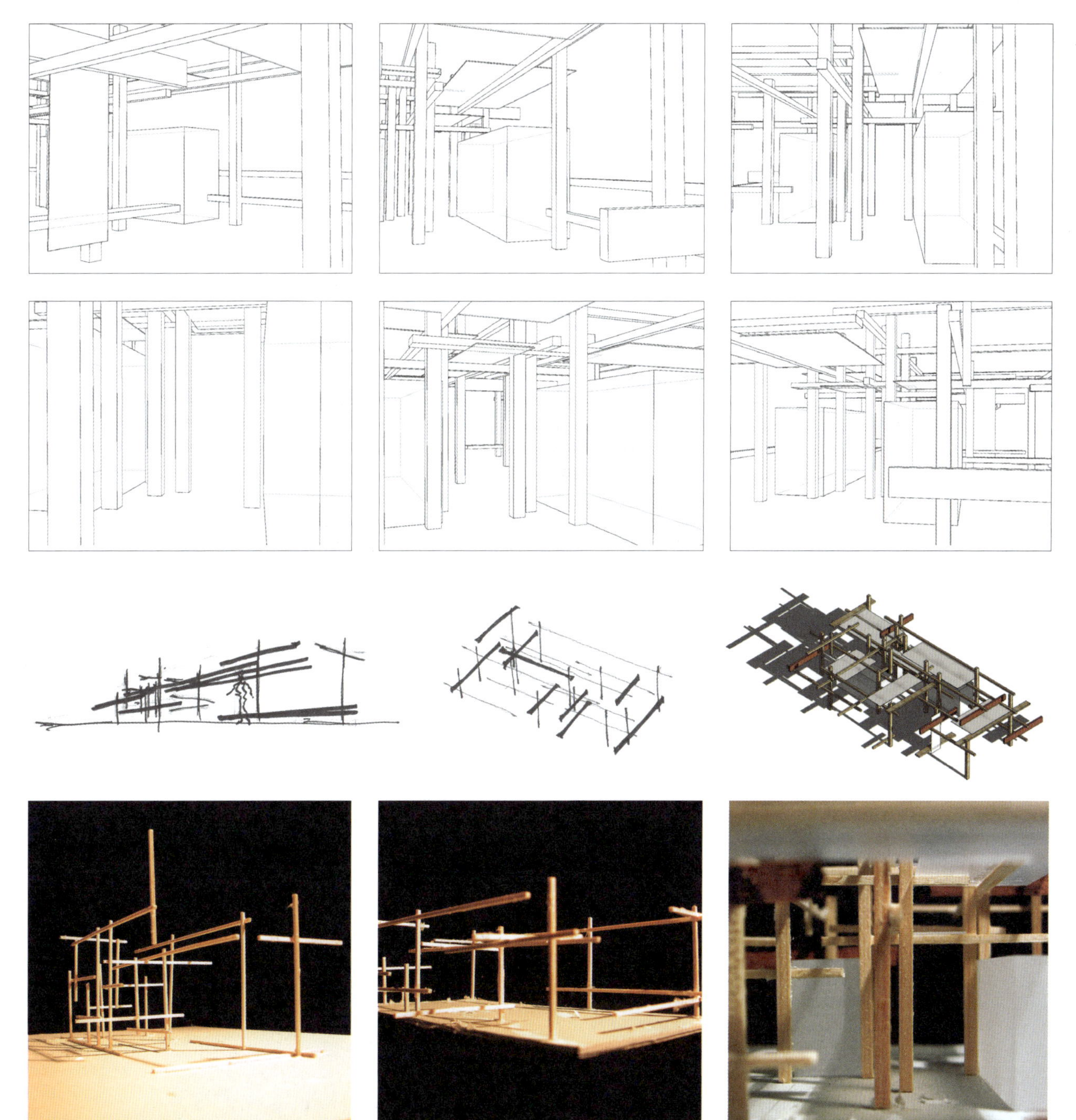

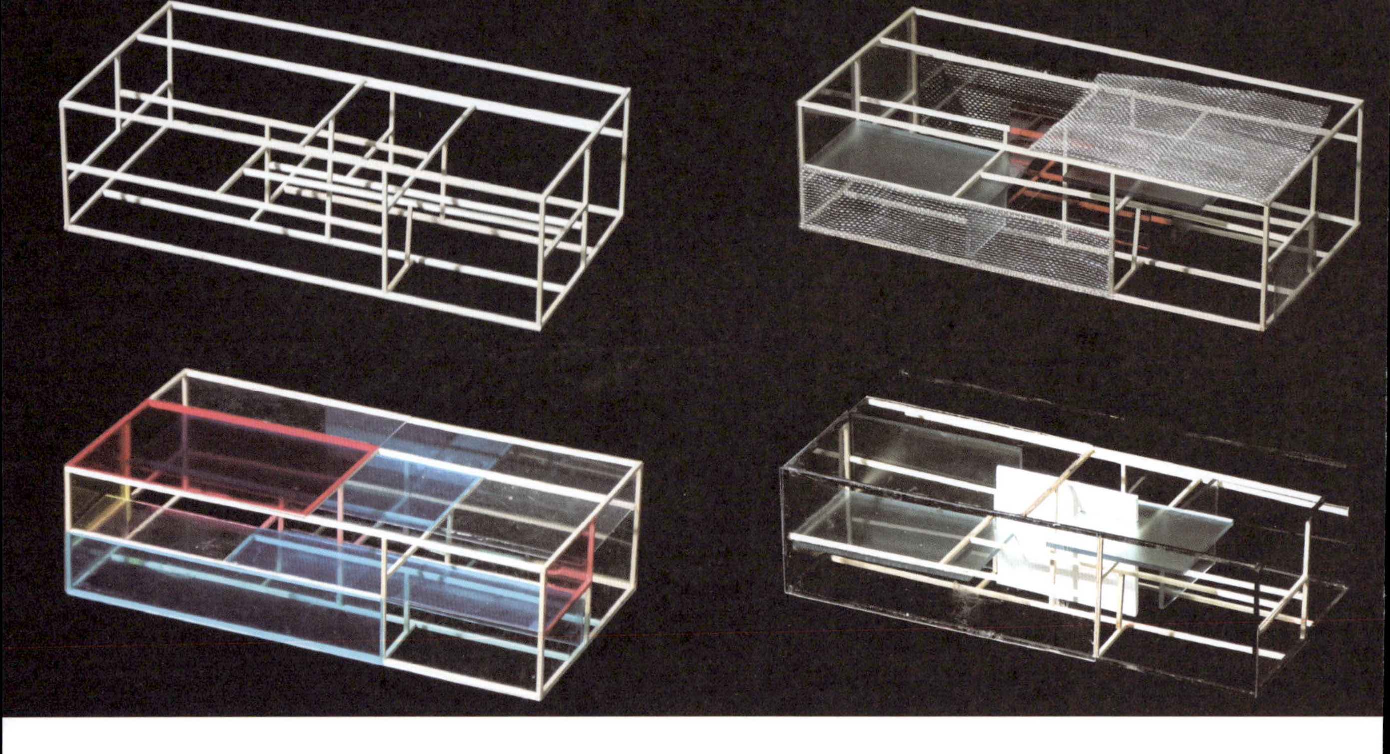

这个设计从研究一种特殊的杆件分叉结构开始，水平和垂直杆件的搭接类似于树木的分叉。设计者总结出两种分叉的方式，一种是在一个平面上的分叉，另一种是在三维空间中的分叉。外框给于整个建筑一个明确的边界，并于内部的分叉结构有机结合。板片的引入不仅为空间的使用提供楼面和墙面，同时也与杆件形成鲜明的对比。

A first operation cuts smaller sticks with a square cross-section. A secondary operation joins twelve sticks axially to form the outline of the object. A tertiary operation inserts repeatedly sticks which join axially in a T-shape. Adjoining nodes can have the sticks in the same plane or in perpendicular planes. The second creates a unique modulation of the space. Some sticks are perceived to outline a frame, and another stick passes through this frame. In the materialisation phase slabs are used for floors and enclosure. They are placed to contrast with the sticks and even strengthen the perception of the sticks inside space.

Martha Tsang Wai-ying, 0506T2Y2

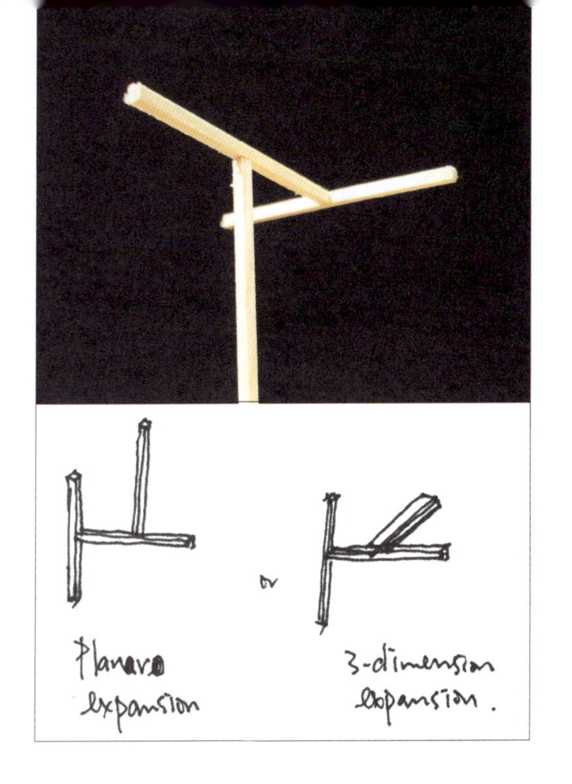
or
Planar expansion
3-dimension expansion.

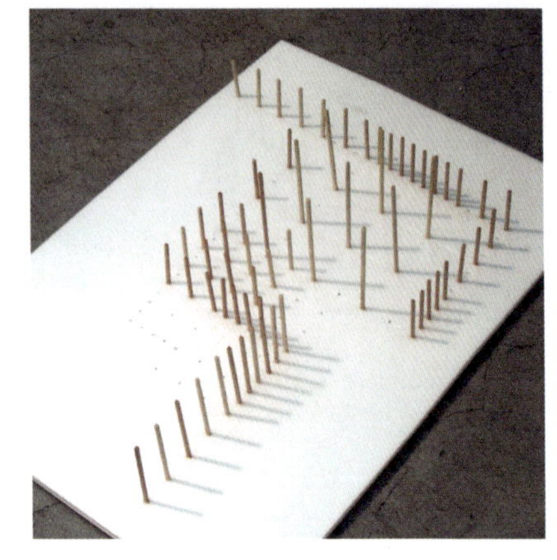

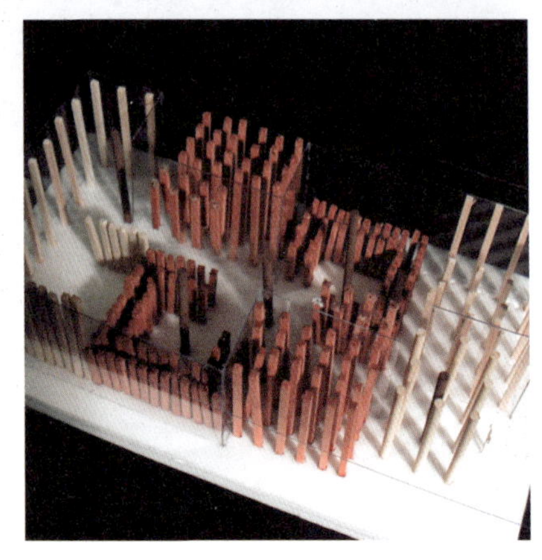

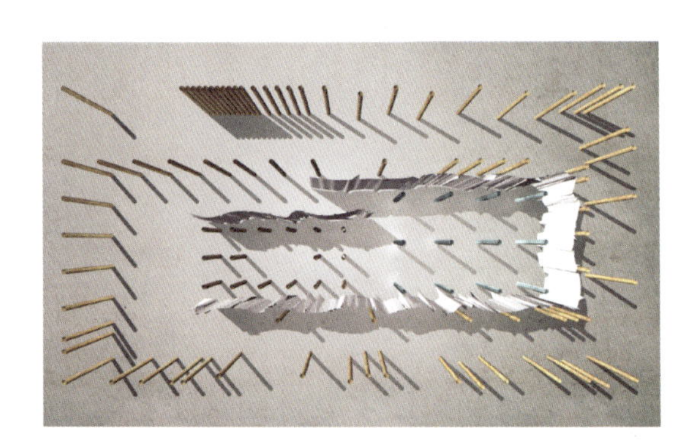

在一块塑料泡沫板上插入杆件，这是一种类似于植树成林的操作。插入杆件的排列方式和密度决定了我们的空间感受。在这个设计中，均匀排列的杆件首先得到一个均质的空间，再在其中拔去一些杆件，空间的密度和排列方式就开始产生变化。杆件的高低、直径大小、材质均可以作为材料研究的变量。

A primary operation cuts identical small sticks from a round longer stick. A secondary operation places these sticks on a grid onto the ground plate. A regularly modulated space like a forest is created. A tertiary operation removes sticks to differentiate the space through the contrast of density. In the materialisation phase the operation is modified, a roof plate is added and curtains are used for enclosure. Sticks are differentiated by material and diameter. Fields of different density are formed, and rows of single sticks modulate the space differently depending on the angle of view – the perception changing from a permeable boundary to a layering of the space in depth.

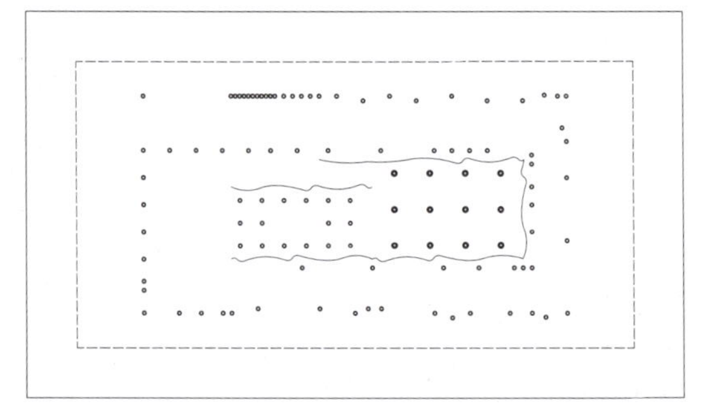

Michelle Tse Hoi-yan, 0304T2M1

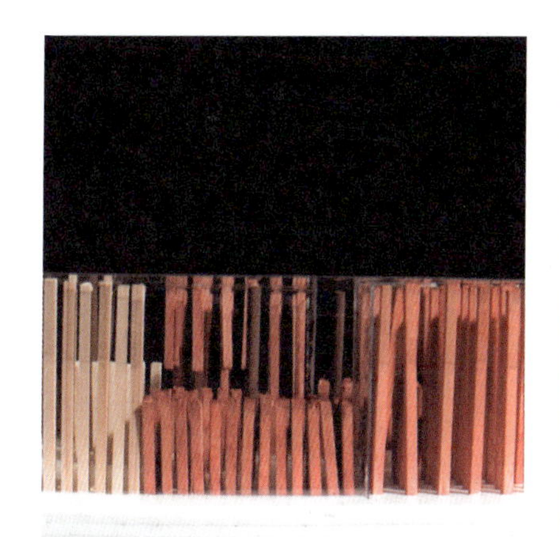

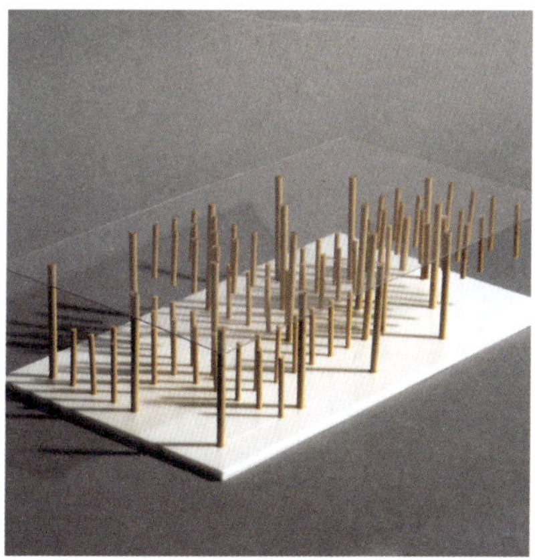

PERSPECTIVE

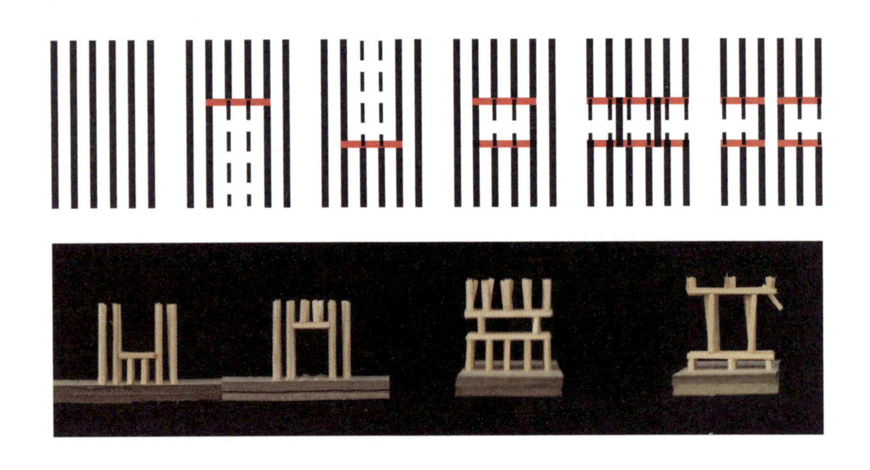

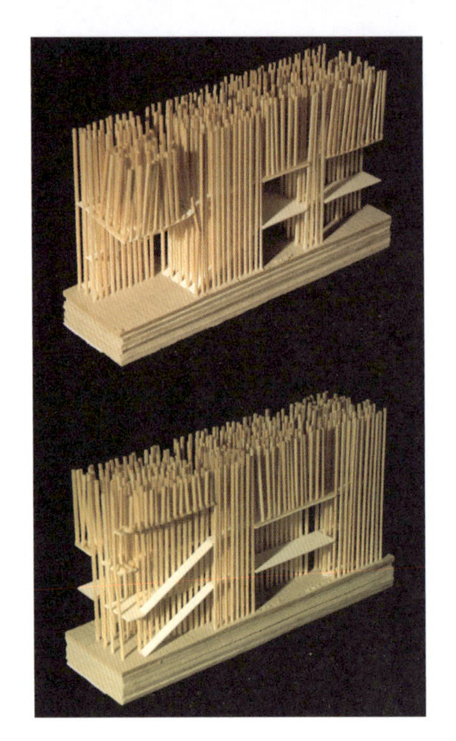

由密实和均匀排列的杆件所构成的一个体积在某种程度上可以被视为一个体块。这个设计在面对这样的一个问题时所采取的操作策略是用水平向的杆件来阻断垂直向的杆件，从而截取出比较大的空间，类似于体块操作中的挖去的方法。

A primary operation cuts identical small sticks from a round longer stick. A secondary operation places these sticks on a dense grid onto the ground plate. A regularly modulated space like a forest is created. A tertiary operation cuts small fields of sticks and adds horizontal sticks to hold the cut ends. A contrast of density is created between spaces with sticks and without sticks, and between the fields of sticks in one direction and the rows of sticks in the other. In certain views, the accumulation of the sticks in depth further differentiates in the perception of the space. In the materialisation phase sleeve-like slabs providing enclosure modify the perception of space.

Larkin Yan Ju-chuan, 0809T1Y3

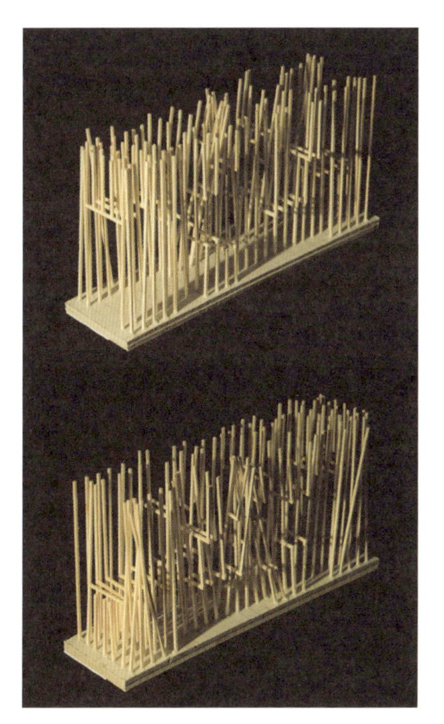

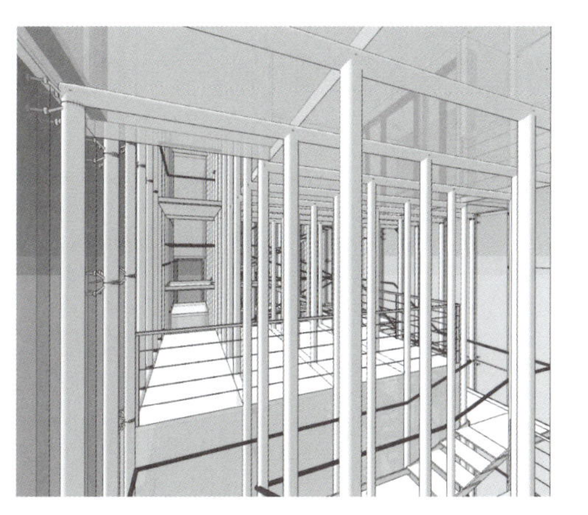

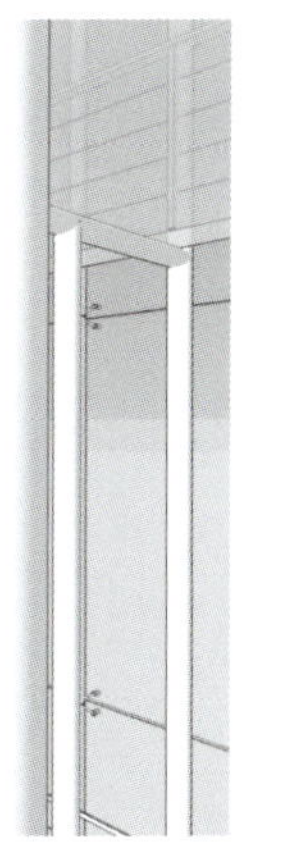

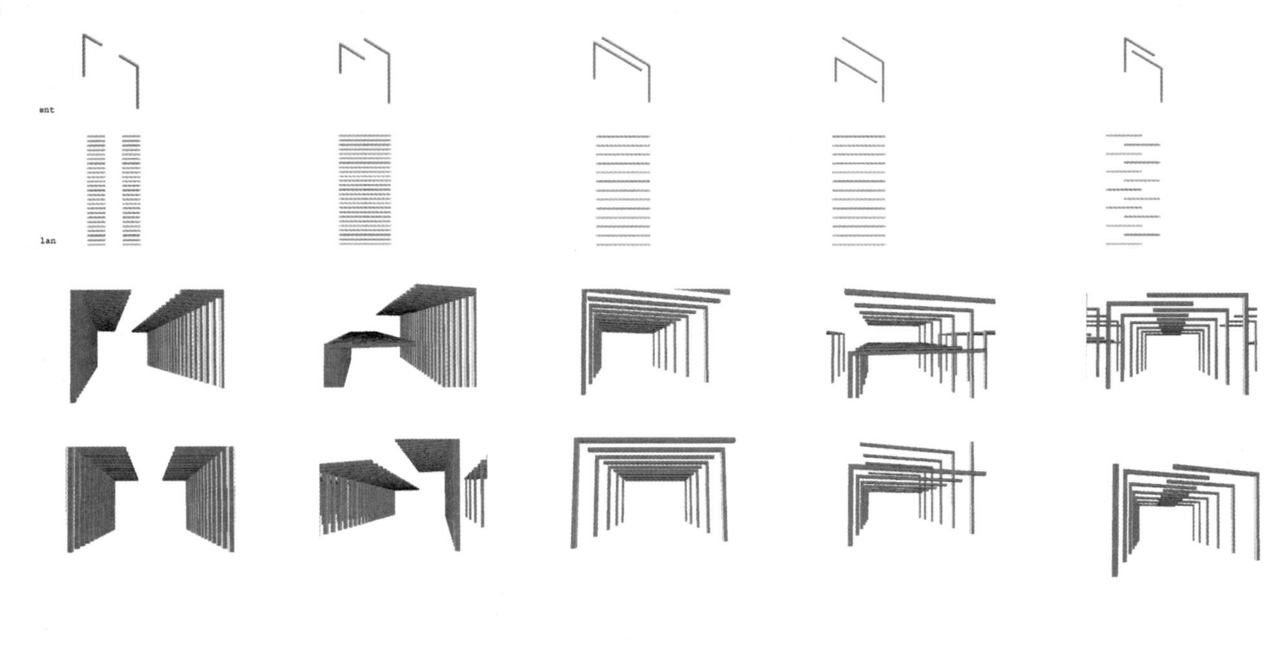

这个设计中反复出现的不是单一的垂直或水平的杆件，而是一个倒“L”形的构件，研究的重点在于如何利用这一同时包含垂直和水平方向的构件来产生空间。构件以单向重复的方式进行排列，产生一个方向上的强烈空间感，同时另一个方向上的半透明屏风。通过高低不同的构件的各种排列方式来制造丰富的空间关系。

A primary operation cuts smaller sections from a long straight wire. A secondary operation bends the smaller wires into an inverted L. A tertiary operation places these on the ground plate in parallel rows with varying overlaps. The differentiated space is perceived depending on the viewing angle as outlines of various densities, as spaces within spaces, as layers in depth, and as accumulations of densities. In the last versions of the abstraction phase, the wire is replaced by wooden sticks of square cross-section. In the materialisation phase, slabs and slabs folded into sleeves are added to provide floors and enclosure.

Iris Wong Hoi-yin, 0304T1Y3

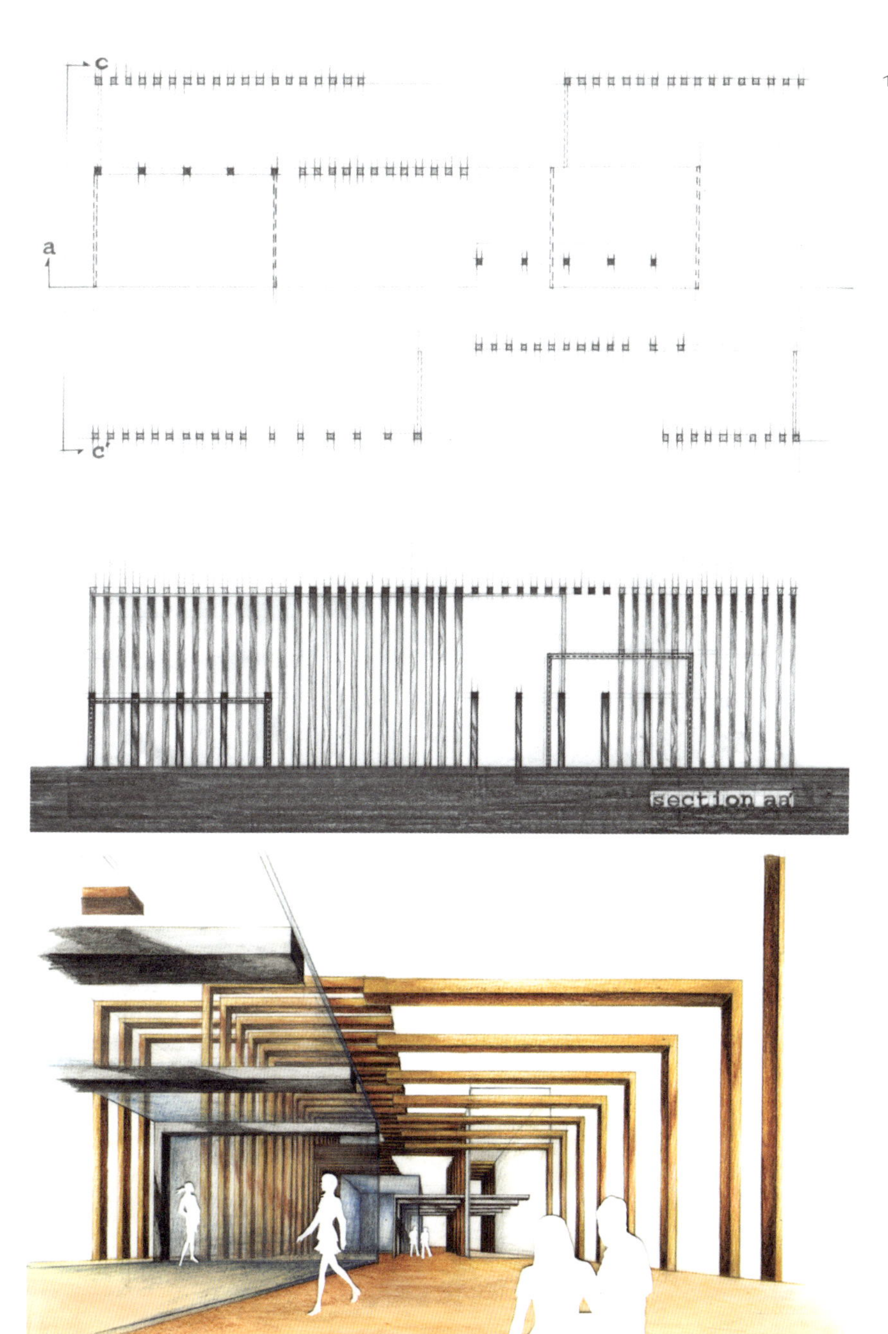
c
a
c'
section aa'

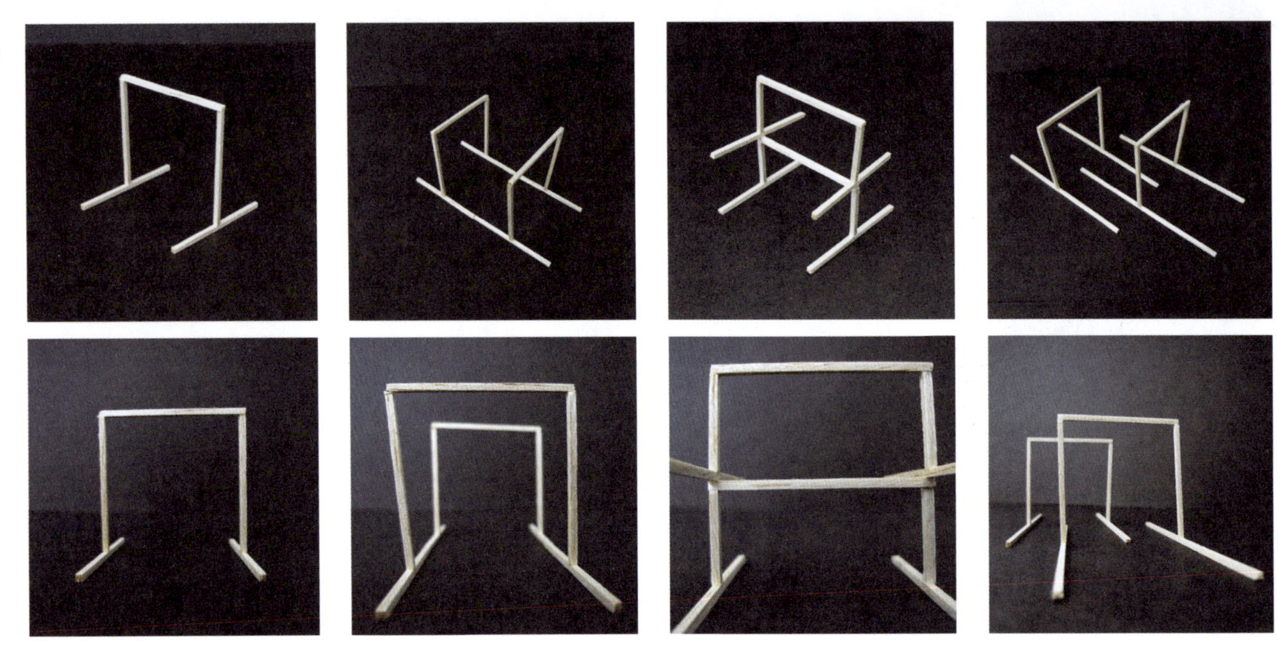

这个设计以跨栏架作为基本的构件，它不仅是一个稳定的结构，而且在两个方向上界定空间。构件的不同的组合方式产生各种空间和结构的可能性。当然，如何用这一构件来构成一个楼房的结构和空间是一个挑战。

A primary operation cuts smaller sticks from a long stick with a square cross-section. A secondary operation connects five sticks each axially to form a hurdle-like frame in various proportions. A tertiary operation forms groups of parallel frames. And a quaternary operation stacks these groups in different directions to form the object. The differentiated space is perceived depending on the viewing angle as outlines of regular densities, as spaces within spaces, as layers in depth and as accumulations of densities. In the materialisation phase slabs are added to provide floors and enclosure with various relationships to the sticks.

Tommy Tam Yung-kin, 0809T1Y2

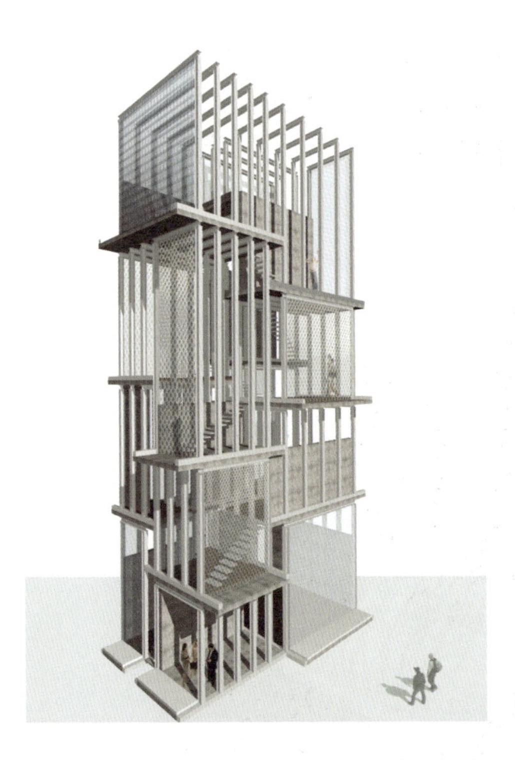

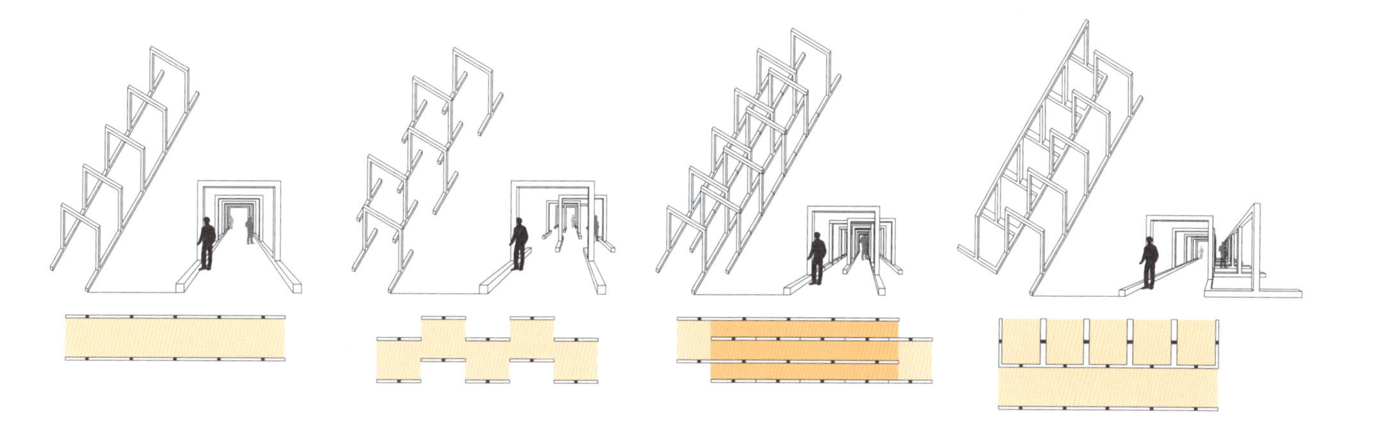

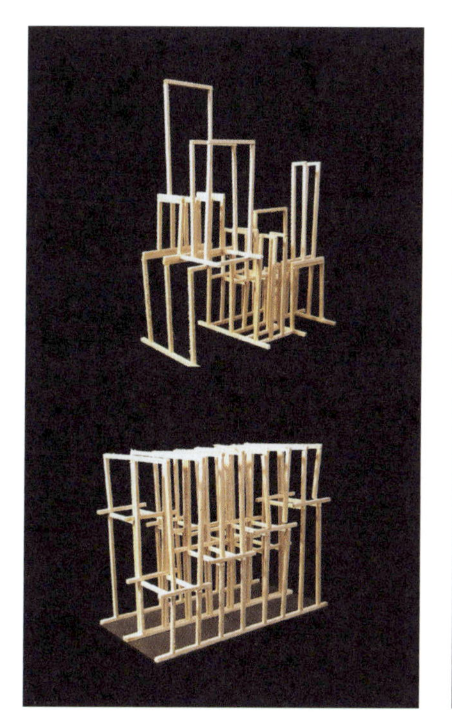

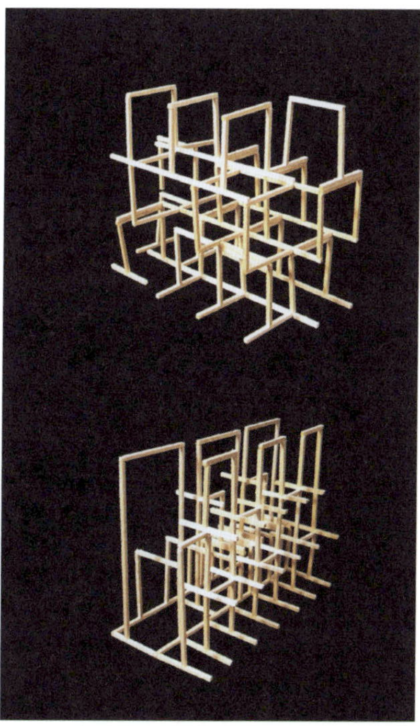

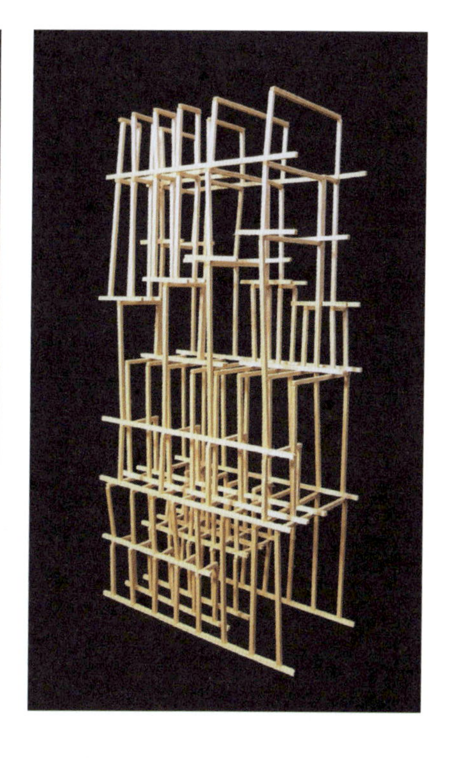

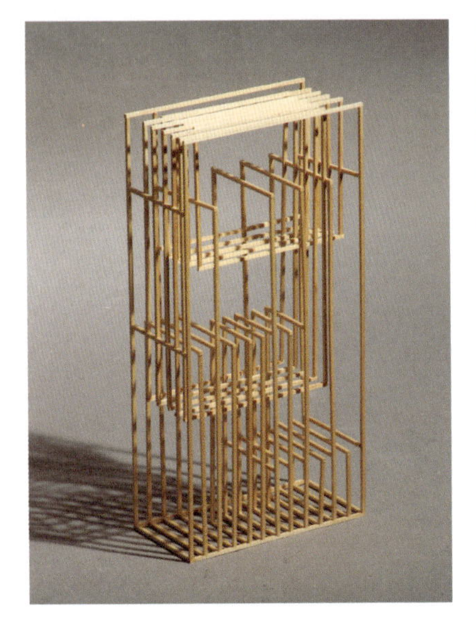

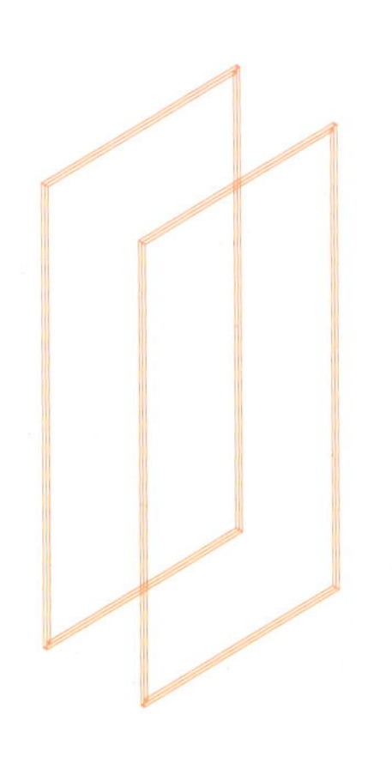

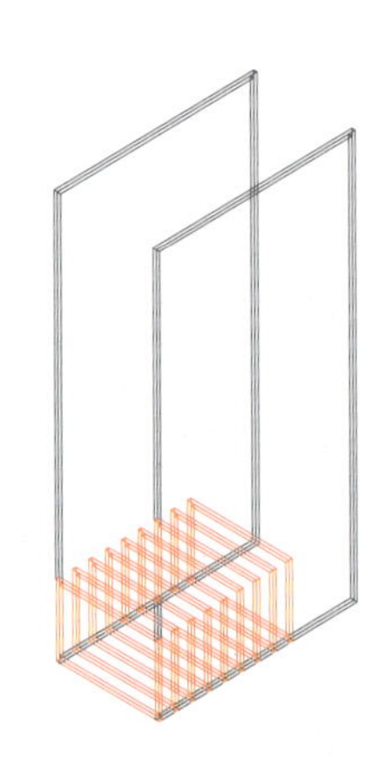

“圈”也是经常被采用的杆件构件。一个由杆件闭合的“圈”界定一个平面空间。“圈”在一个方向上的重复形成空间的容积，该空间具有明确的方向性。这个设计通过不同大小和方向的由“圈”构件单向重复产生的空间的相互咬合形成复杂的空间套叠关系。

A primary operation cuts a long square stick into shorter sections. A secondary operation bends these into loops of various proportion. This gives a minimal definition of space through an outline of edges. A tertiary operation forms groups of parallel loops with regular distances. This defines space through outline, permeable boundary and accumulation of layers in depth. A quaternary operation places these groups into the tall object, letting groups overlap in various ways. Through this the space is further modulated, strengthening the perception of space through accumulated layers, and adding other types of perception – space inside space and sticks inside space.

Lo Wan-lok, 0405T1Y3

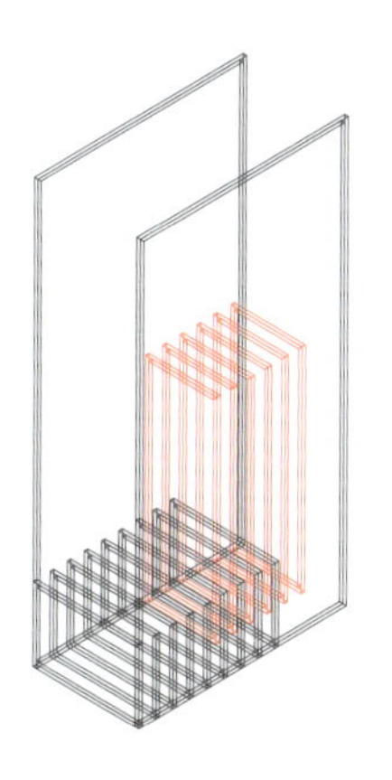
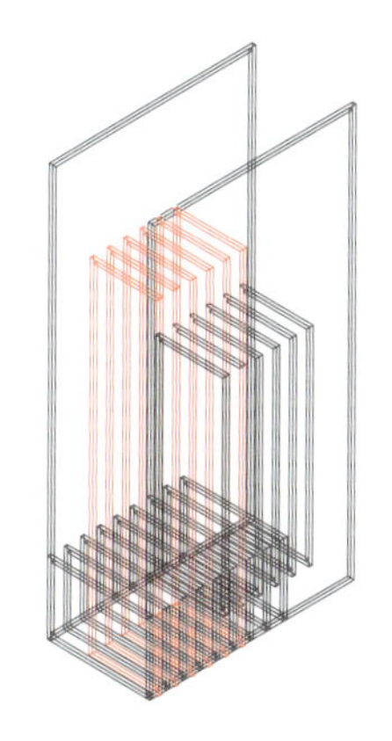

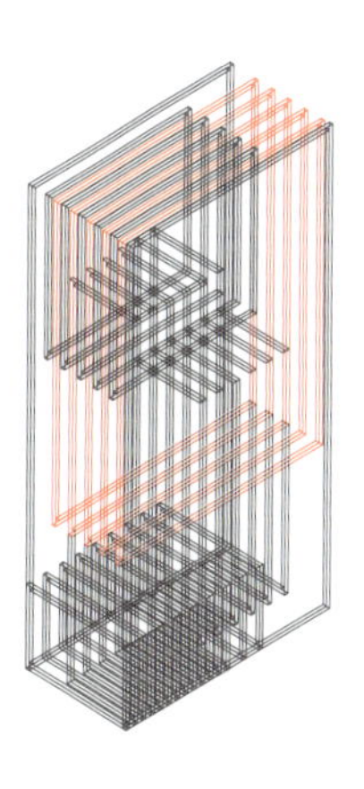

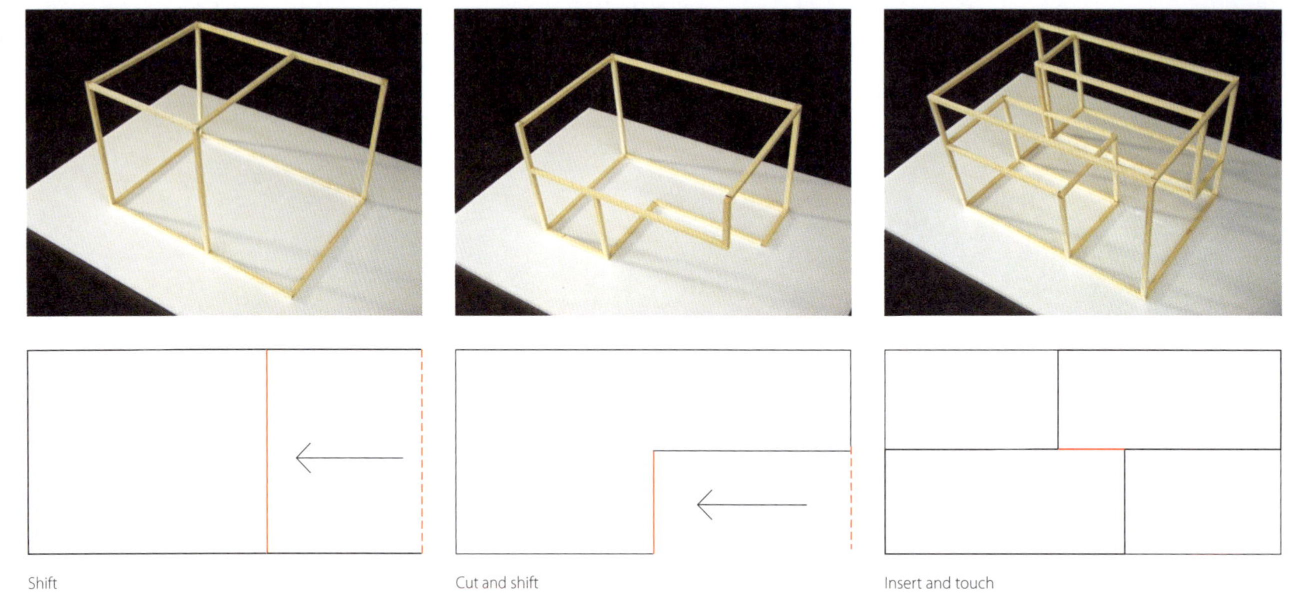

一个由六根杆件构成的立方体成为空间和结构的基本单元，杆件界定出一个空间的容积以及结构框架。设计者研究的兴趣在于如何通过特定的操作来改变该空间的特性，比如在原来的杆件上作位移，或加入新的空间框架。最后的楼房由八个空间框架单元来构成，内部则是若干个插入的空间框架相互作用给于空间进一步的界定。

A primary operation cuts a square stick into smaller sticks in a few lengths. A secondary operation forms a regular grid with one size of sticks by joining them axially. With this the space becomes uniformly modulated. A tertiary operation inserts into each cube more sticks forming a smaller cuboid. With this the space becomes differentiated. The space is mainly perceived through the outlines of its edges, but also through the accumulation of sticks in depth, depending on the viewing angle. In the materialisation phase, planes are inserted into the frames to create floors and enclosure.

Matthew Leung Kwan-yin, 0809T1Y3

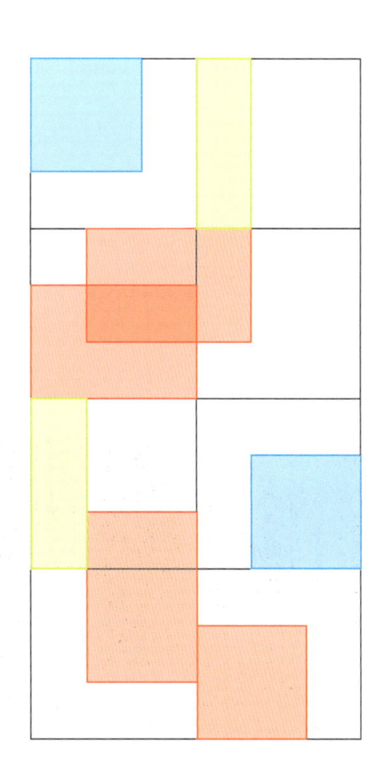

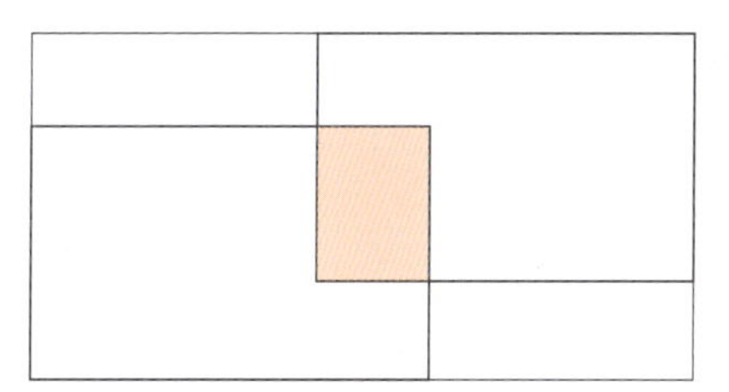

Insert and interlock

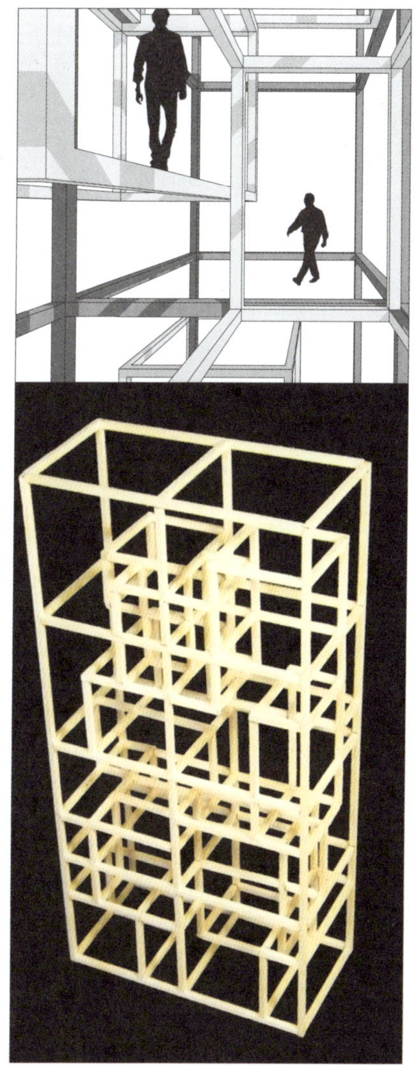

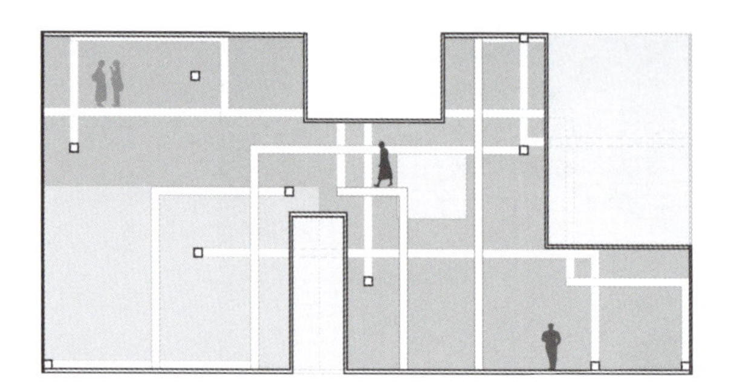

连续的杆件通过弯折在空间中伸展。杆件的走向引导着我们的眼睛在空间中移动，或向上，或向下，或向前，或向后。杆件的密度根据空间的意图来调节，或密集，或稀疏。这是一种充满运动感的空间。而杆件空间的感觉因为两片板片的围合而更加强烈，这说明杆件的调节空间是发生在空间之内。

A primary operation cuts a long square stick into shorter sections. A secondary operation bends these by 90°, changing direction each time, starting and ending on the surfaces of the object. Space is perceived, depending on the direction of view, as partially outlined on its edges, as surrounding sticks, through the contrast of partial frames and inserted sticks, and the accumulation of sticks in depth. In the materialisation phase, blocks are used to provide enclosure and the object itself is reinterpreted as block. The materials differentiation then contrasts the lines of the sticks on the surfaces and the lines of the sticks through space.

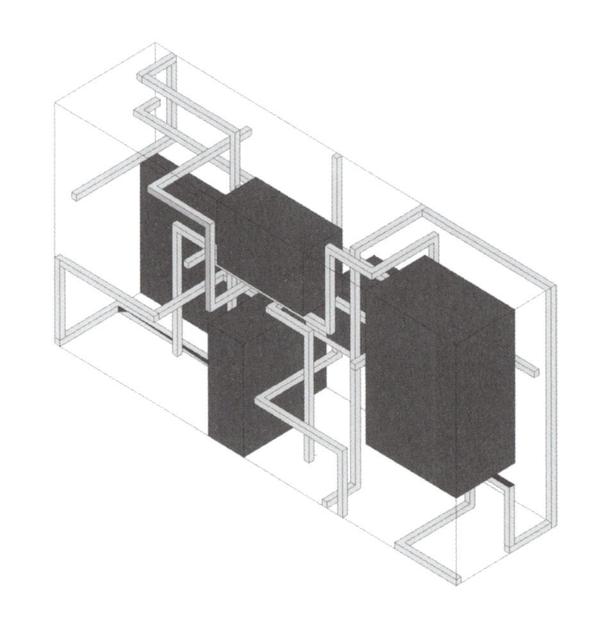

Raymond Wong Yun-hong, 0506T2Y3

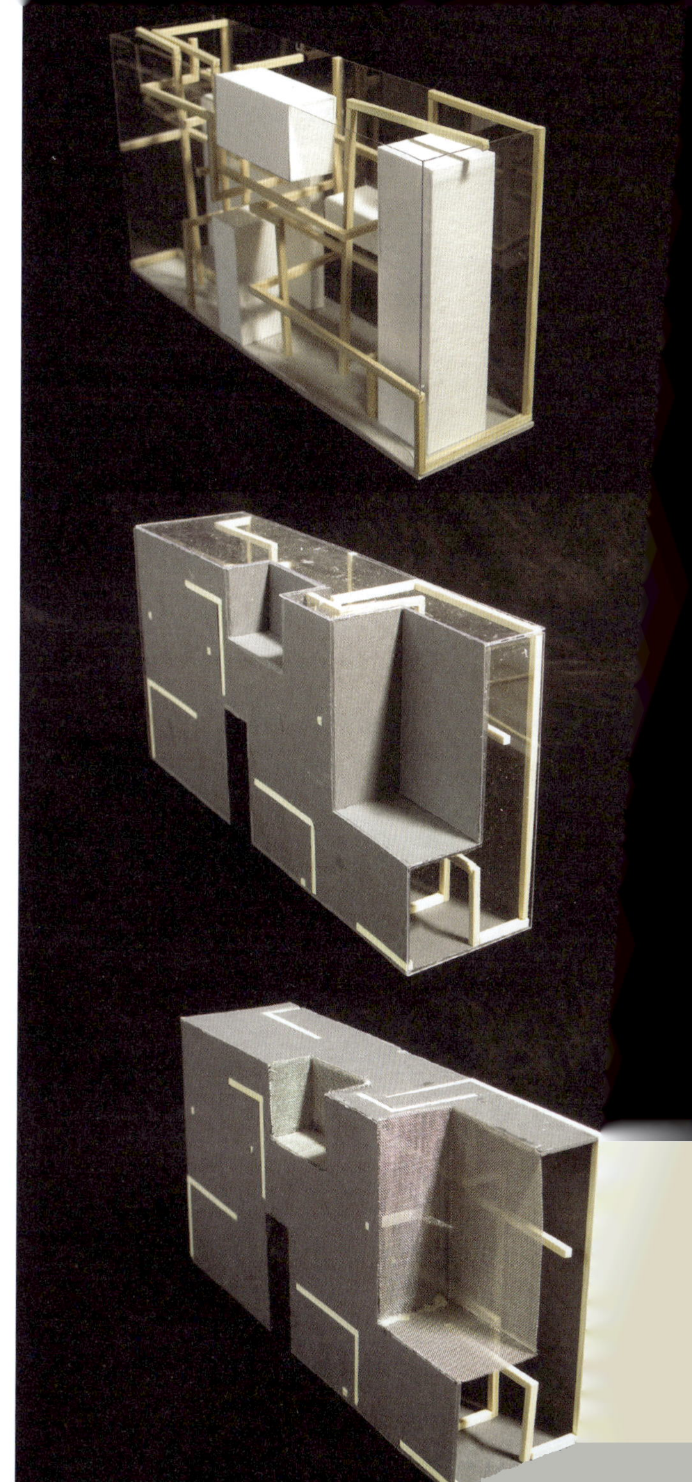

5 设计 | design

练习加方案　exercise + project

练习即方案　exercise = project

方案练习化　exercise ⊂ project

从设计练习到设计方案 | from design exercise to design project

在这个课程中我们将工作室里的各种设计课题分为“练习”和“方案”两个基本的类型。所谓的练习是指它的抽象性，即将设计的条件进行了必要的简化和抽象，以便于对某些问题作专门的研究。而所谓的方案是指一般的建筑设计课题，它的特点是综合性，包含了设计的任务、场地等基本条件。一个设计工作室教学的最终目的是使得学生能够处理复杂的建筑设计方案，但是综合性的问题不利于特定问题的专门研究。好比任何一个运动项目，运动员除了参加运动项目的比赛外，必须要先有各种专项的训练，即练习。所以，练习和方案在一定的程度上可以理解为手段和目的的关系。

这本书前面的第二、三和四章，我们主要讲述了“建构”练习的三个方面：基本的概念、方法和手段，练习的操作过程，以及练习研究的内容。建构练习的抽象性主要体现在对三种空间限定要素以及各自对应的空间特点的定义，以及对练习过程的四个基本阶段的设定。为了研究和学习的方便，我们把建筑的具体使用功能抽象化为人在空间中的体验，把建筑的体量简化为抽象的板房、平房和楼房，并把建筑的场地问题抽取掉等。但是必须在这里指出的是，我们所说的抽象，并不是“非设计”的抽象形式训练，是一种对设计问题和方法的抽象，本质是一种设计训练和设计研究。

如何从抽象的练习转换到综合性的建筑方案的设计？这是本章所要讨论的问题。根据全系的设计专题工作室的基本设置，一个学期的设计教学被划分为两个部分，前面的一个较短的部分（通常是四到五周）是每个平行进行的工作室自行决定的专题练习，后面的一个较长的部分（通常是七到八周）则是由学系来共同决定的设计方案。所以，从2001－2002学年开始直到2005－2006学年，我们完全以“练习加方案”这一基本模式来运作。从2006－2007学年开始，我们对练习和方案的关系作了三种不同的探索，主要是想解决过往教学中出现的一些问题。如“练习即方案”将功能和场地问题结合进练习；“练习即练习”是将方案设计排除在外，专门集中于练习的教学；最后是“方案练习化”则是把练习的设计过程融入方案的设计过程。关于“练习即练习”的模式已经在前面的“练习”章节中专门讨论过，这里就只介绍练习＋方案，练习即方案，以及设计练习化三种模式。学生作业的选取尽量体现体块、板片和杆件三种基本要素的研究和运用。

我们的最终目的是要发展一种具有可操作性的建筑设计方法。

In this chapter we discuss the exercise-project relationship and show different examples with student work. In the exercises, we work with a certain degree of abstraction and with a reduced set of issues. In the project, we deal with less abstraction, with an expanded set of issues and closer to a real project. We also make a distinction between the process within an exercise which somehow can seem artificial, and a more natural design process that we would expect in a project.

The previous chapters are about the exercises as abstractions of projects. We strongly emphasise the role of the space defining elements with the three distinct types: block, slab and stick. We structure the process into steps with articulated questions to explore. And we reduce the number of issues to consider by either eliminating or drastically reducing the number of questions relating to use and site. But we don't reduce the issues to a form-making exercise. We try to deal with form and expression, but the emphasis is on the relationship between the physically built form and the perception of the otherwise invisible space through it.

Exercises are made for a purpose. They are opportunities to explore issues, to practise skills, to develop a vocabulary for discussion, and to acquire working habits. What was achieved in exercises should be usable in the design of a project. Over the years, we have experimented with different exercise-project relationships. Initially, the organisation of the studio teaching into thematic studios provided a possible structure by articulating two distinct projects each term, a four to five-week studio project and an eight-week school project. Later other relationships were tested – to treat the exercises more like an abstract project, to expand the exercises to the whole term or to embed the exercises into a project.

from 2001/02 to 2005/06

exercise + project — studio and school project

from 2006/07 to 2008/09

exercise = project — the exercise as a project

exercise = exercise — the exercise as an exercise

exercise $\subset$ project — the exercise is imbedded in the project

As the exercises are described in Chapter 3, we discuss and illustrate the other three possibilities on the following pages.

1 练习加方案 | exercise + project

在整个学系的设计教学框架中，由各个专题设计工作室自行决定的练习也叫做“工作室课题”或“自选课题”，由学系共同商定的设计方案也叫做“学系课题”或“规定课题”。学系课题只是规定一些建筑使用的基本特点，以及适应不同年级的设计方案的规模，而具体的建筑类型则由各个工作室来定。这样的课题有：学习的场所，居住的场所，祈祷的场所，工作的场所等。每个学期，学系共同决定一个课题。这一课程设计的意图是在推动各个专题工作室的各自研究方向的同时，在全系的层面上同一年级之间仍然有一个共同的评判的基准。

就建构工作室的特定课程设计而言，这一先练习后方案，先抽象后综合的教学，其实是有不少重叠的部分的。因为，我们的练习并不是单纯的分析性研究，或抽象的形式训练，它本身也是设计训练。而我们感到这种课程结构的最主要的特点是前面的练习中发展出的建构概念对后面的设计方案的构思有直接的影响。从一个方面来说，这体现了训练的特点和优势，从另一个方面来说，这也有形式先导之嫌，设计过程不够自然。

In the studio project, each studio clarifies its architectural position and approach. In the school project, all studios work on similar projects. A common topic, relating to a situation of life, is shared, for example a place for learning. And each studio formulates projects according to the circumstances and differing in the number of persons involved; it is appropriate for undergraduate year 2, year 3 and graduate year 1 students. When the work is discussed in the final review it is expected that the differences of approach will become apparent.

We observed several issues. Naturally the students tended to carry the idea directly from the studio project to the school project. This was in some cases very successful; in others it raised questions of appropriateness or became a limitation. The school project didn't trigger enough variation, so that to a certain degree, there was a repetition of certain processes in the two projects, which sometimes resulted in the school project not being strong enough as a project, but rather another version of the exercises.

城市住宅 | an apartment building

Helen Lam Wai-yin, 0304T1Y3

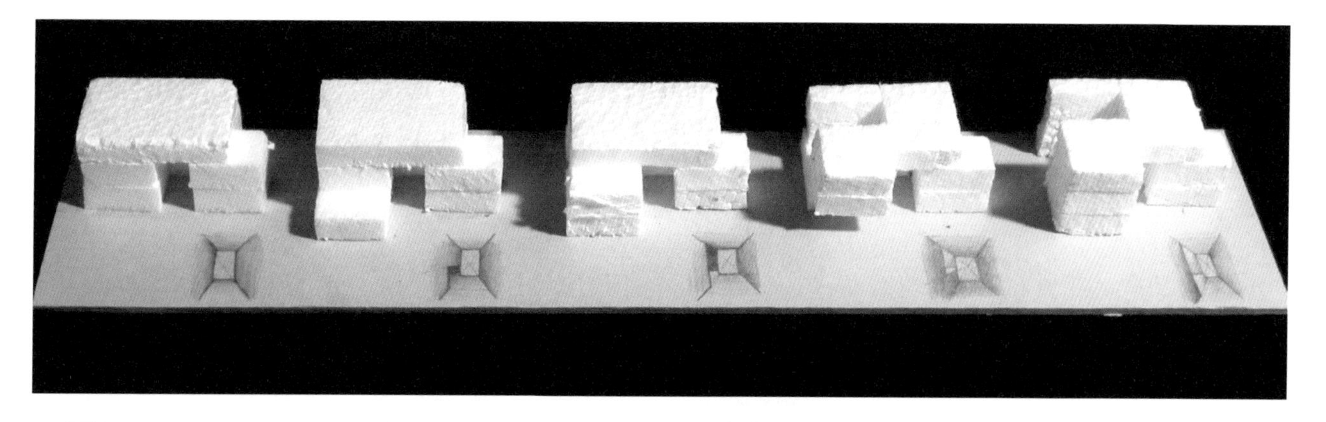

BUILDING AS A WHOLE BLOCK

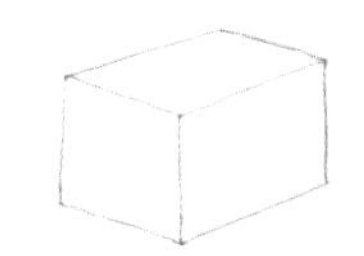

THEN USE "CARVE" TO DESIGN SPACE

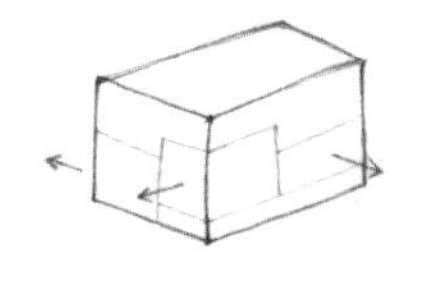

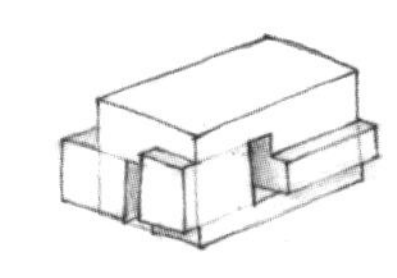

THE ELEMENT JUST MOVE HORIZONTALLY

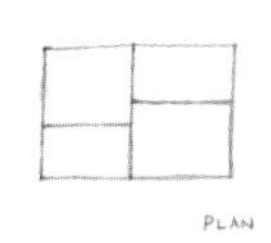

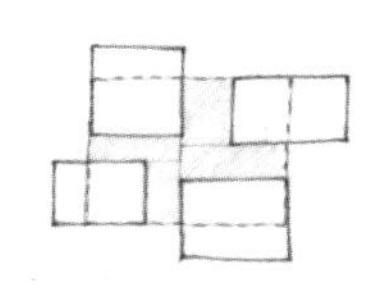

PLAN

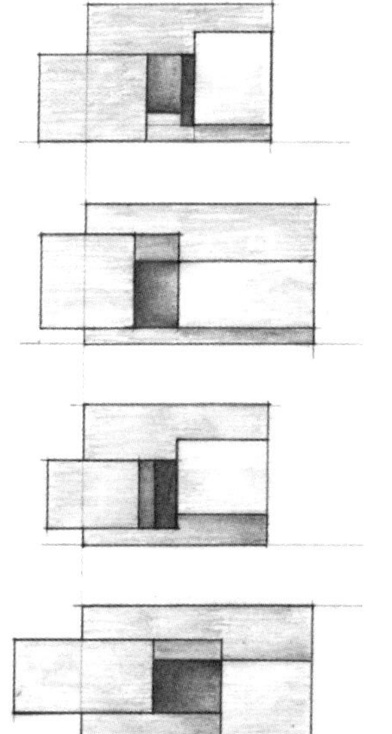

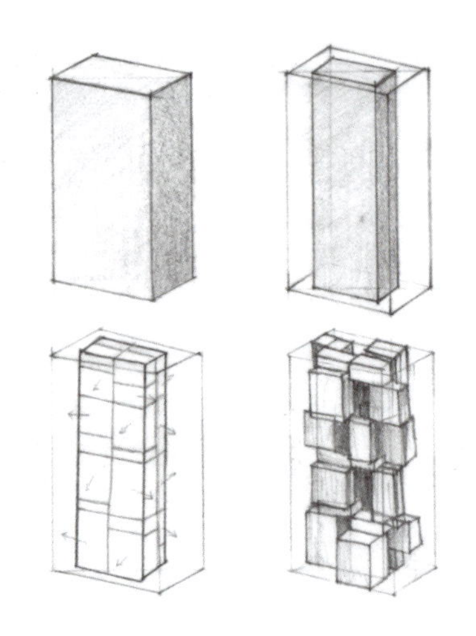

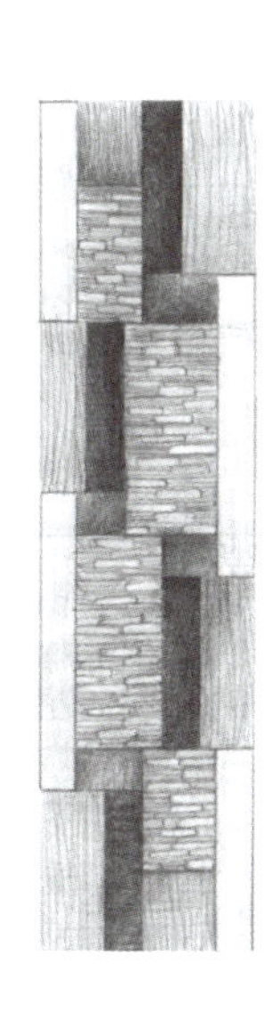

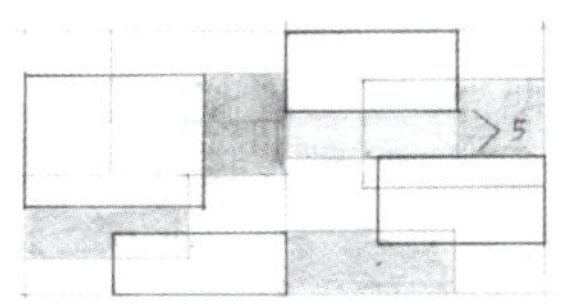
5

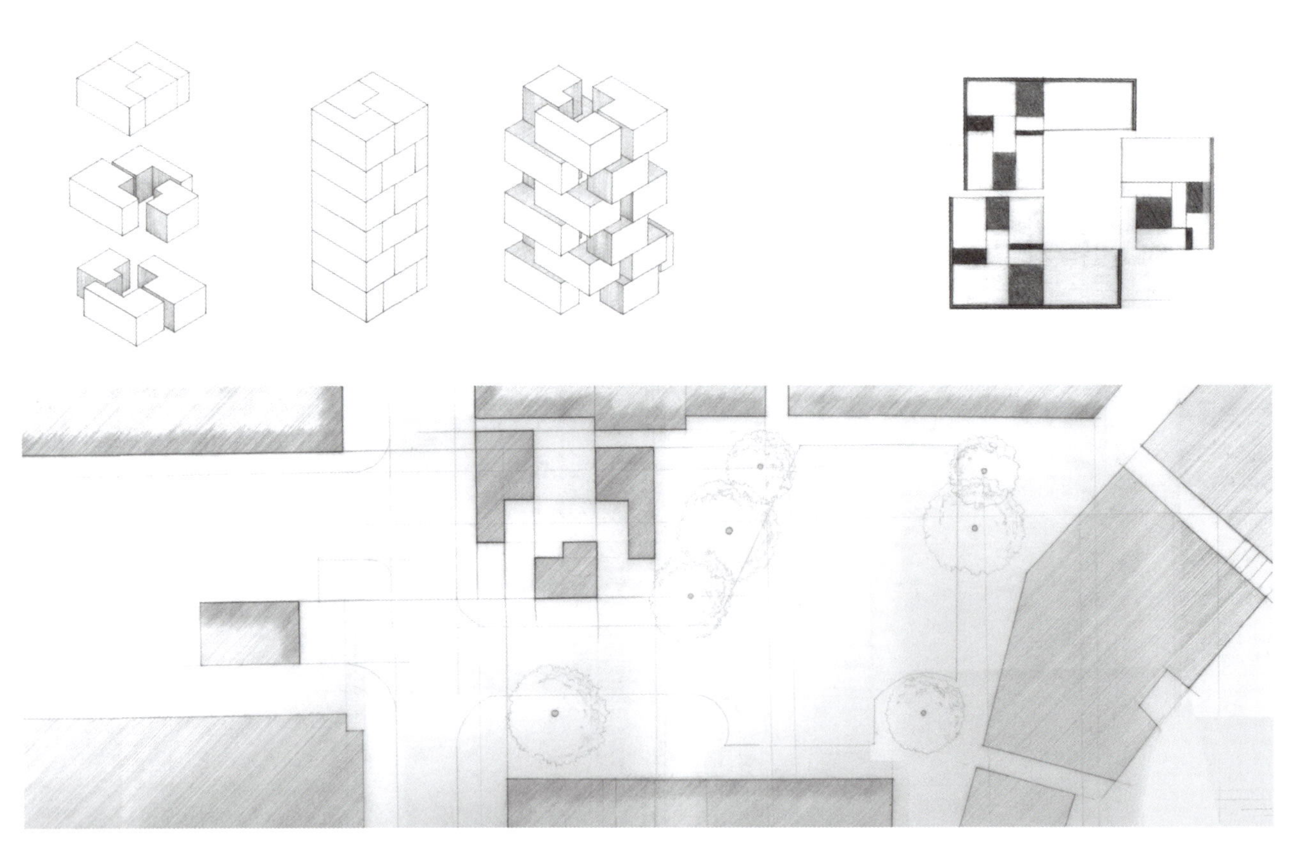

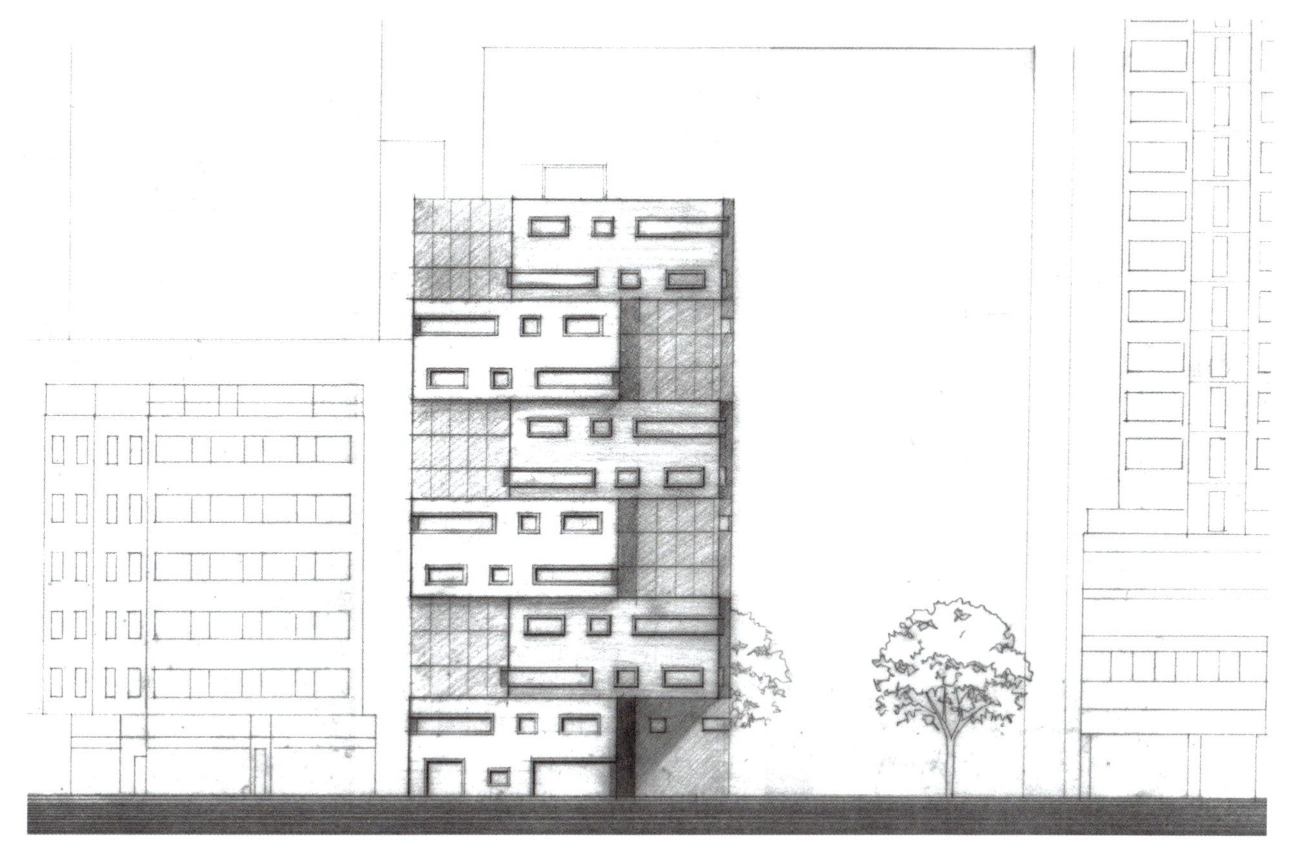

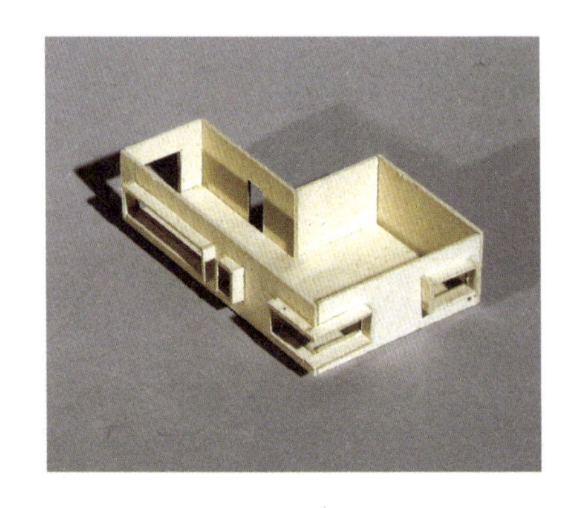

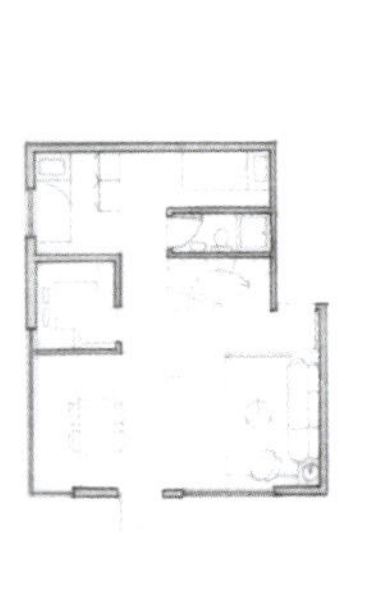

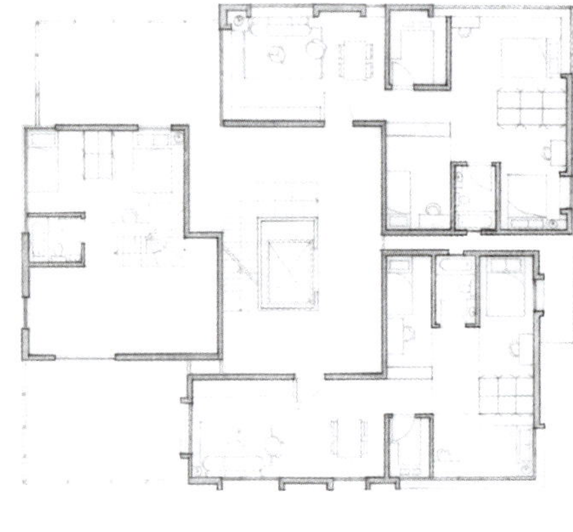

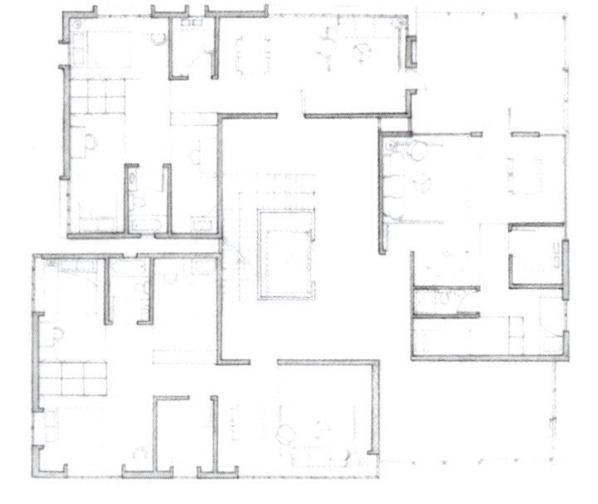

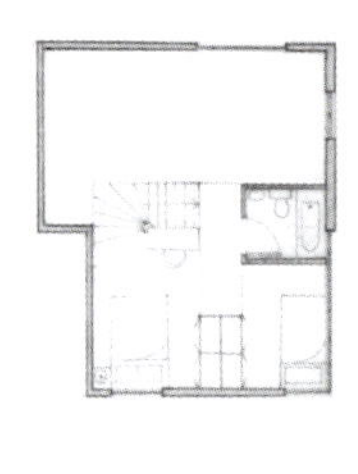

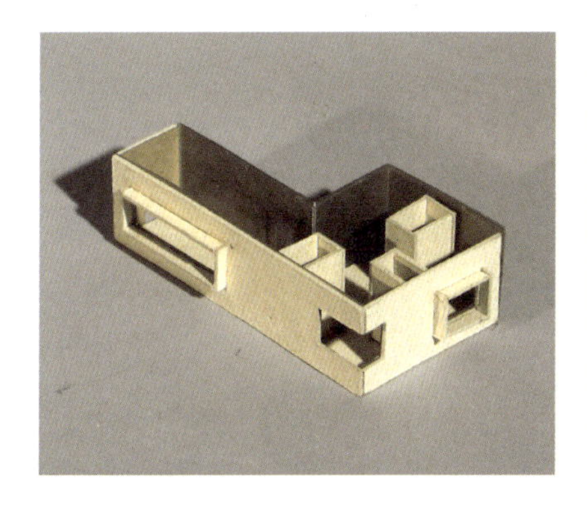

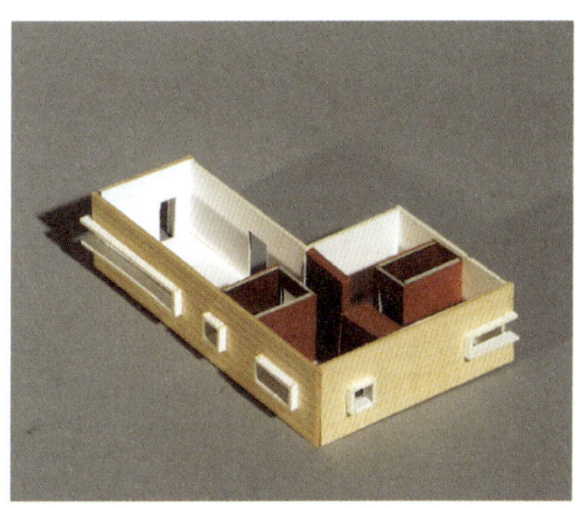

村落民宅 | a village house

Lam Tat, 0304T1Y2

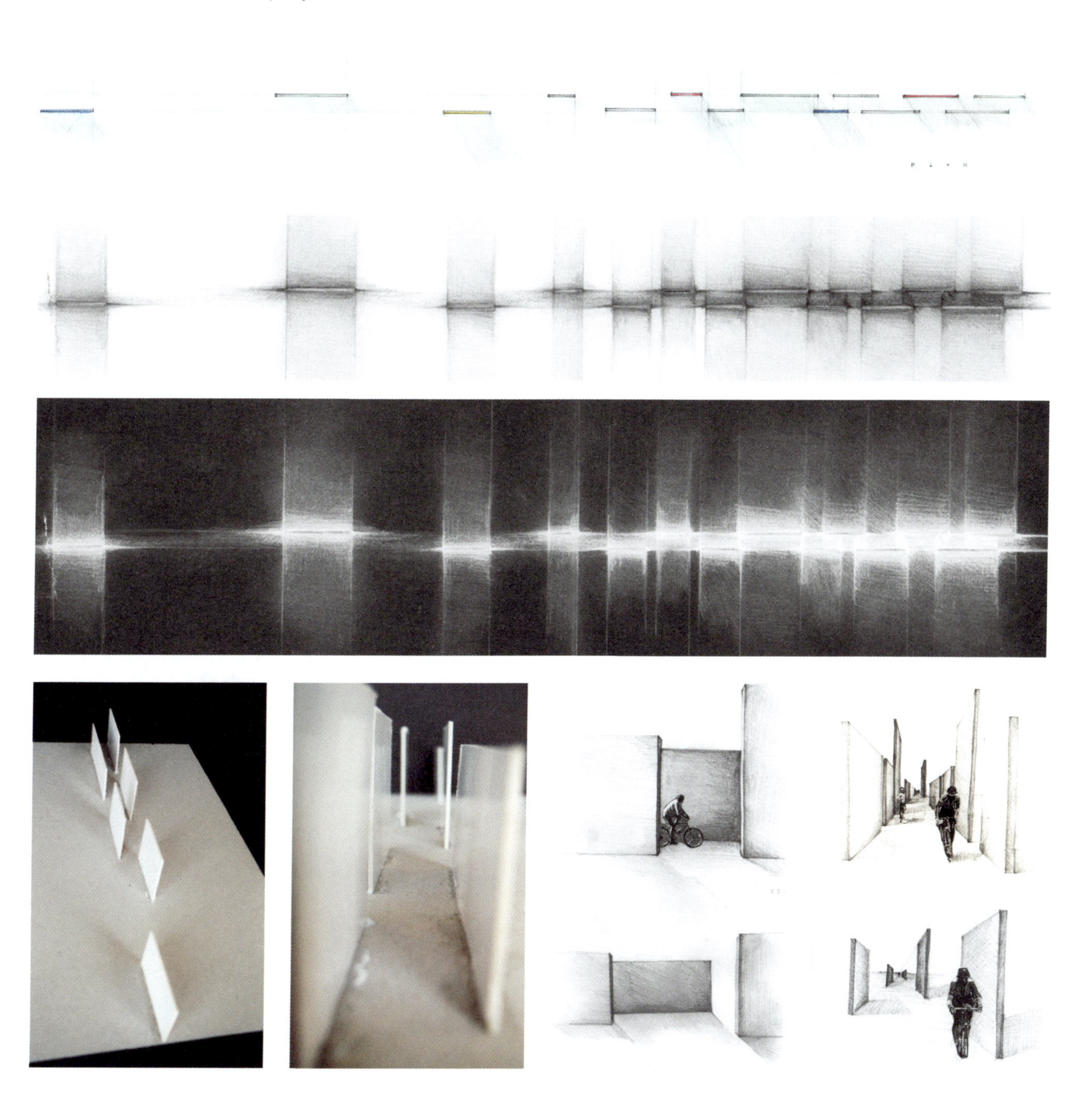
PLAN

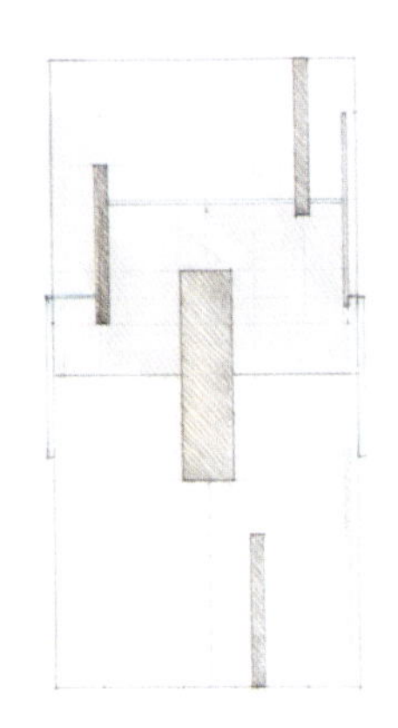

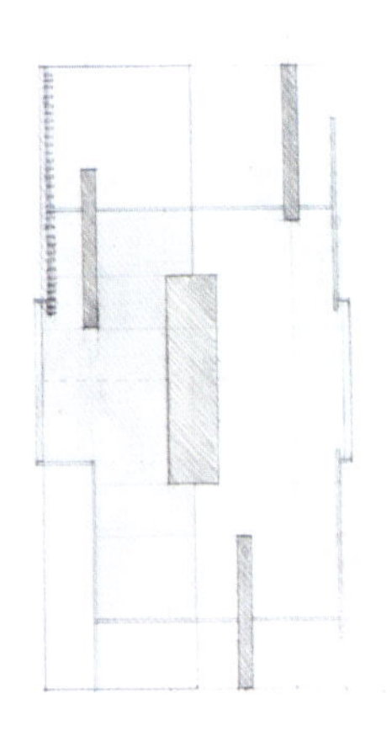

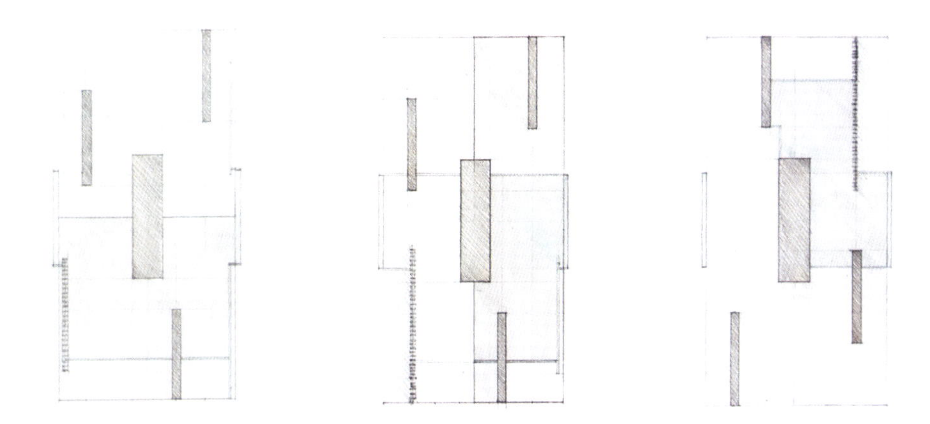

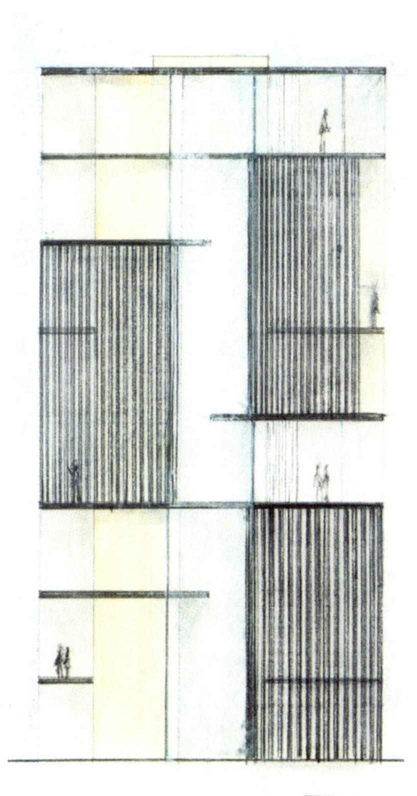

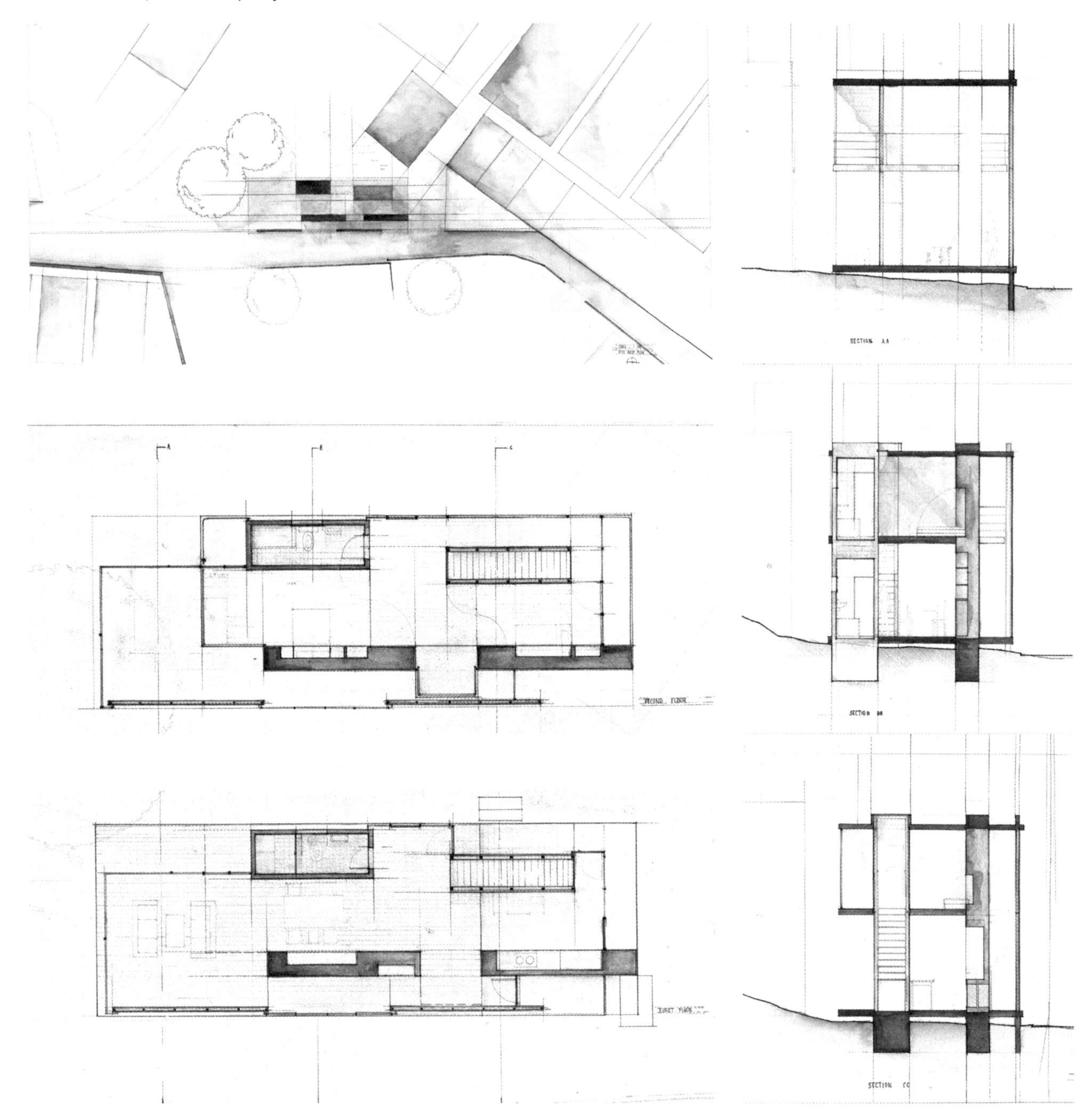
SECTION AA
SECOND FLOOR
SECTION BB
FIRST FLOOR
SECTION CC

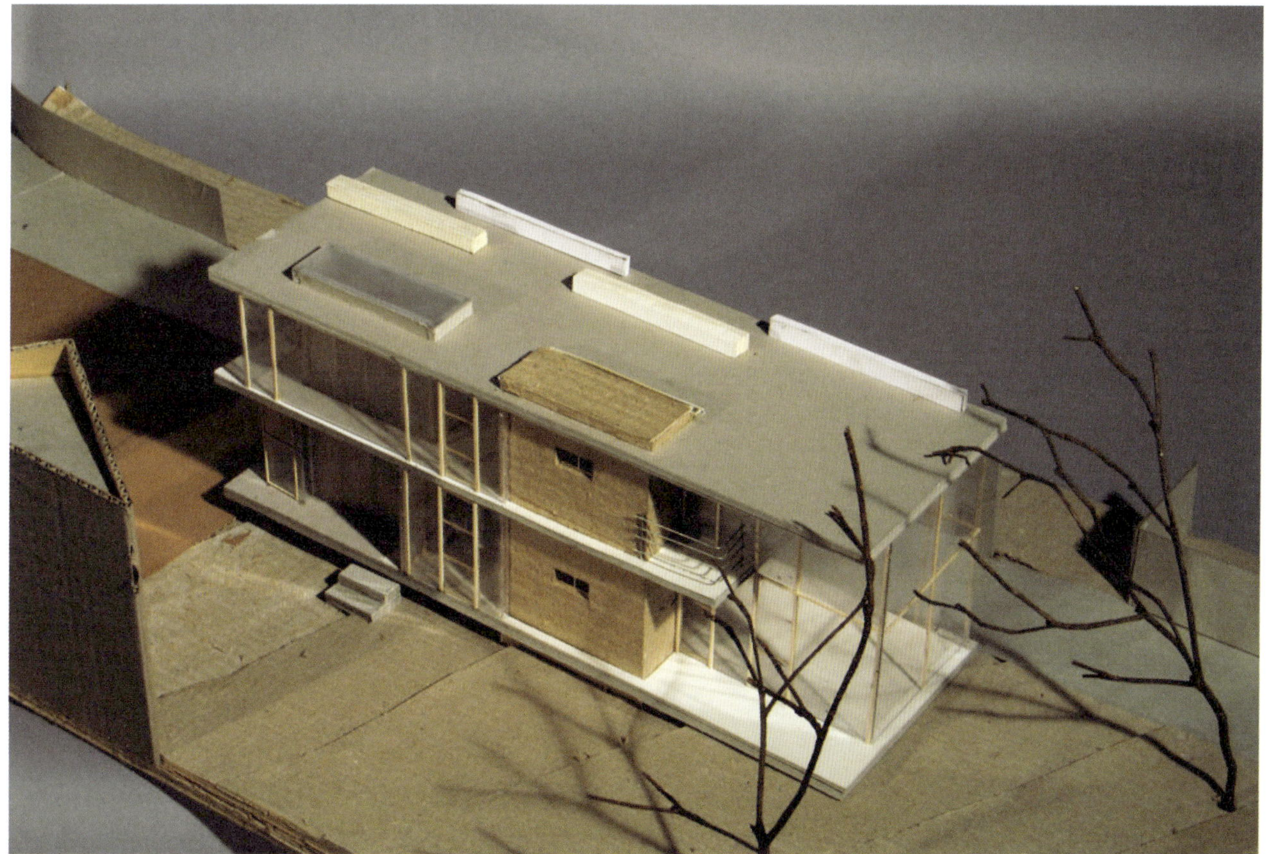

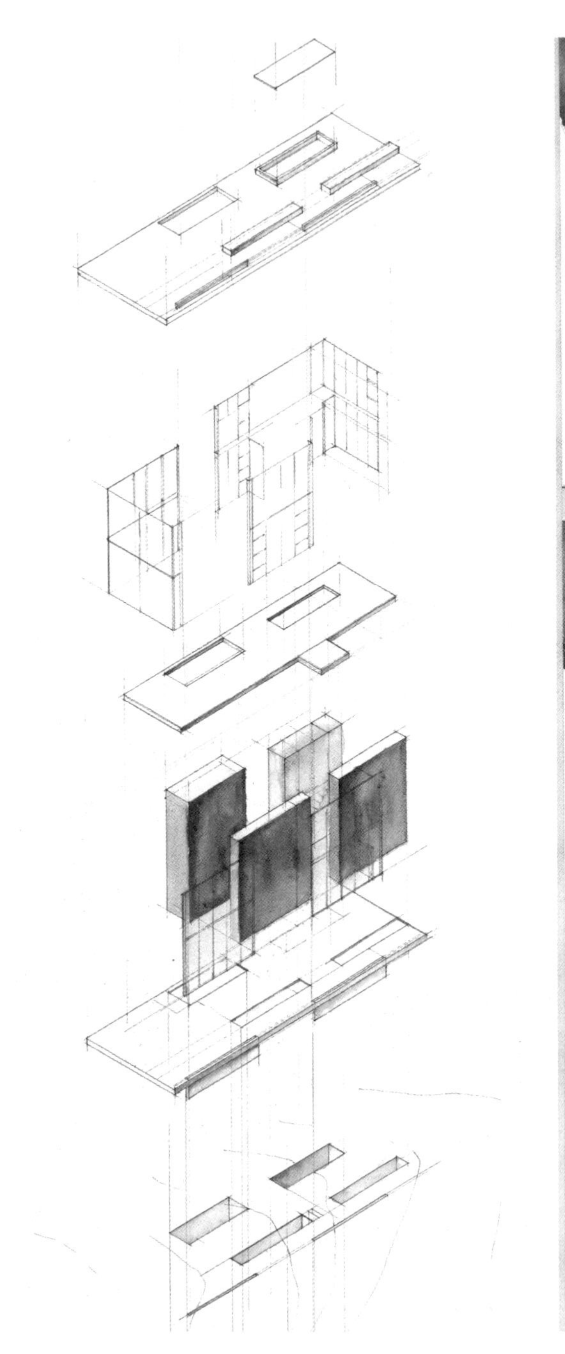

书店 | a bookstore

Catherine Pei Tin-wan, 0405T1Y2

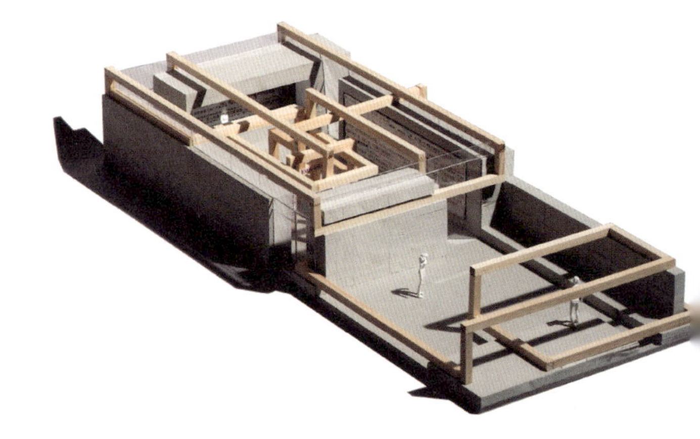

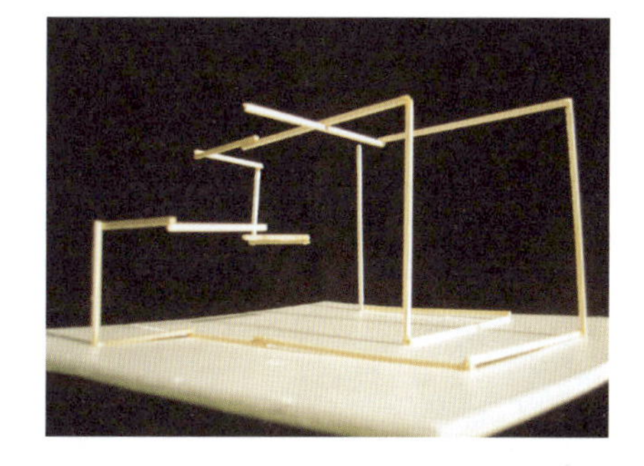

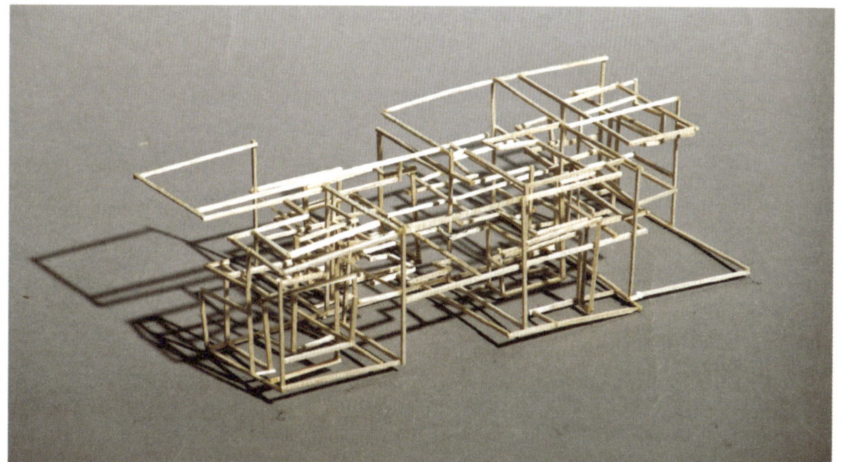

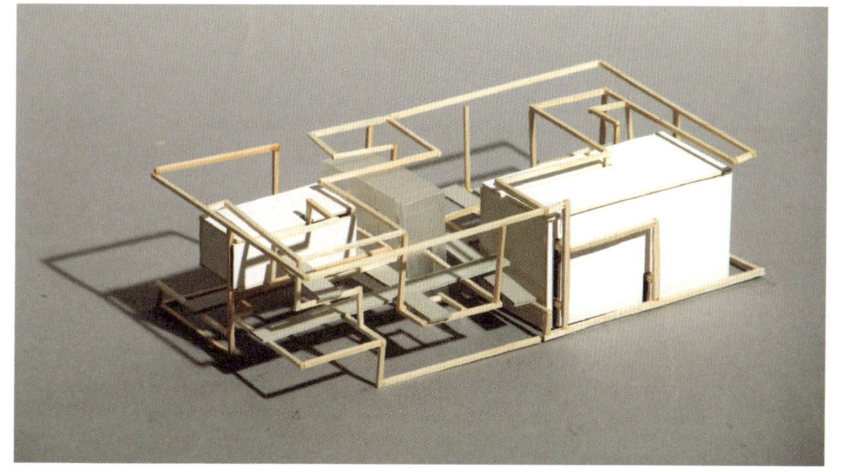

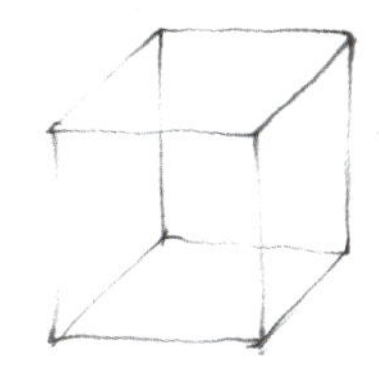

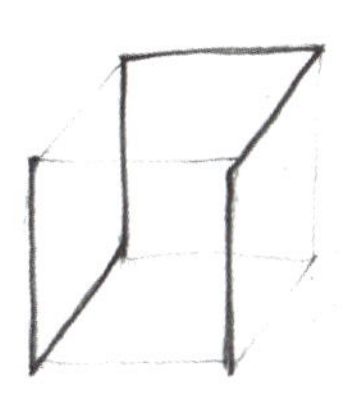

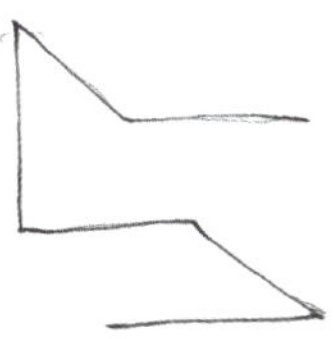

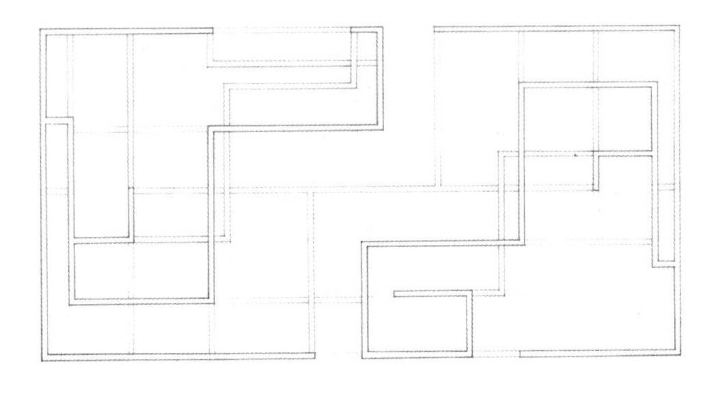

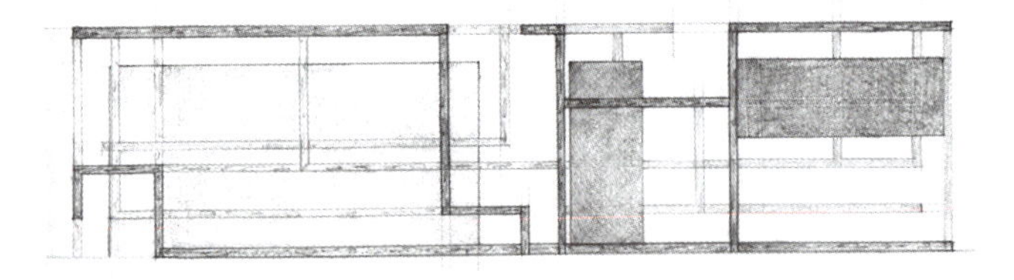

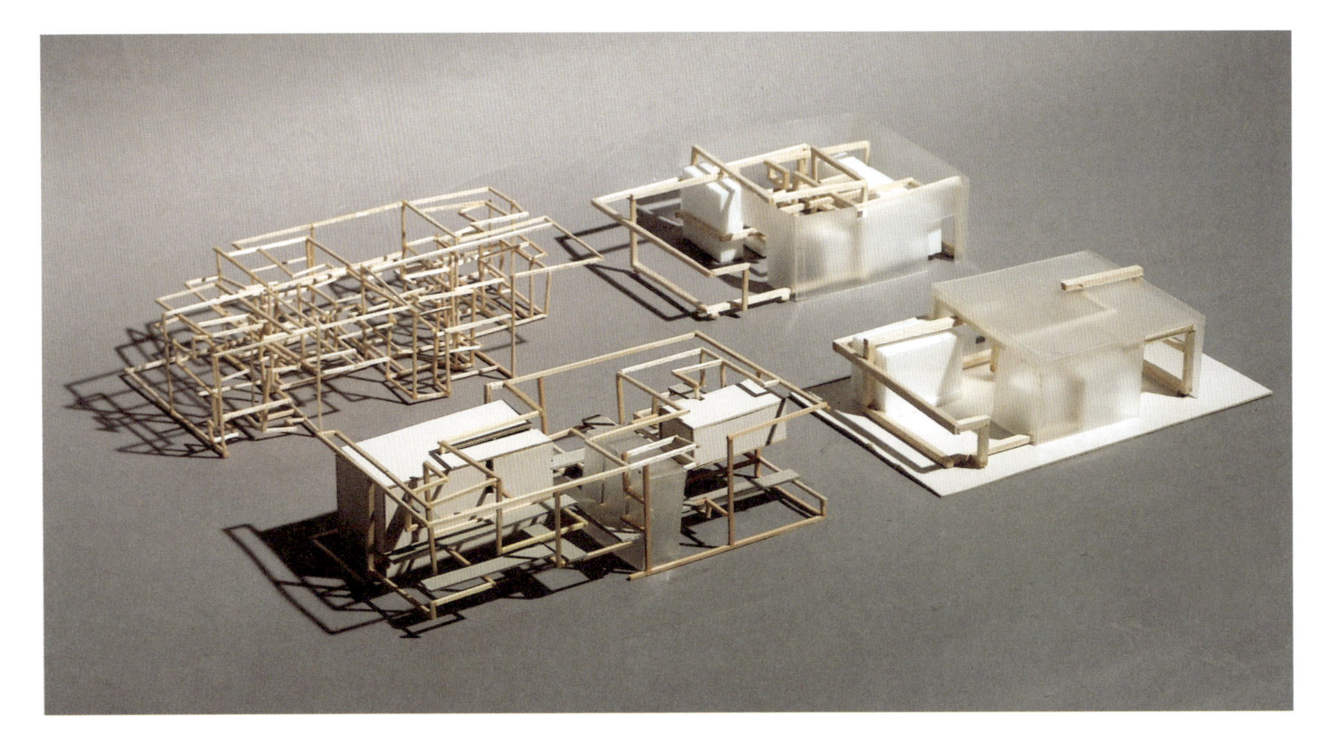

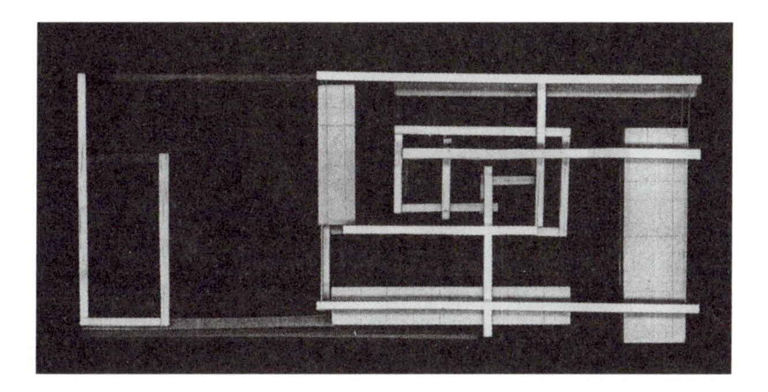

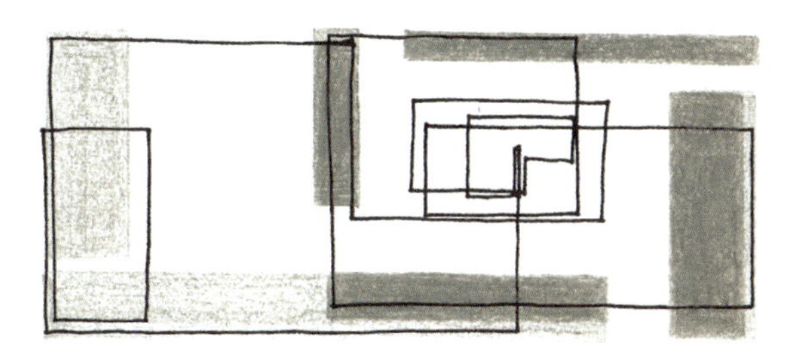

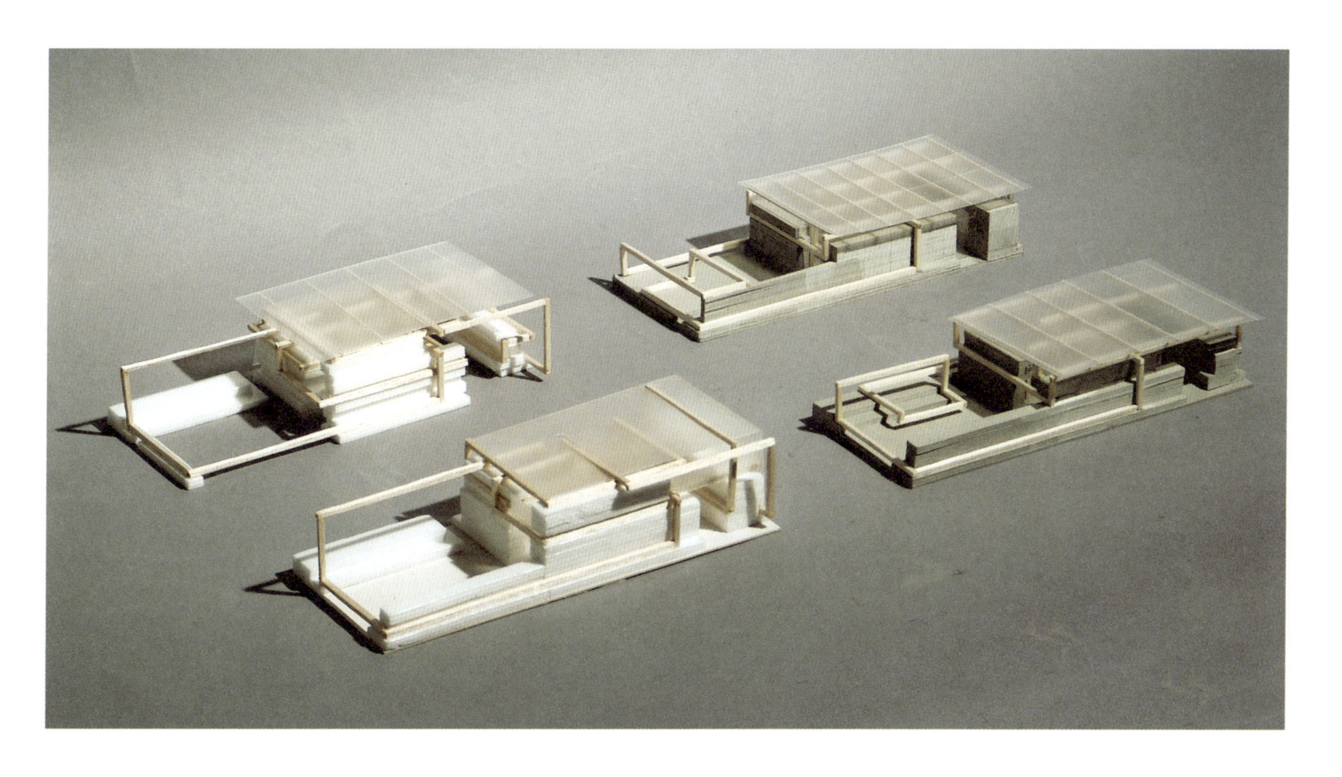

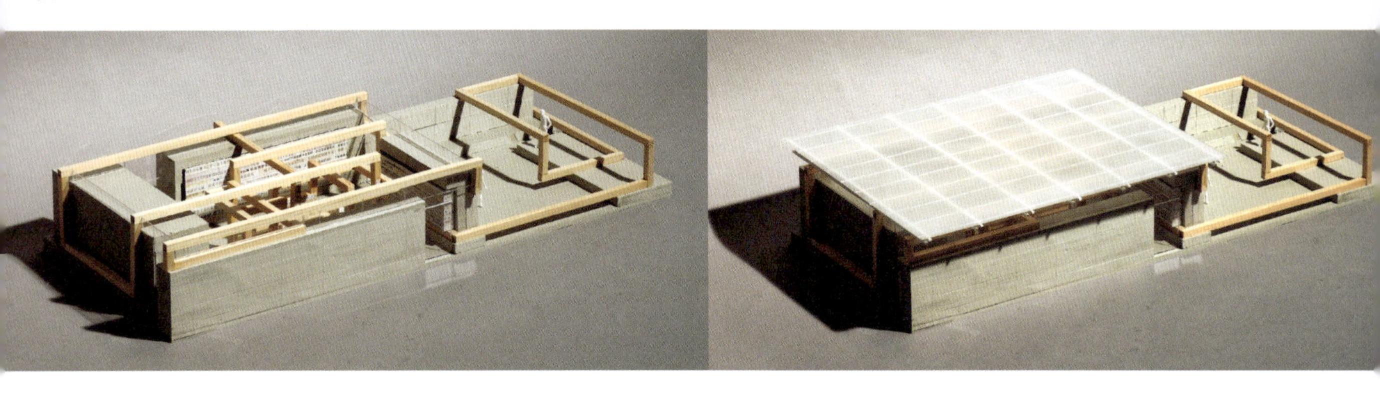

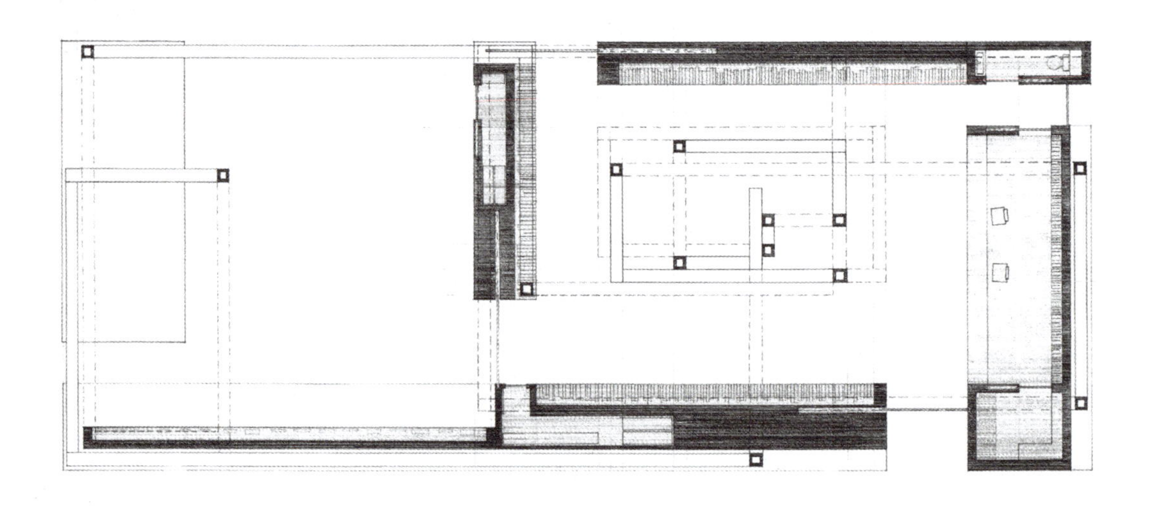

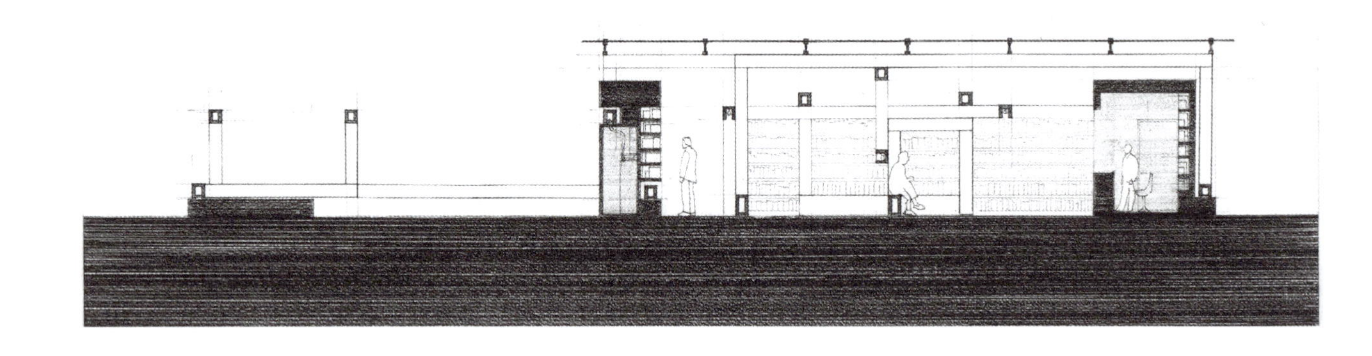

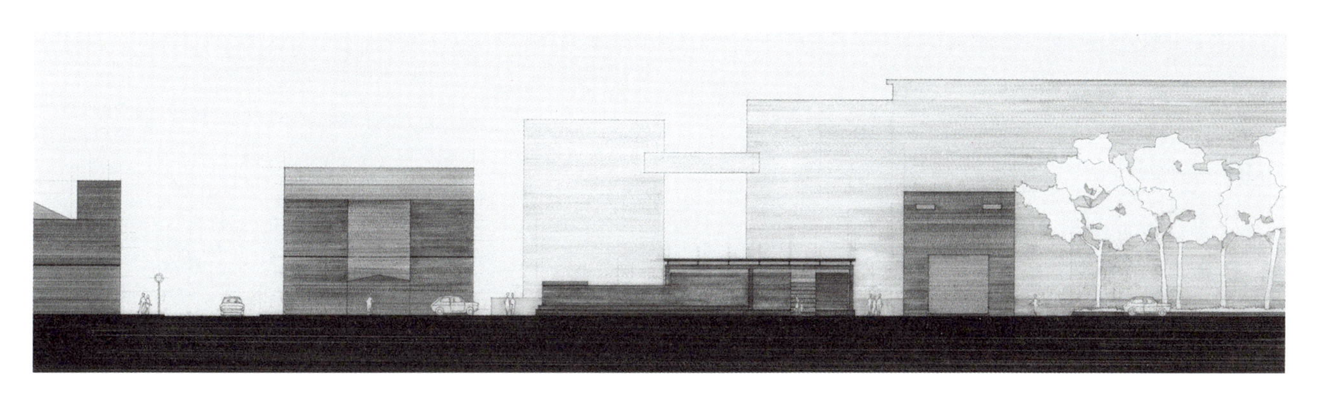

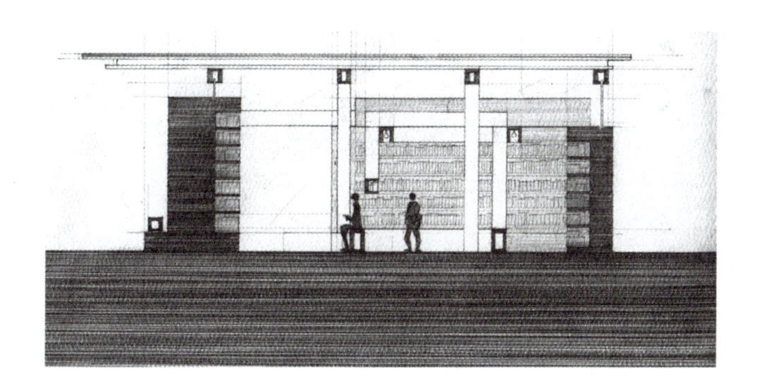

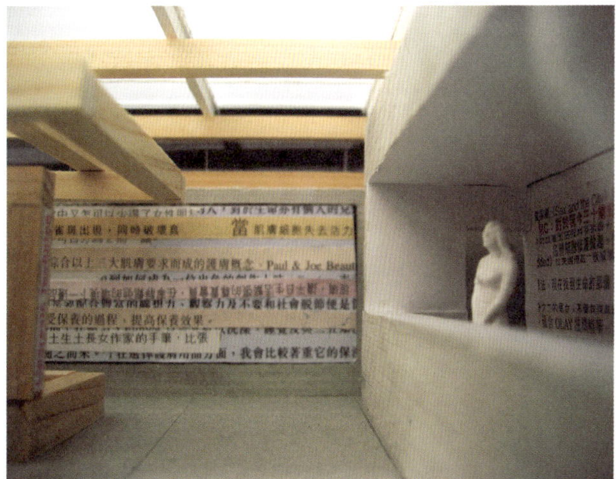
受保養的過程，提高保養效果。
土生土長女作家的手筆，比張
我會比較著重它的保

2 练习即方案 | exercise = project

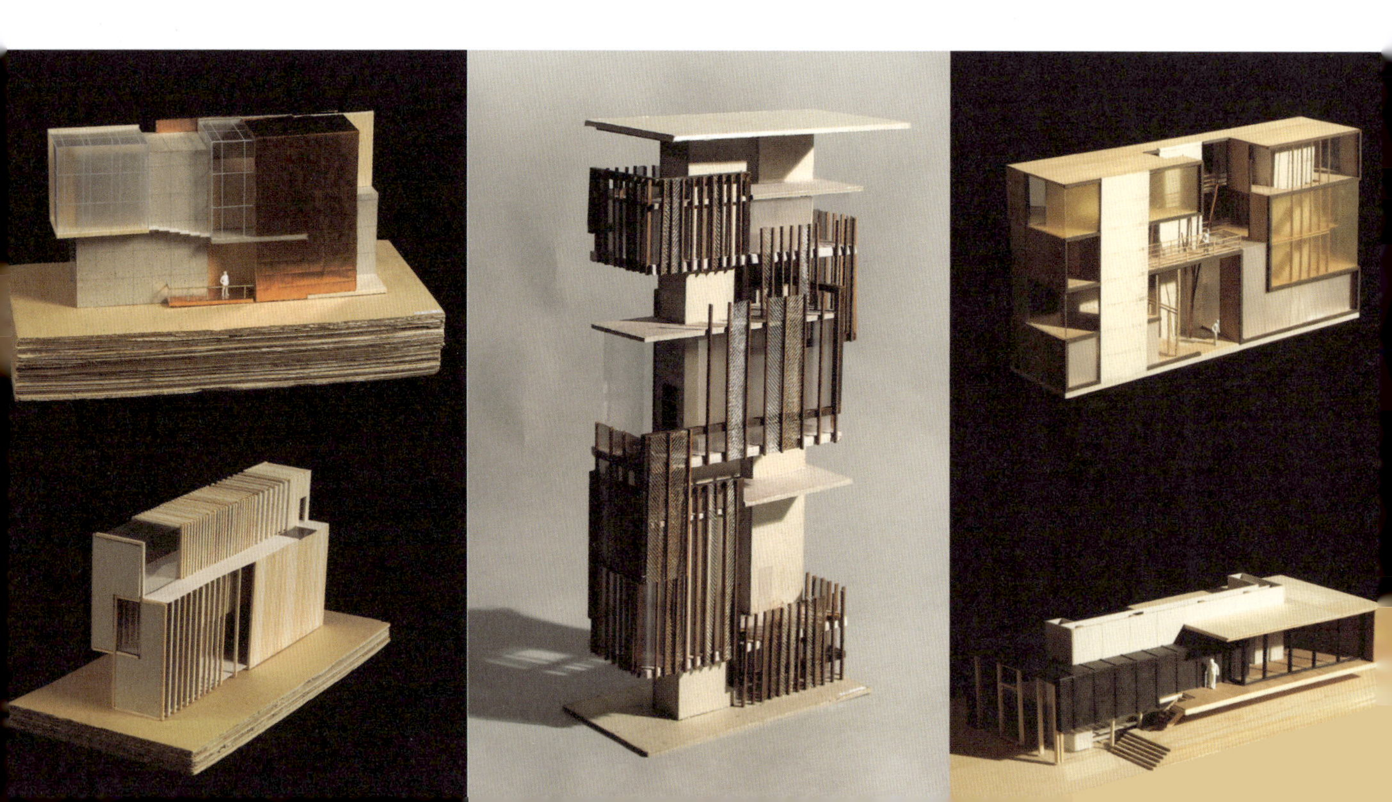

“练习加方案”的课程结构的另一个问题是两种训练的时间均不足够，特别是前面的专题练习的部分，有很多的问题可以深入研究，有很多的方法问题需要特别处理。所以，在2006－2007学年我们对课程的结构作了一个改变，即将设计方案的部分融入前面的练习中，这就是“练习即方案”。根据练习的平房、板房和楼房的特点，我们将建筑统一规定为居住加另外的一个功能，如画廊加住宅，商店加住宅，作坊加住宅等。此外，在原来的四个设计阶段中，我们加入了一个“场地”的阶段。在完成第二阶段“抽象”的练习后，一个小组的设计共同组织在一个抽象的场地上。这个场地的练习其实是最初建构练习的一个部分，后来被简化掉，现在又重新加入。总的来说，“练习即方案”的模式依然体现了问题设定的抽象性。

这一改变的好处是有比较充裕的时间来发展每个阶段的练习，特别是作图的环节。此外，我们还在过程的最后阶段加入了一个1:50的模型，作为设计研究的结束。

After finding that neither during the studio project nor the school project there was enough time to develop either the exercise or the project to a desired level, we tried to turn the exercise into a project for the whole term. It was the first time an abstract site had been introduced, but it was later dropped again. A simple programme of use was given as a living space for two people and additional spaces for working, selling or exhibiting.

This gave more time for each phase of the exercise which allowed for more depth in dealing with the issues. And the programme supported the imagination for the use of the space. The extra time also provided the chance to introduce a concluding phase with comparative drawings of the phases, key views of the final stage and a presentation model in the scale of 1:50.

画廊兼住宅 | a gallery with apartment

Justin Law Chun-wai, 0607T2Y3

概念 | concept

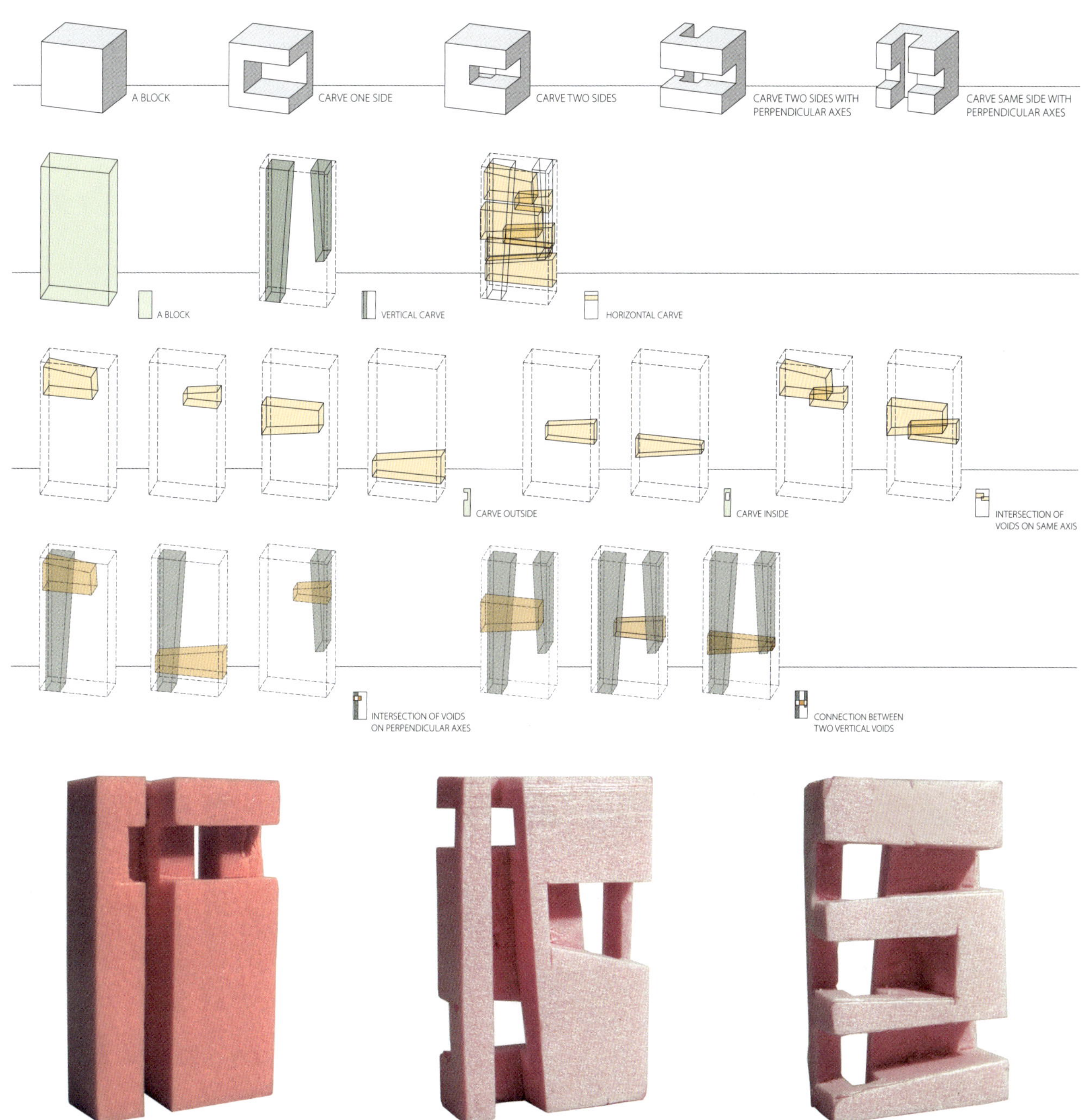
A BLOCK
CARVE ONE SIDE
CARVE TWO SIDES
CARVE TWO SIDES WITH PERPENDICULAR AXES
CARVE SAME SIDE WITH PERPENDICULAR AXES
A BLOCK
VERTICAL CARVE
HORIZONTAL CARVE
CARVE OUTSIDE
CARVE INSIDE
INTERSECTION OF VOIDS ON SAME AXIS
INTERSECTION OF VOIDS ON PERPENDICULAR AXES
CONNECTION BETWEEN TWO VERTICAL VOIDS

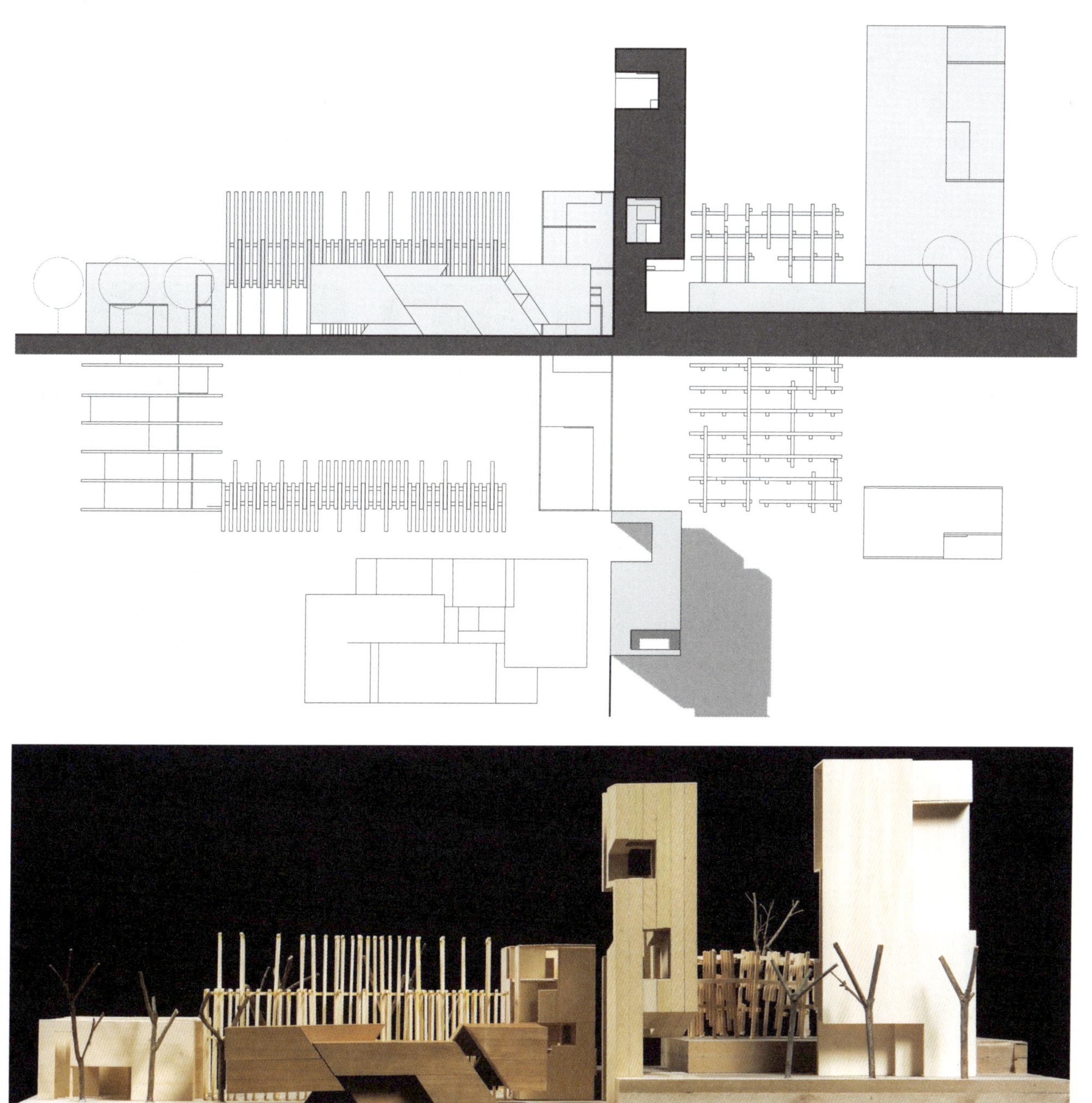

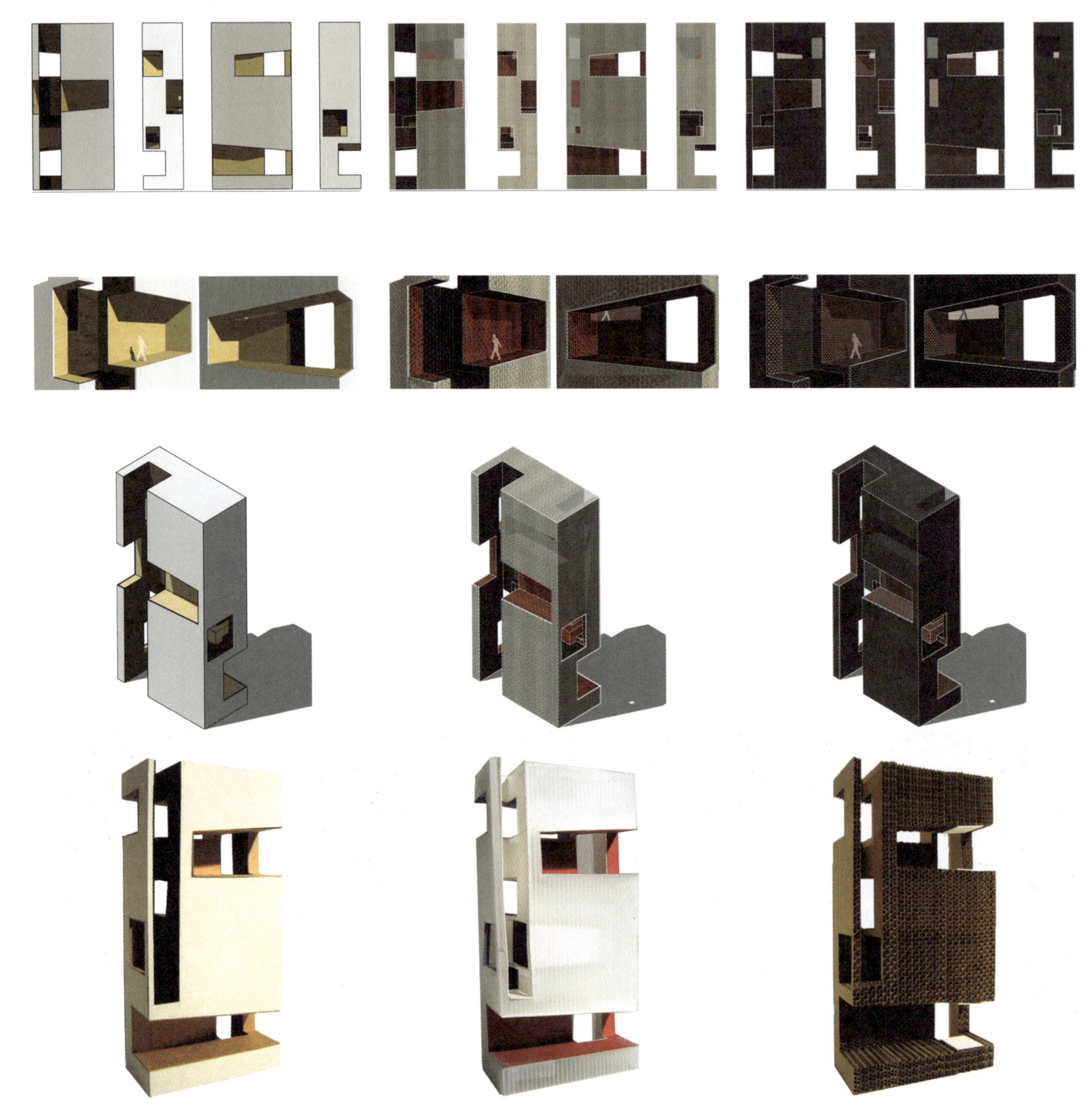

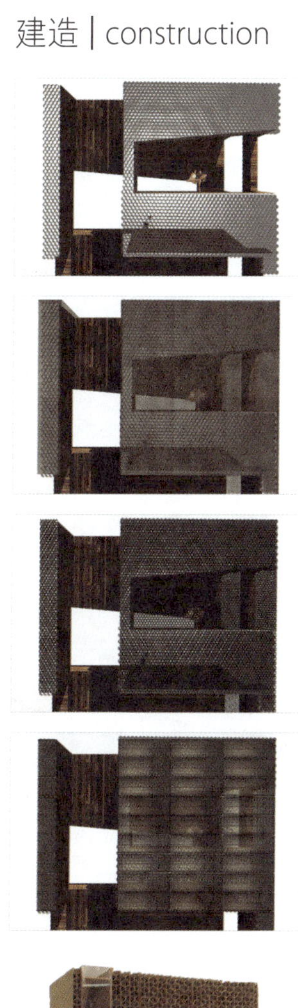

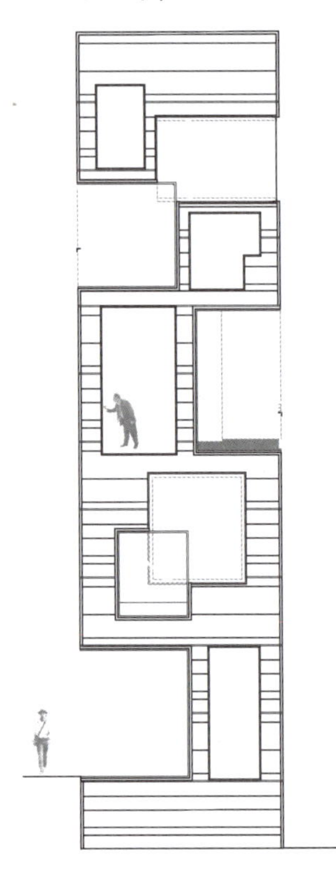

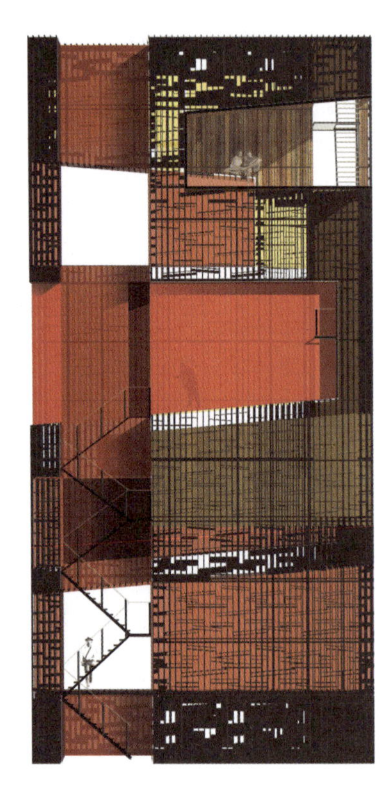

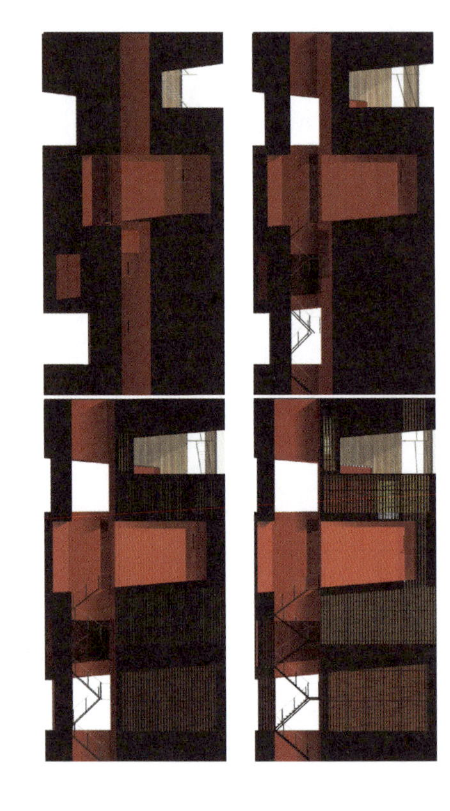

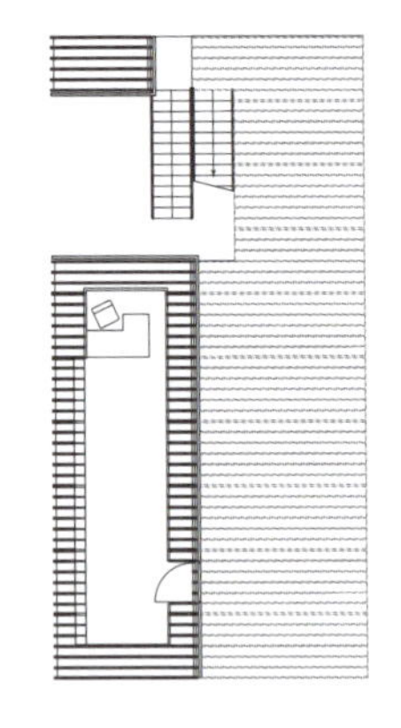

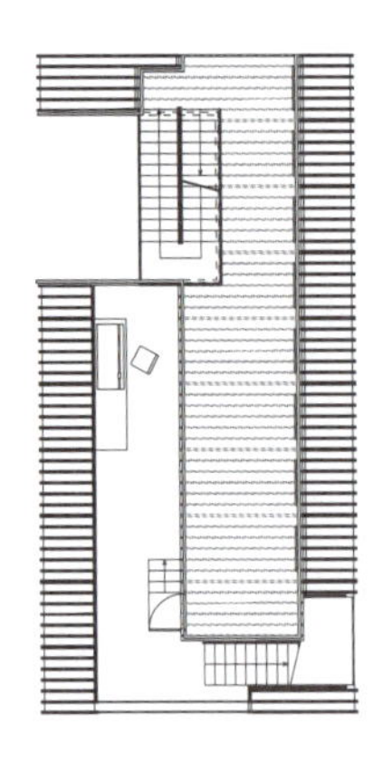

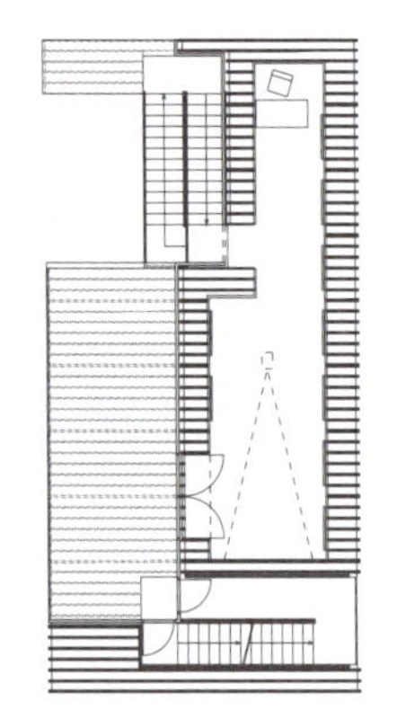

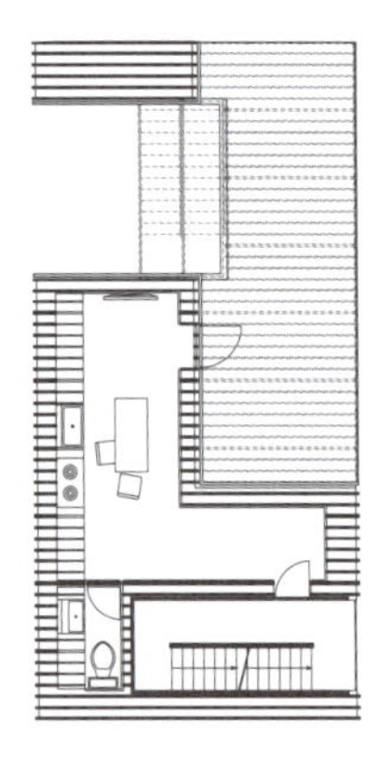

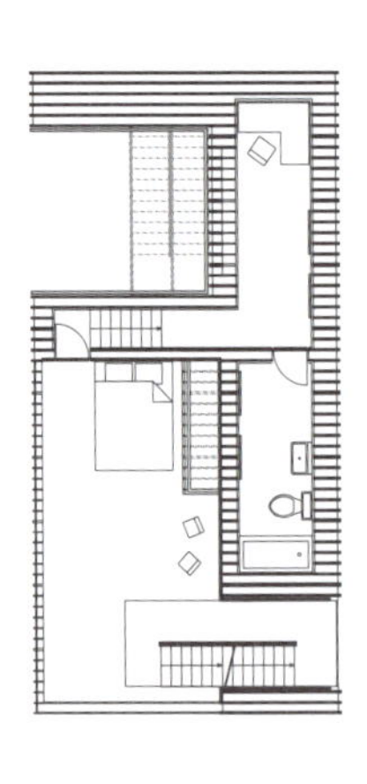

时装店兼住宅 | a boutique with apartment

Kit Chung Yuk-ching, 0607T2Y3

概念 | concept

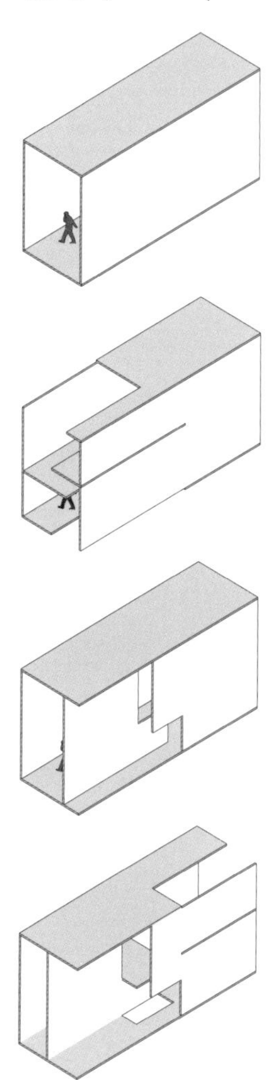

抽象 | abstraction

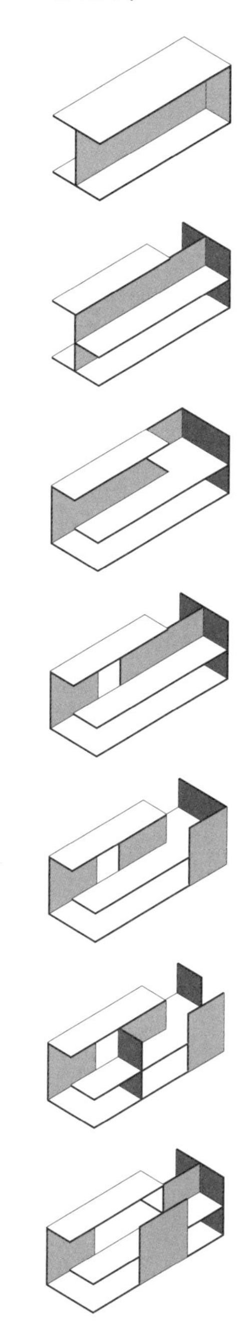

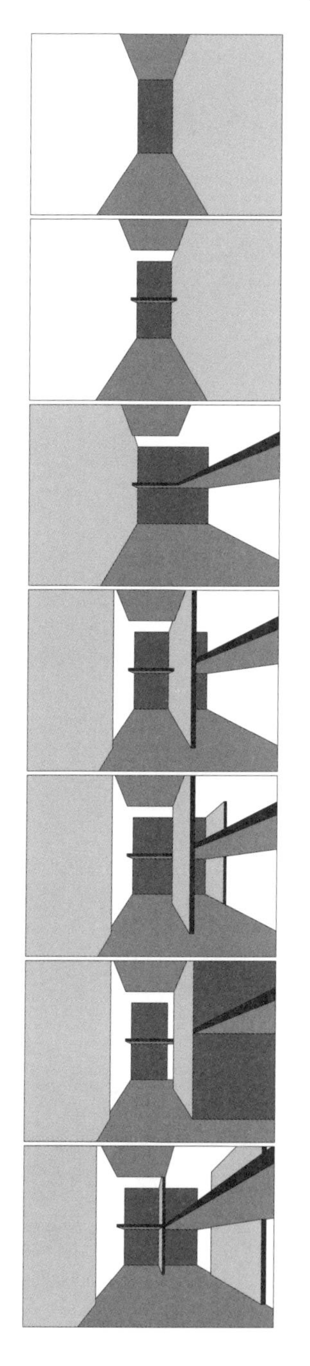

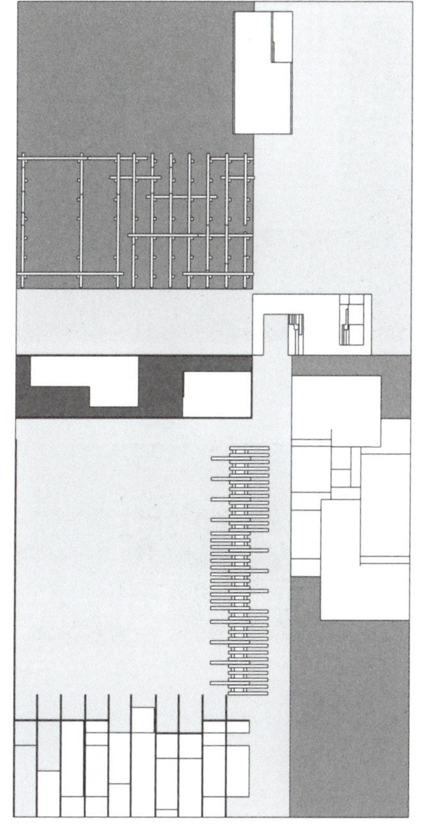

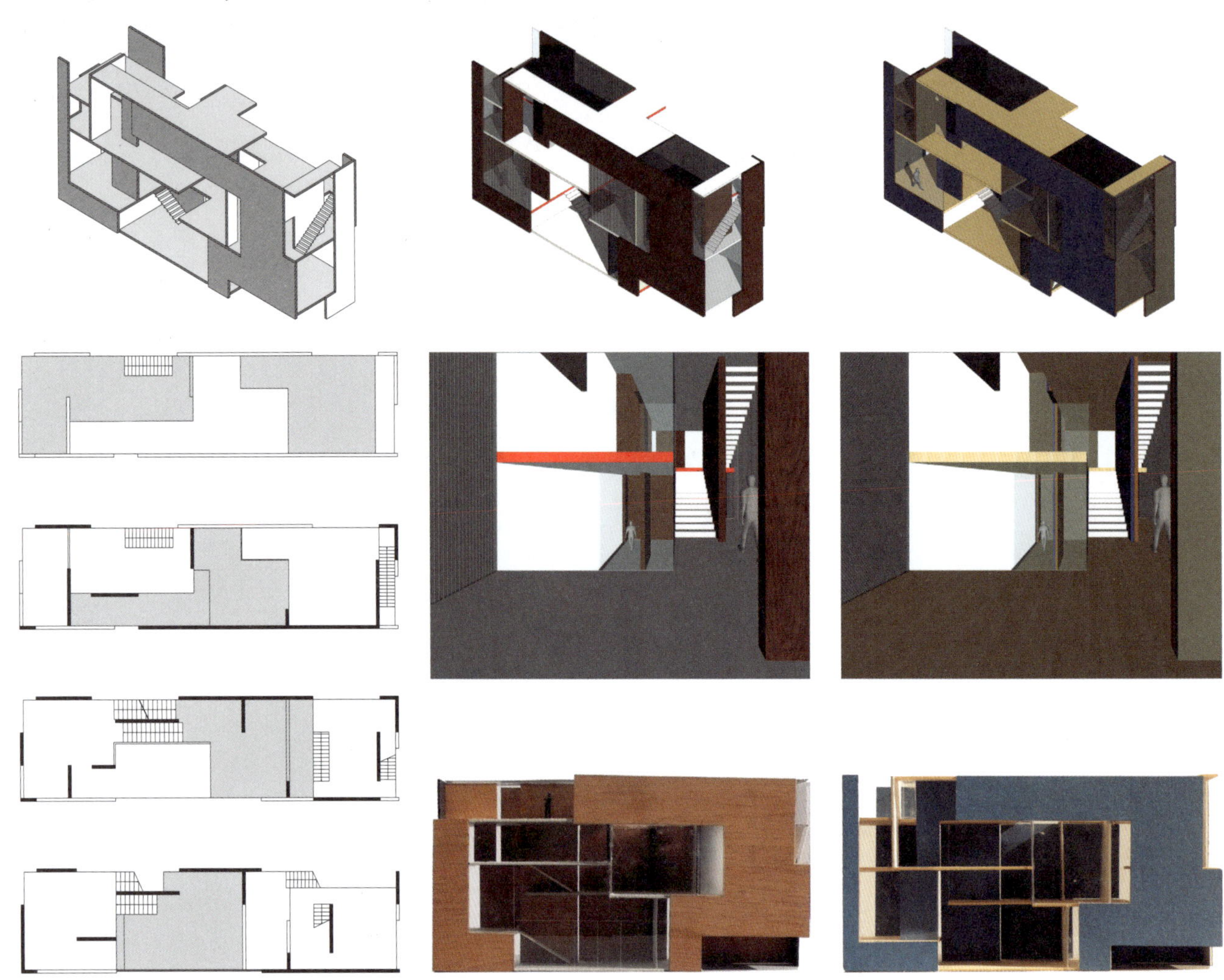

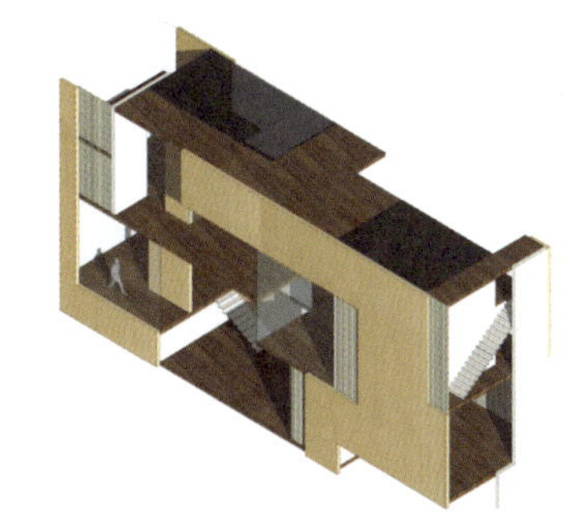

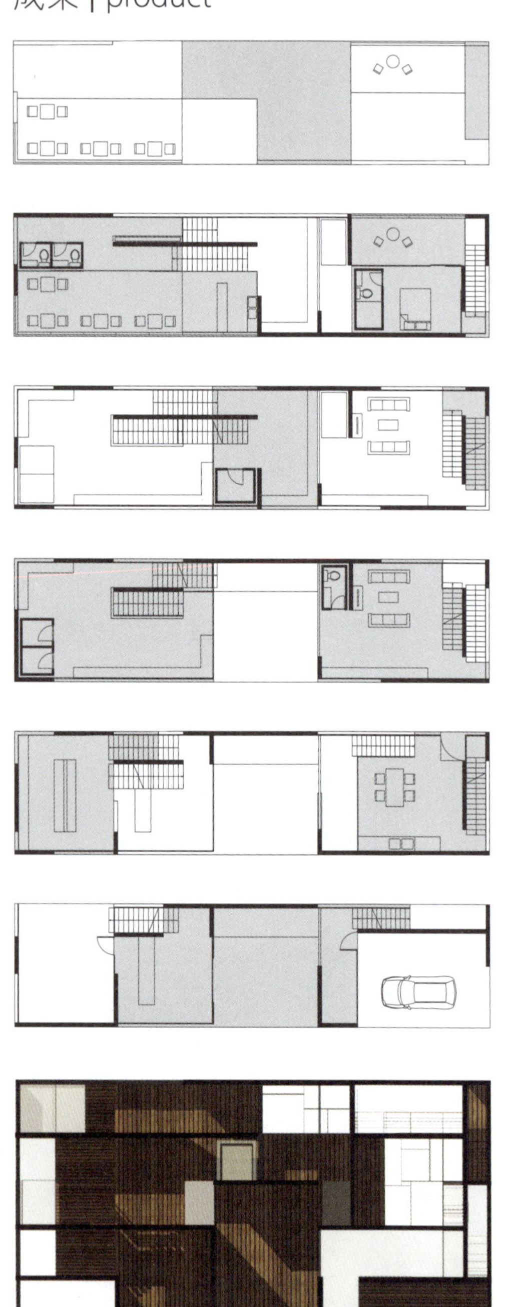

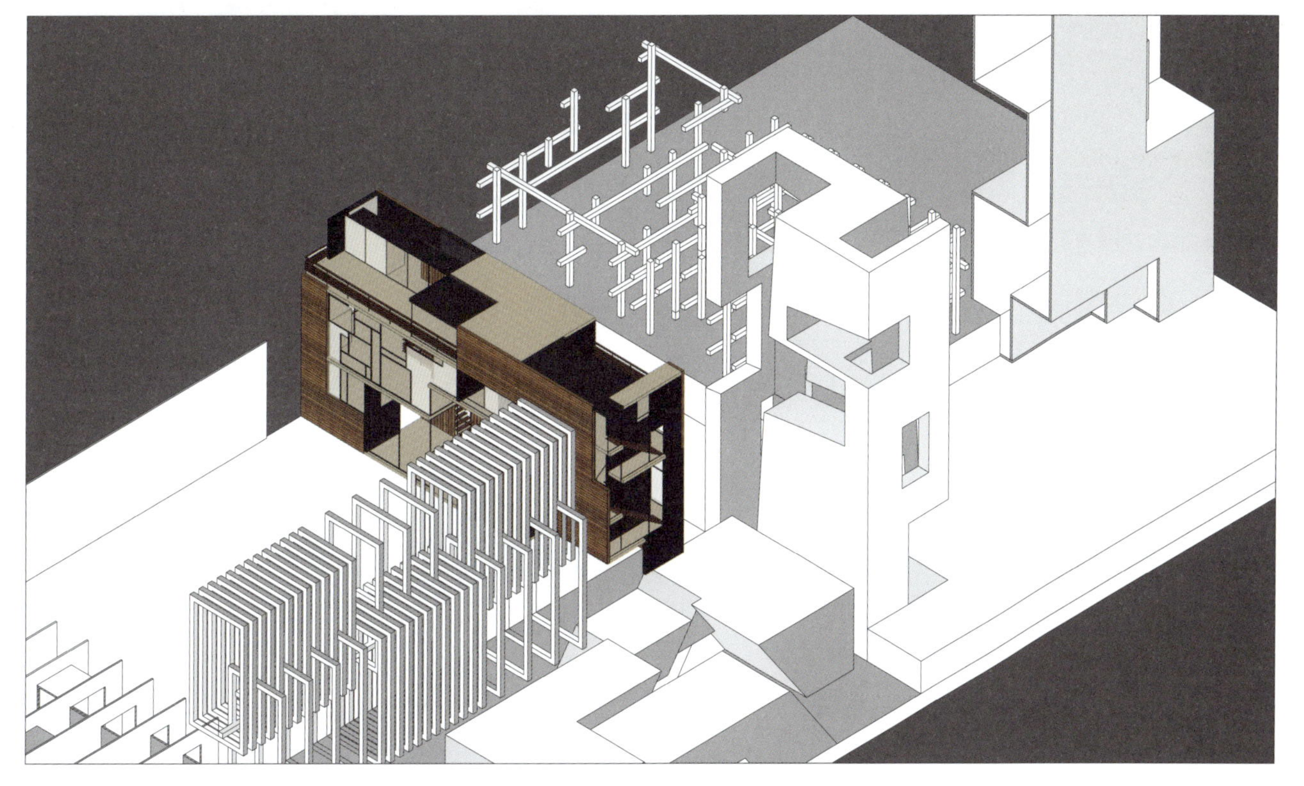

书店兼住宅 | a bookstore and apartment

Tracy Yip Chui-chui, 0607T1Y3

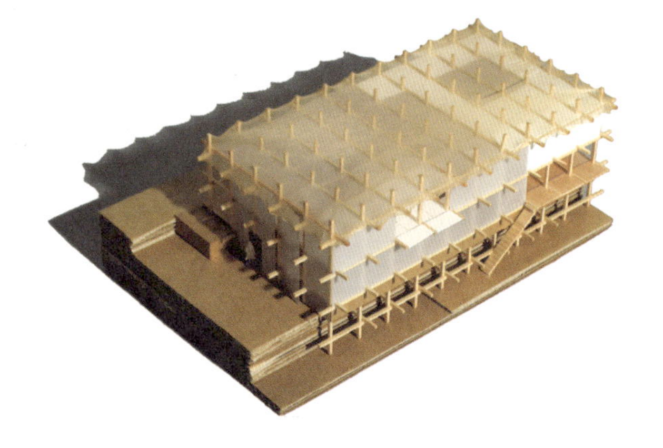

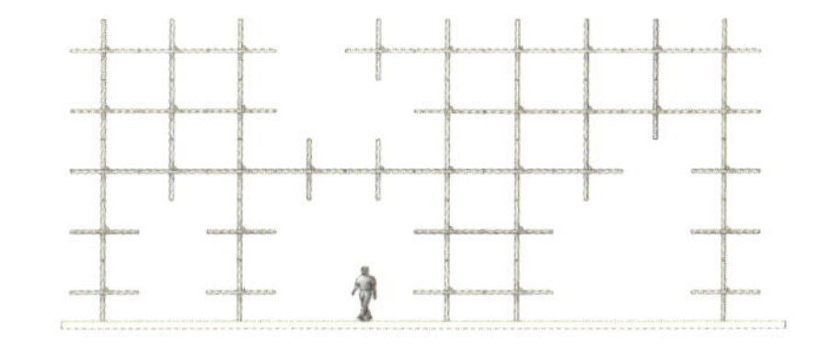

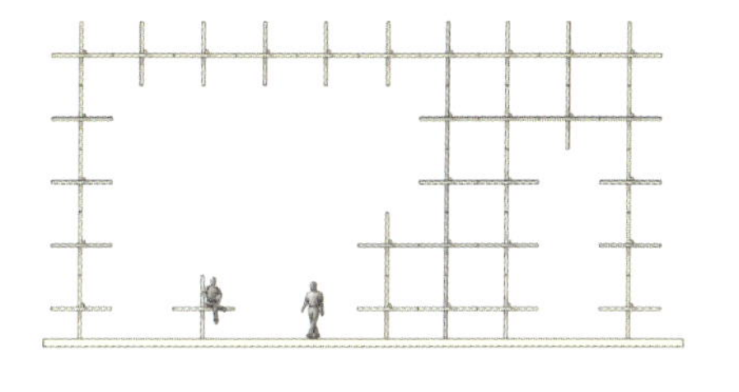

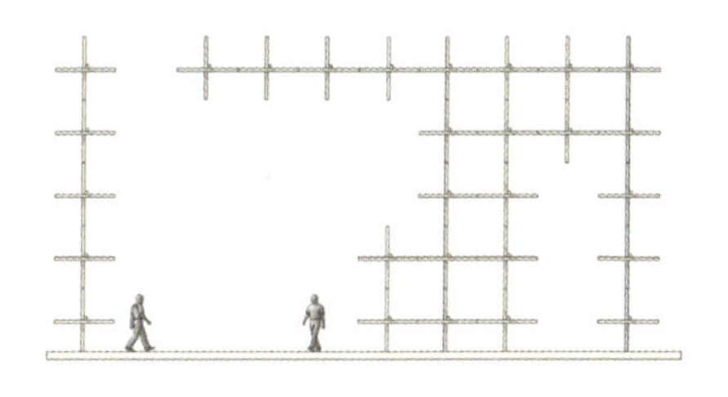

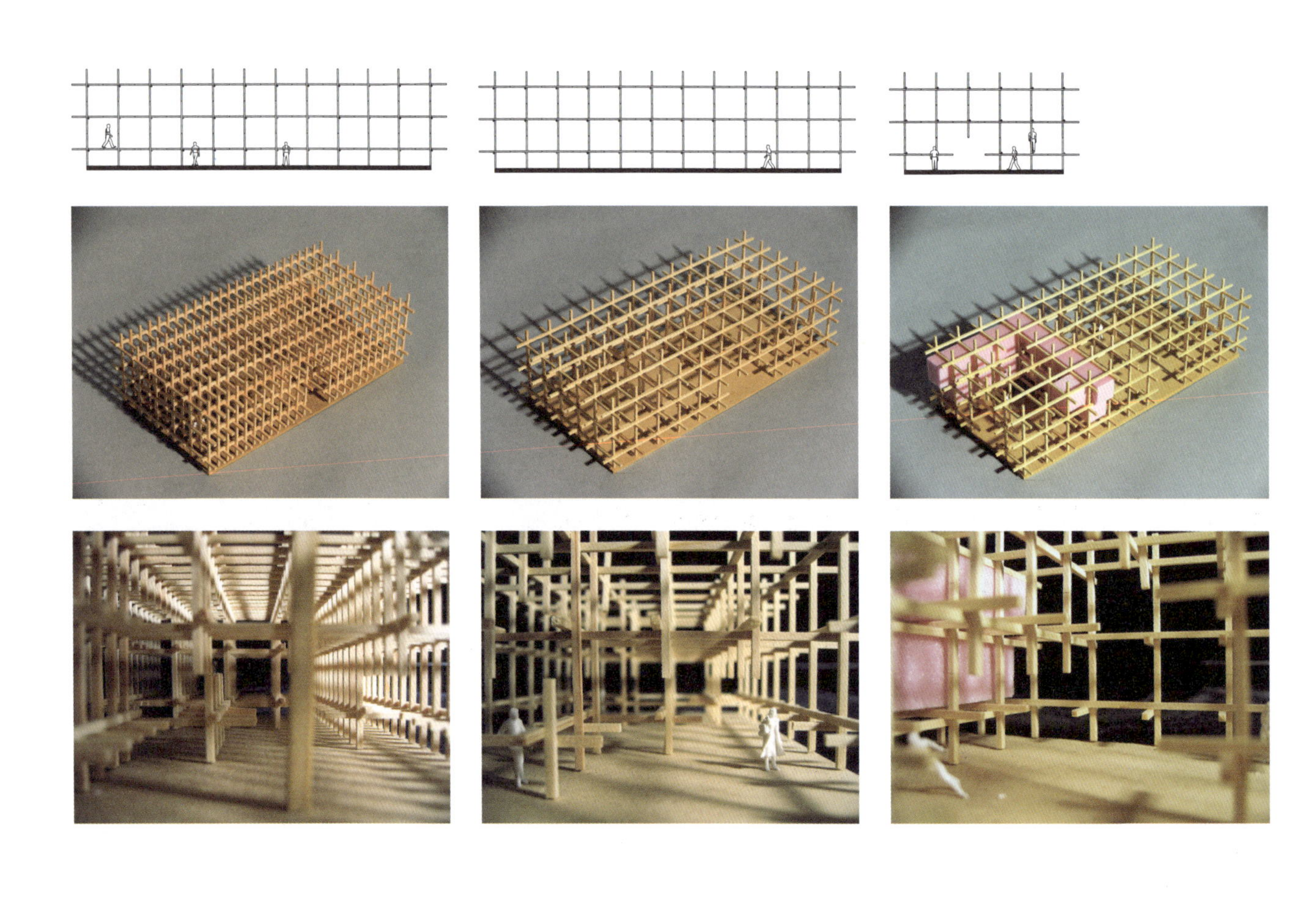

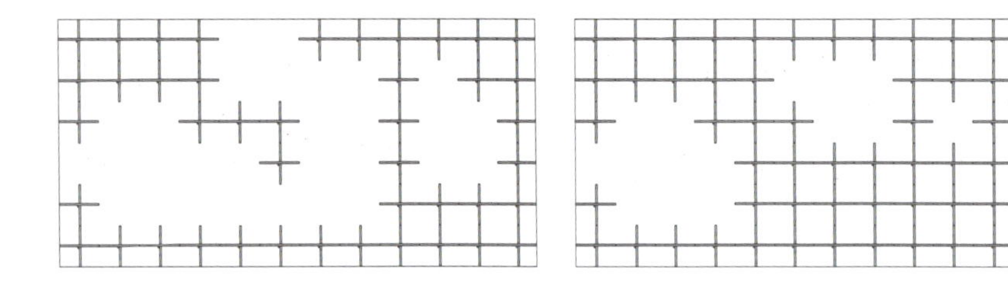

环境 | context

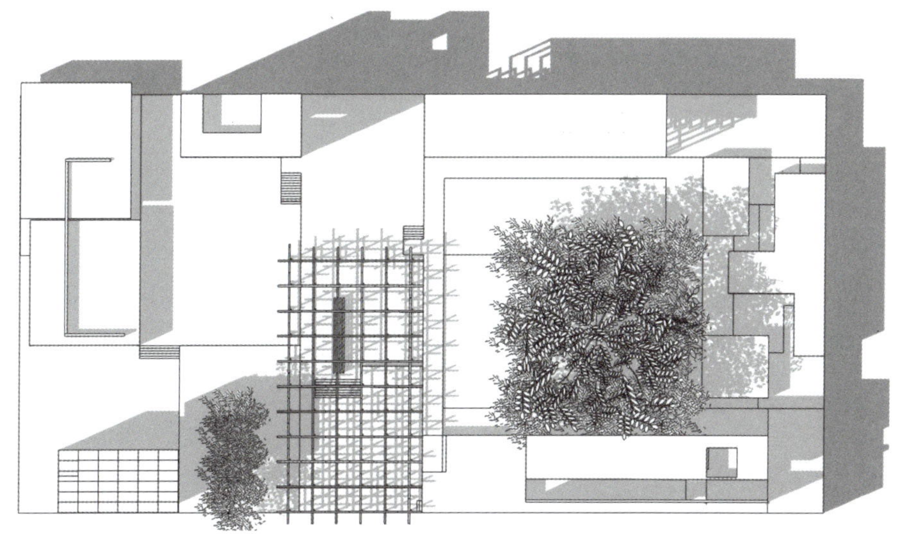

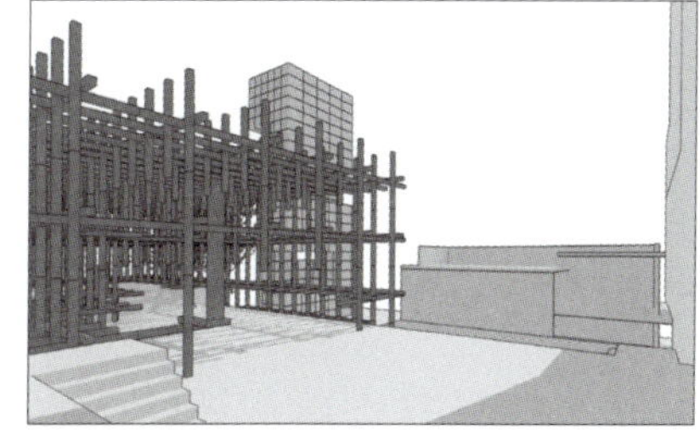

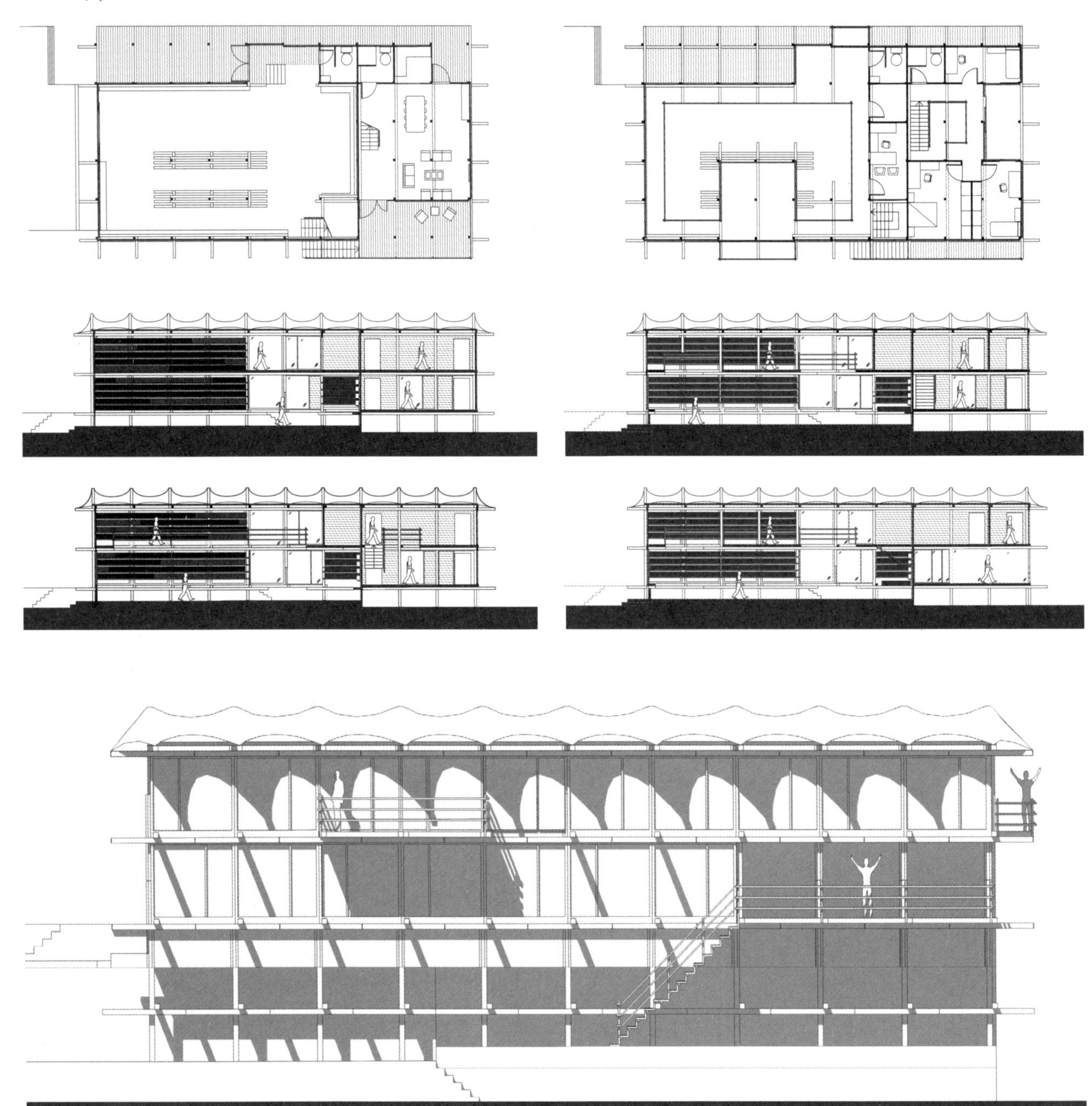

3 方案练习化 | exercise ⊂ project

在“练习即方案”的模式之后，我们还尝试了“练习即练习”的模式，即重新回到练习的抽象本质。2008－2009学年的第二学期，我们又探讨了“方案练习化”的模式，即如何使得对建构形式和空间的追求成为一个设计过程的有机组成部分，一种更为自然的设计方法。与前面的几种教学模式相比，“方案练习化”的模式有两个重要的改动。一是在设计开始的“场地”阶段，通过对特定场地的形体和空间关系的研究来确定建筑的体量。也就是说，我们终于放弃了平房、板房和楼房的概念。二是在“组织”阶段，如何用不同的建构和空间概念来对特定的功能要求作各种诠释，如体块的诠释、板片的诠释、杆件的诠释，或综合的运用。也就是说，不是先有一个形式，而后再看看它能做什么，而是根据功能来作相应地“建构”诠释。“材料”和“建造”两个阶段则基本上延续了原先的方法。

The results of the change were again mixed. Seen as exercises, it was clearly an improvement, but as a project, the process still felt artificial. The strength of the exercises, and remarks by the external examiner, made us put the question of the project aside, and concentrate on the exercises as exercises – basically extending the original studio project to the whole term. But we still were interested to see how we could integrate these exercises in a more natural design process. So in the last term of the year 2008/09 we tried a completely new approach, embedding the exercises in the design process of a project.

The main issues of the exercises were kept, but studied in connection with other considerations. The mass of the building was determined in the study of the site. The mass was differentiated and organised based on the study of the programme. The definition of space was studied as an interpretation of the space for the programme. One interpretation was further articulated through the study of material differentiation. And finally the transformation of model material to building material was done through the study of construction.

训练中心 | a training centre

Cathy Tang Nga-chi, 0809T2Y2

体积 · 场地 | massing · site

Variation 1

Variation 2

Variation 3

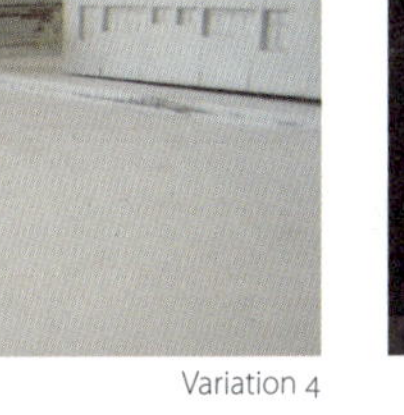

Variation 4

组织 · 功能 | organisation · programme

Final Massing

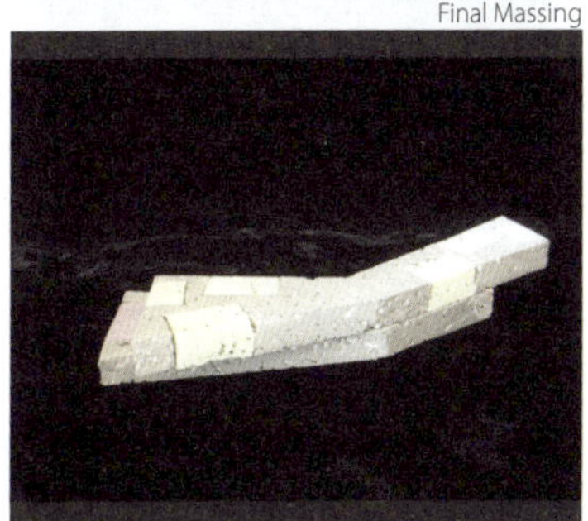

Variation 1

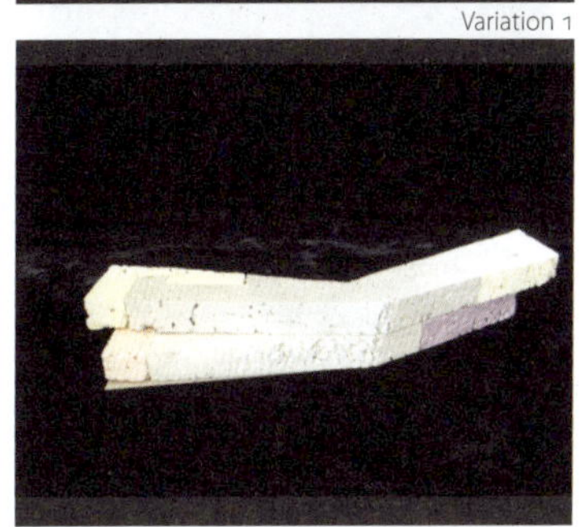

Variation 2

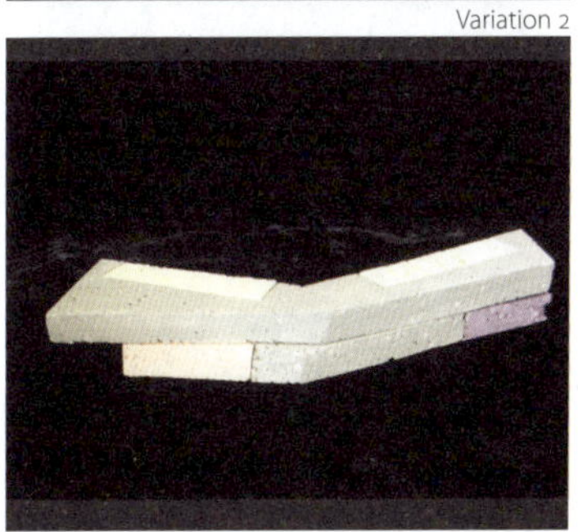

Variation 3

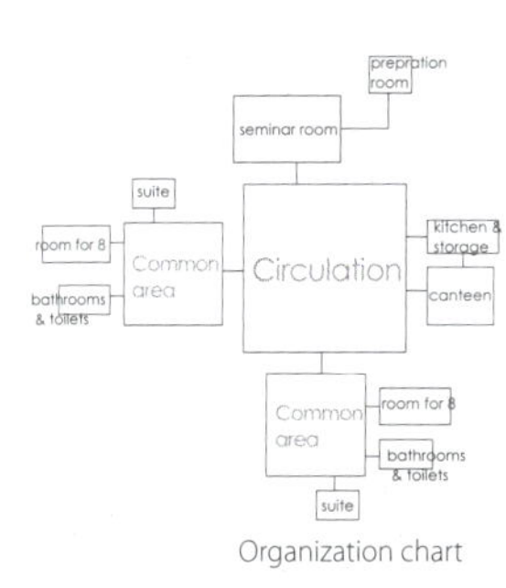

Organization chart

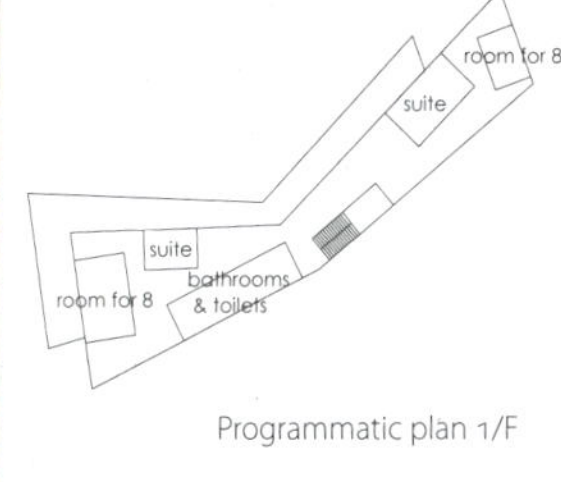

Programmatic plan 1/F

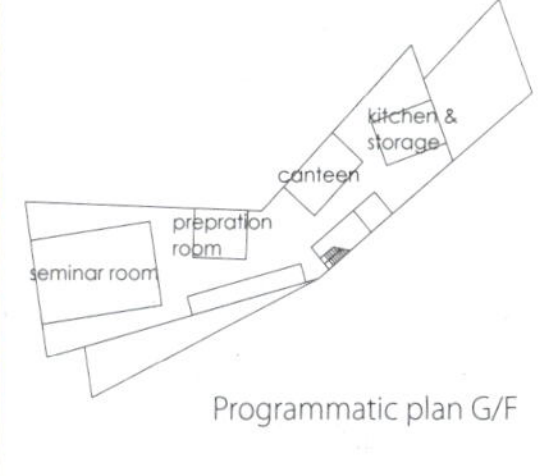

Programmatic plan G/F

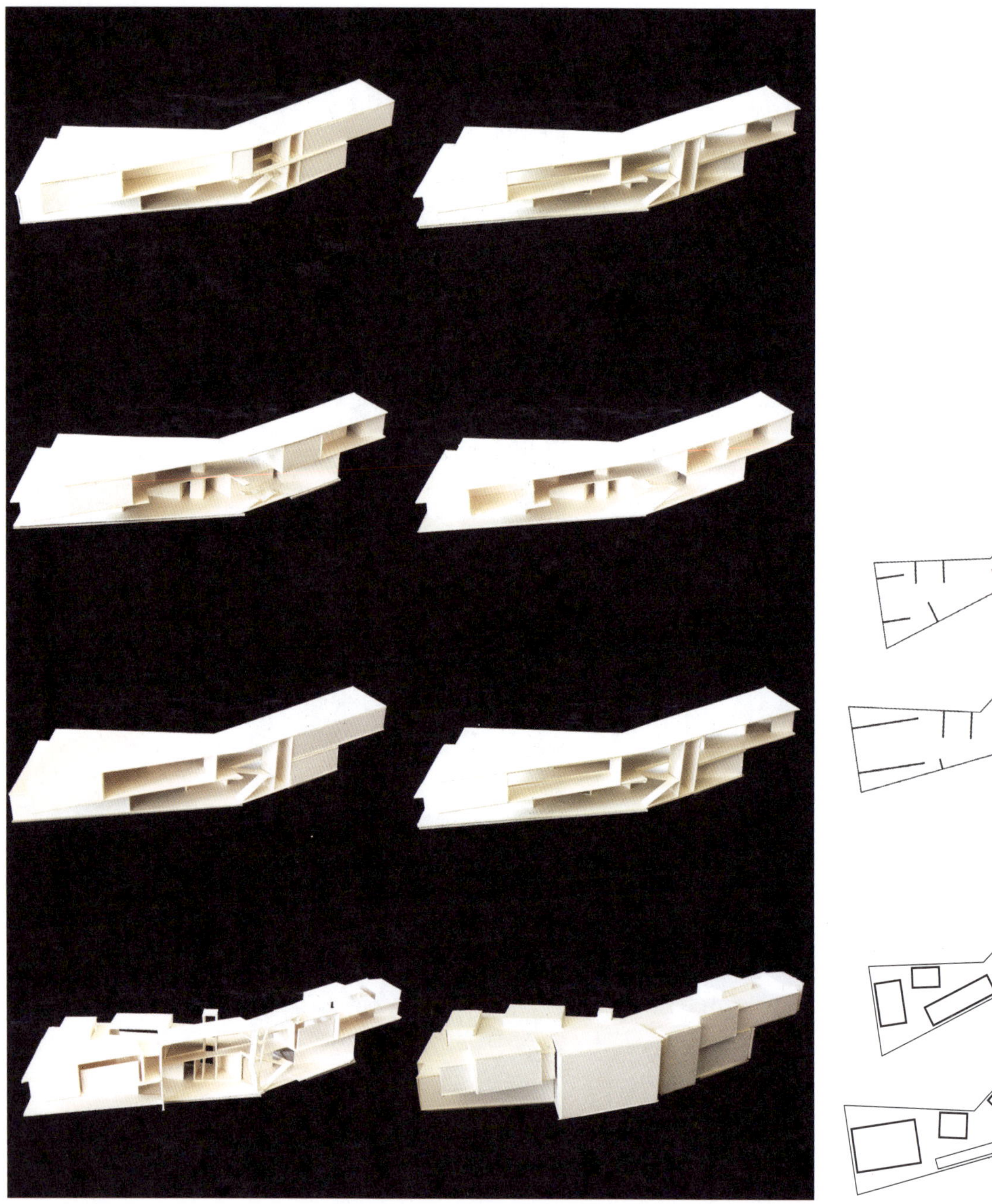

Mesh as facade - translucent

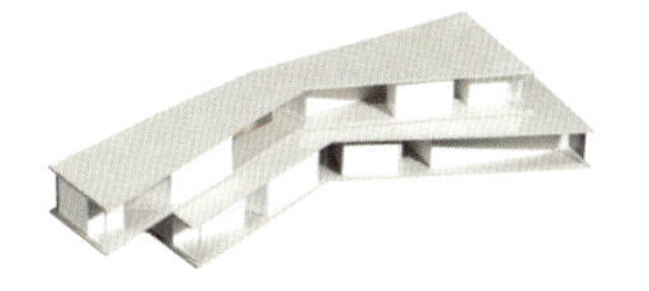

Glass as facade - transparent

Glass with screen as facade - nearly opaque

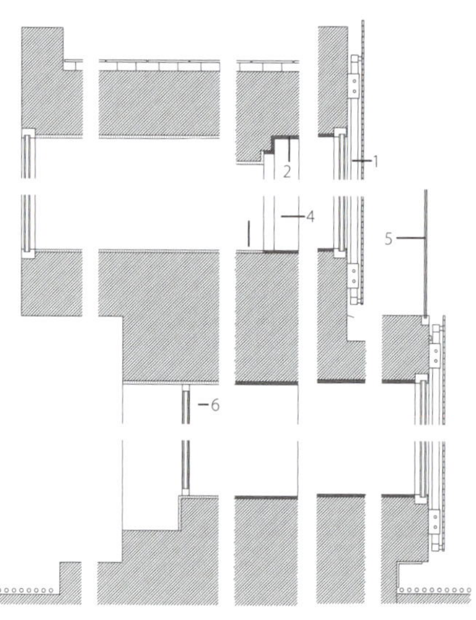
1
2
4
5
6

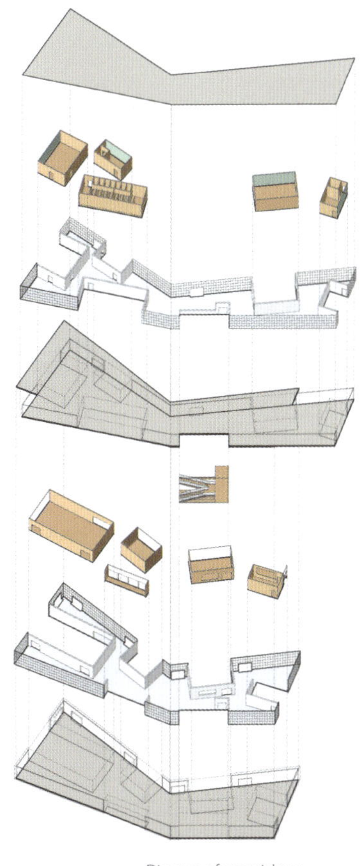

Wood

White paving

Structural Concrete

Mesh

Diagram of material use

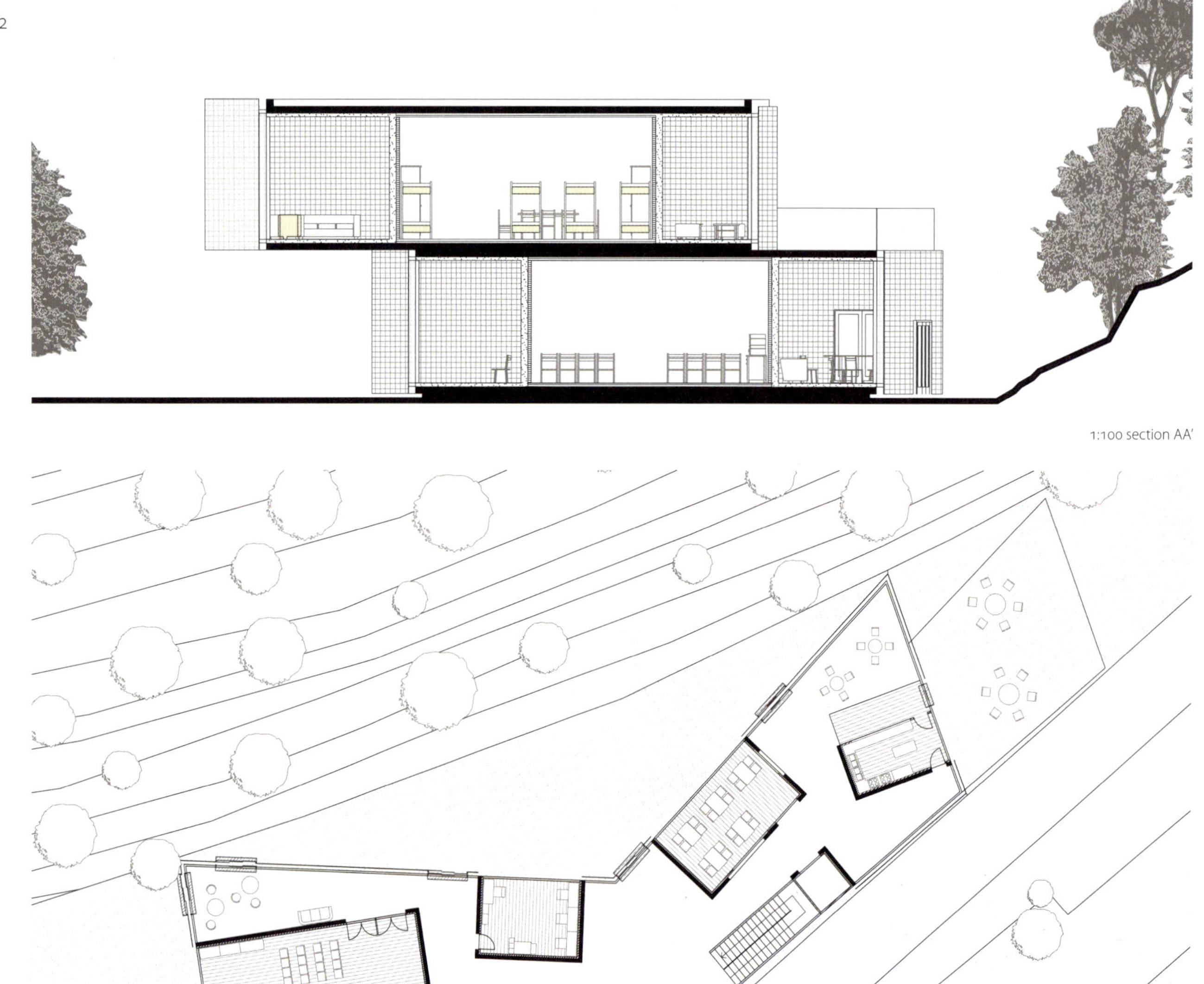

1:100 section AA′

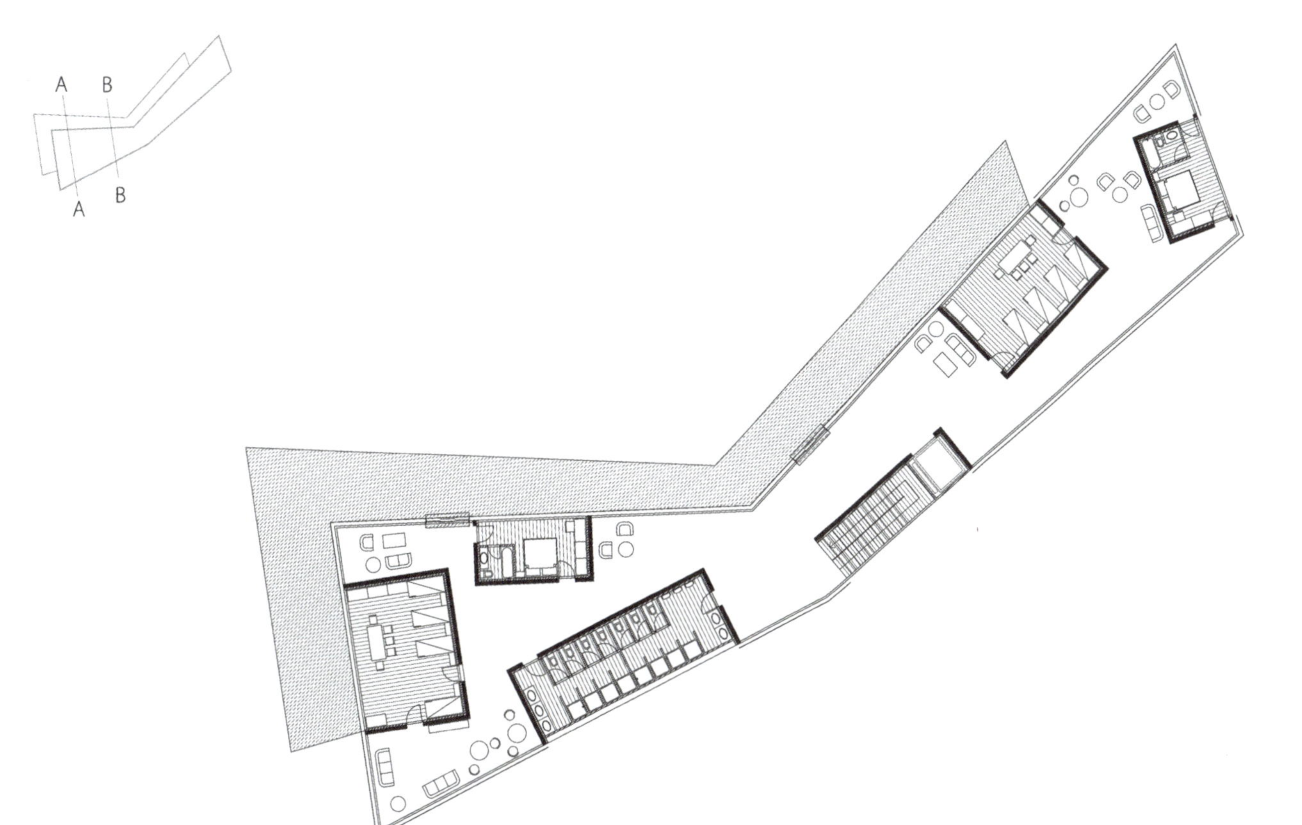
A
B
A
B

训练中心 | a training centre

Tsoi Yun-fung, 0809T2Y2

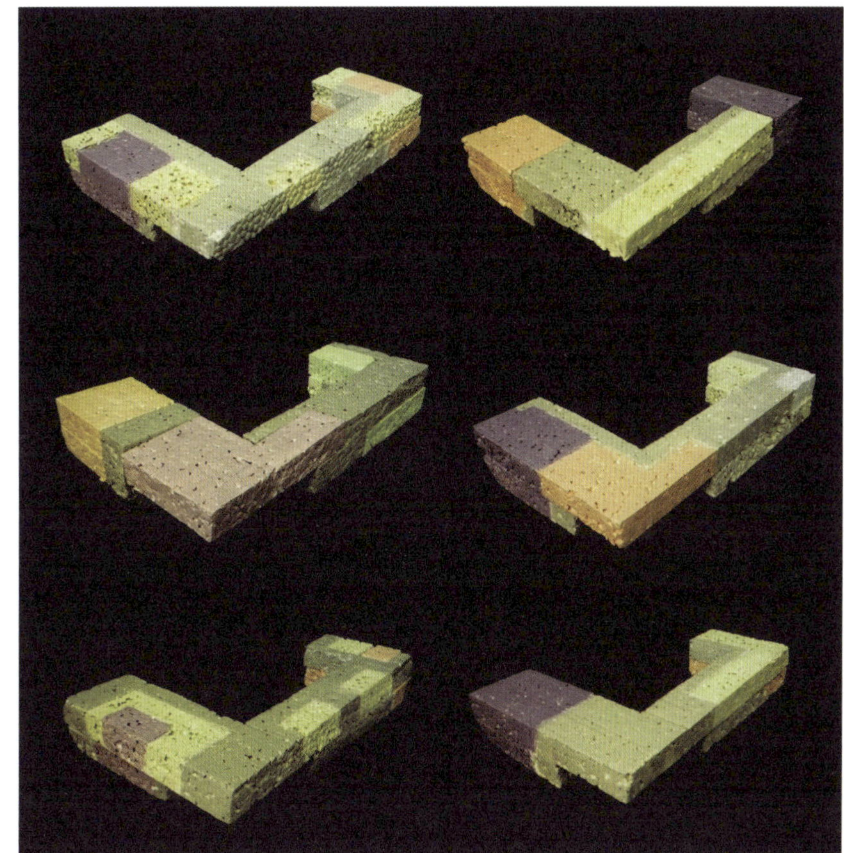

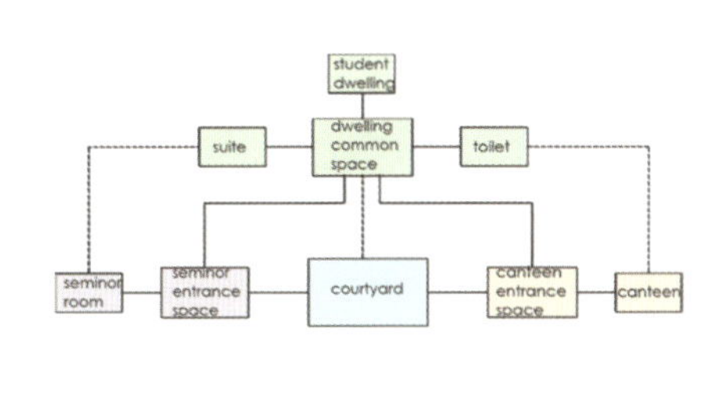

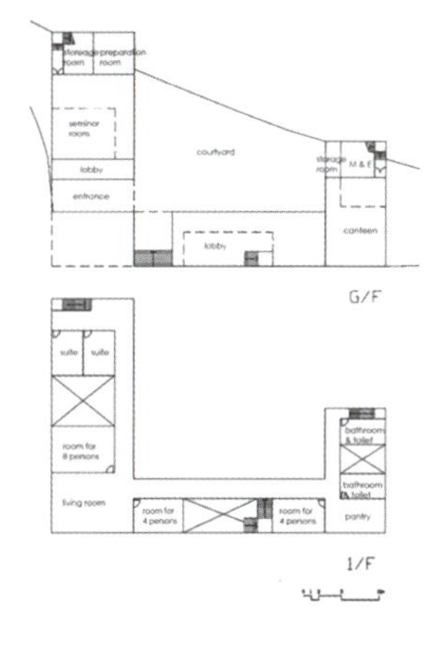

Organisation chart

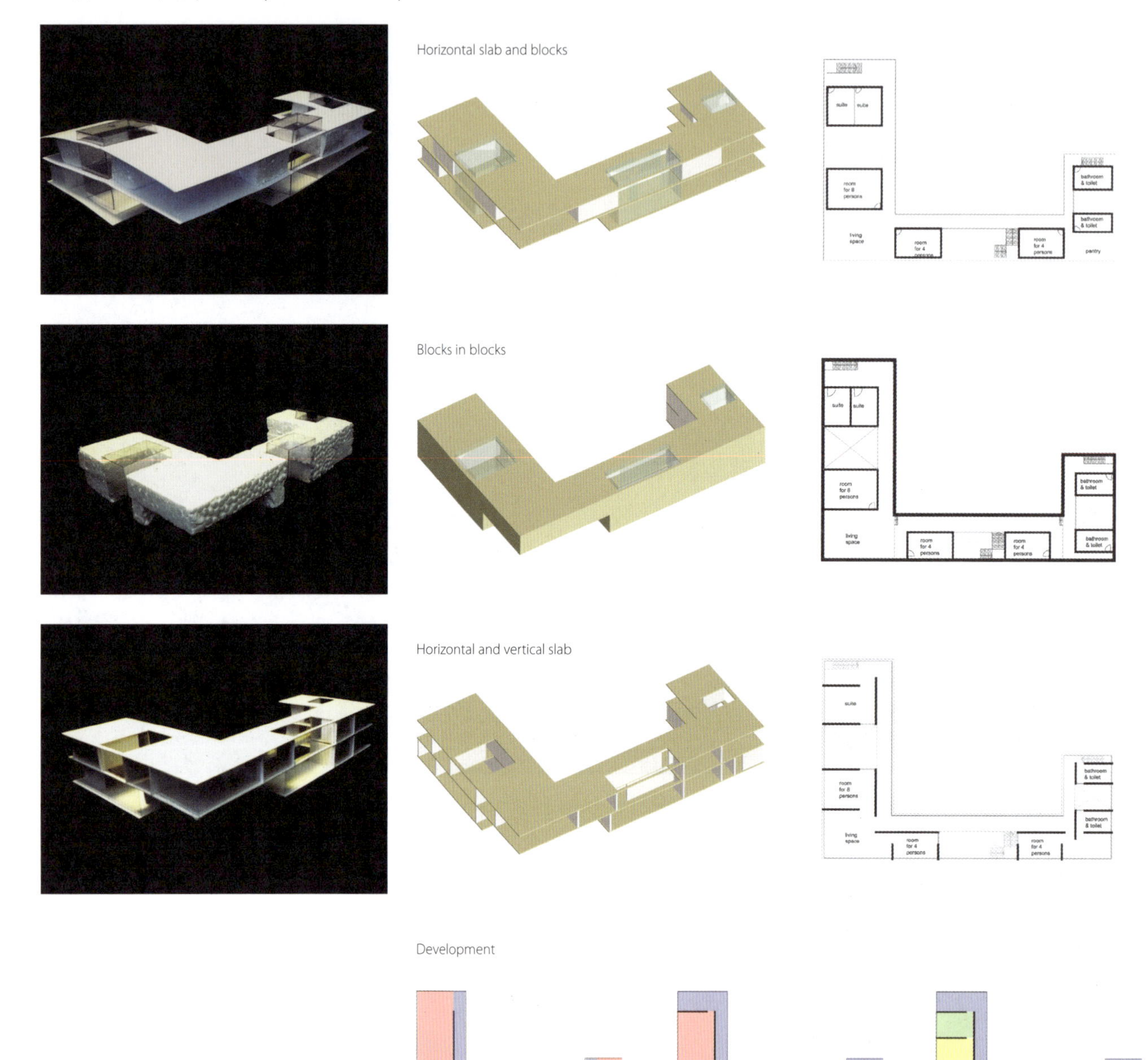
Horizontal slab and blocks
Blocks in blocks
Horizontal and vertical slab
Development

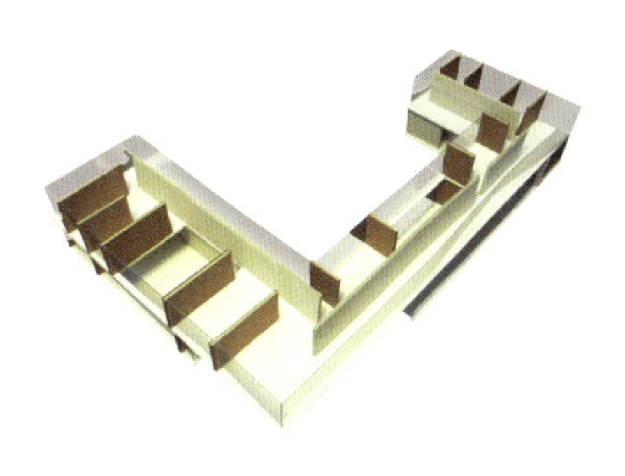

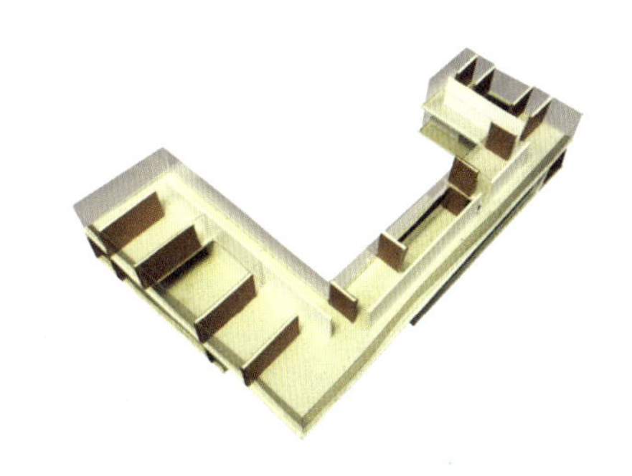

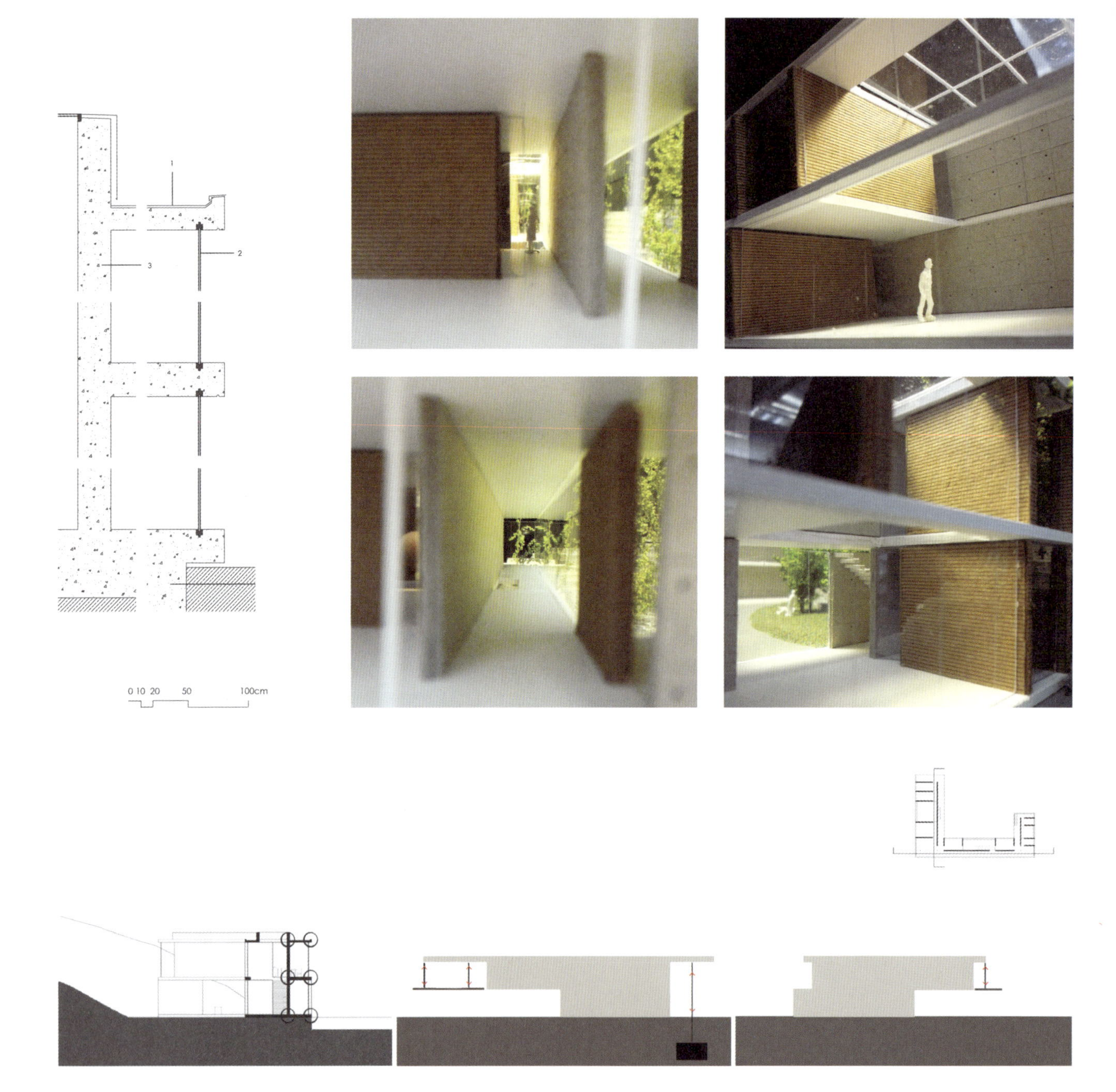
1
2
3
0 10 20 50 100cm

成果 | product

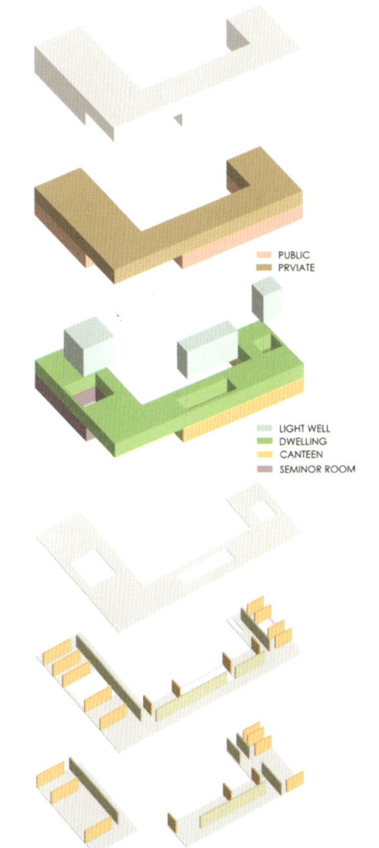

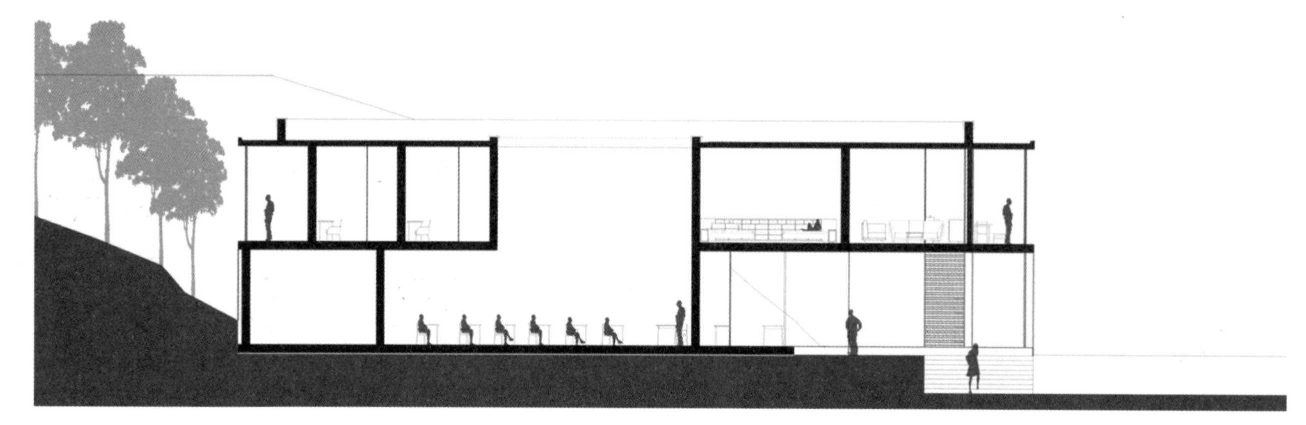

Section

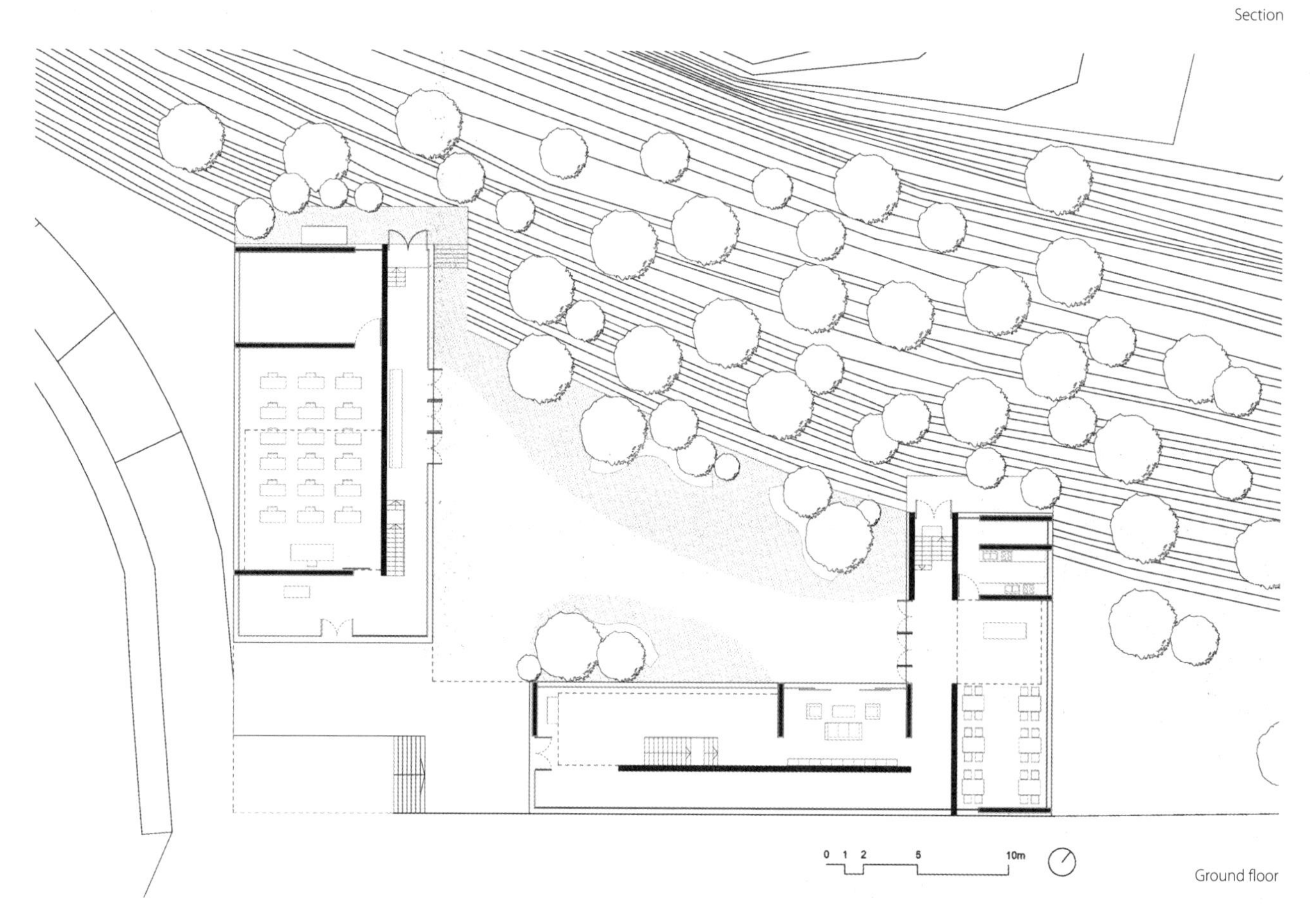

Ground floor

Corridor

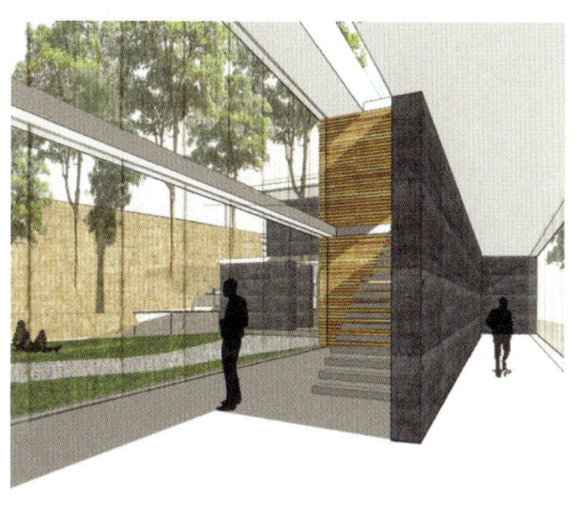
Lobby

Courtyard

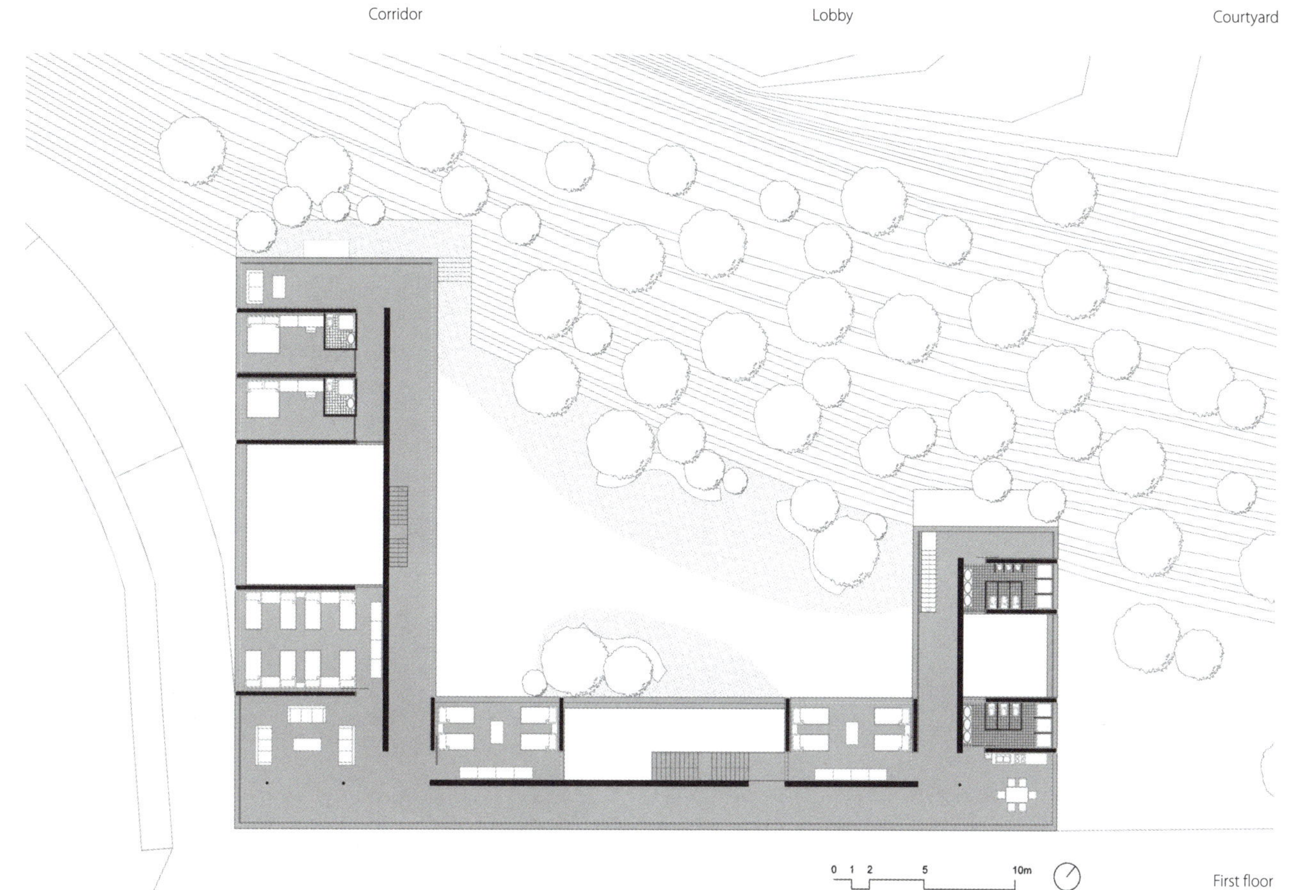

First floor

训练中心 | a training centre

Tse Kwun-lok, 0809T2Y2

体积 · 场地 | massing · site

组织 · 功能 | organisation · programme

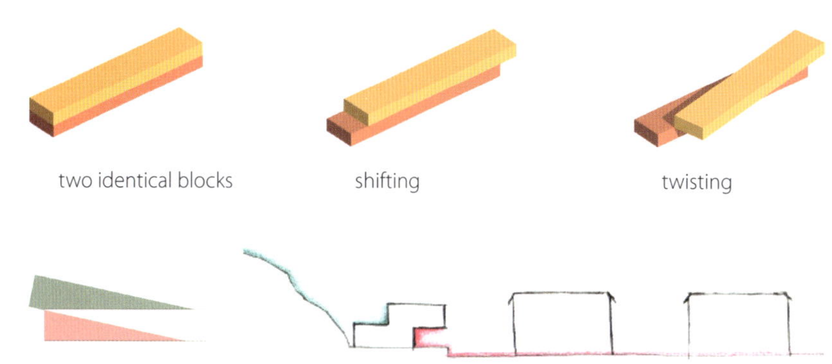

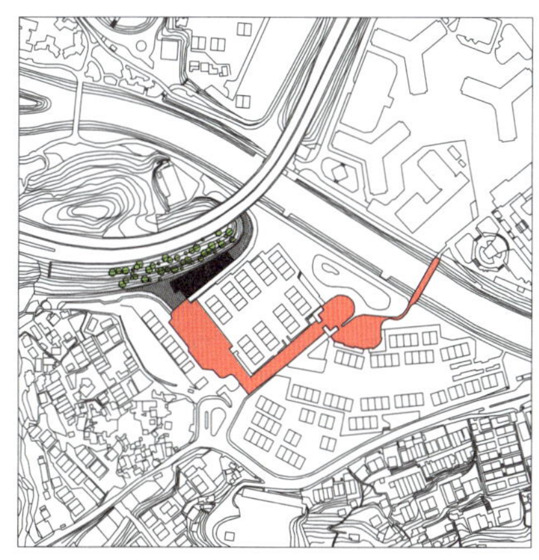

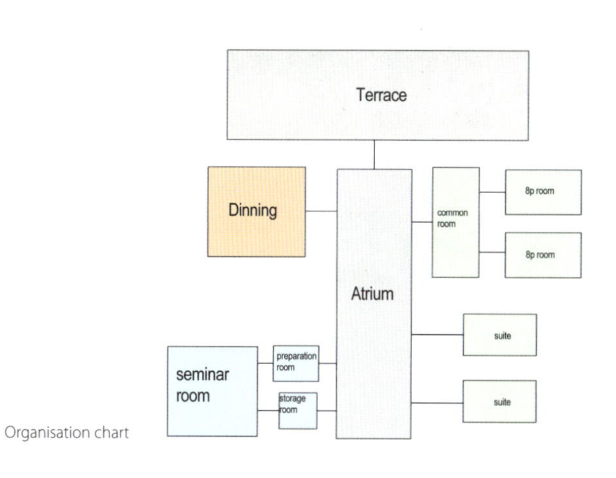

Organisation chart

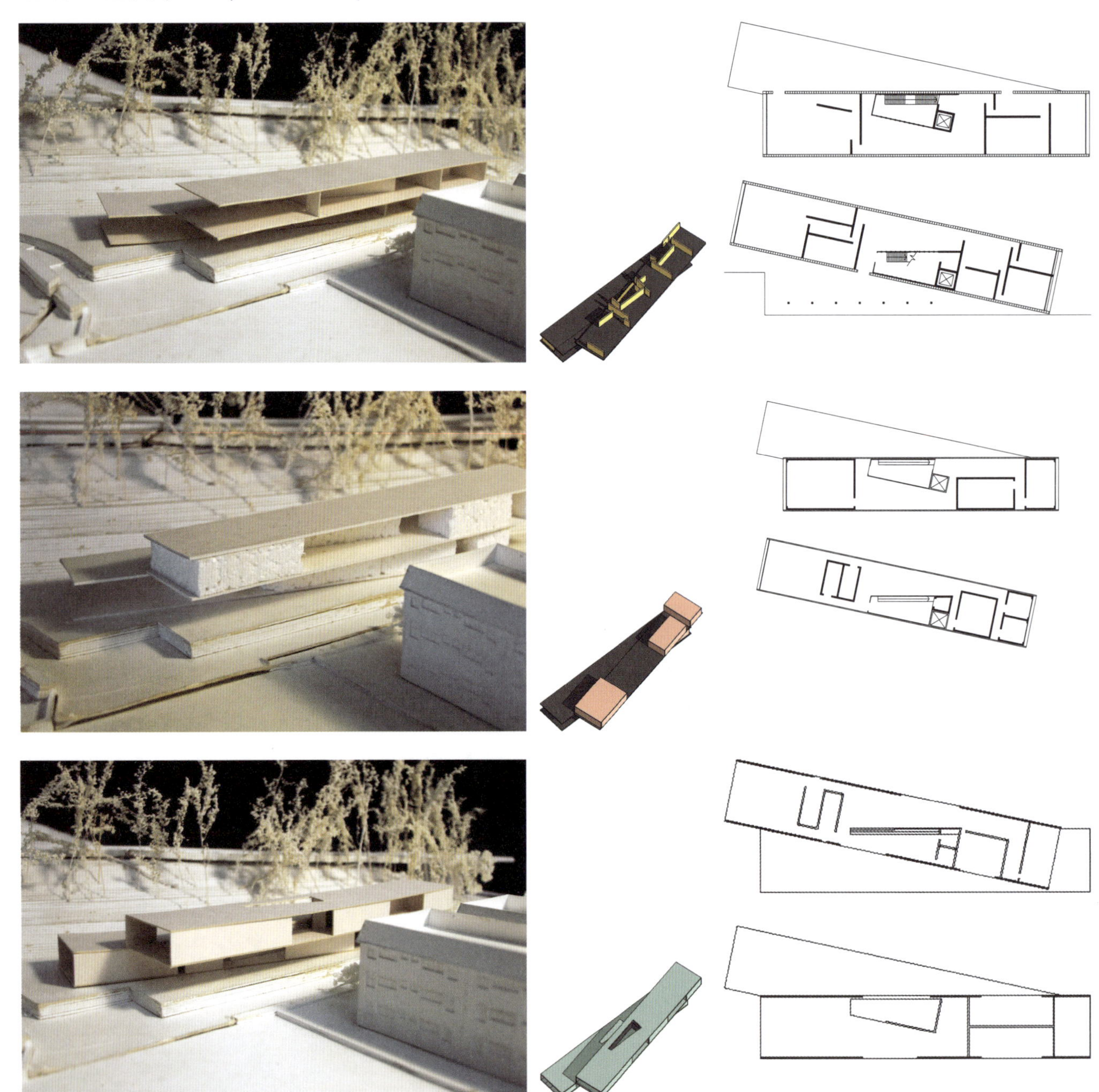

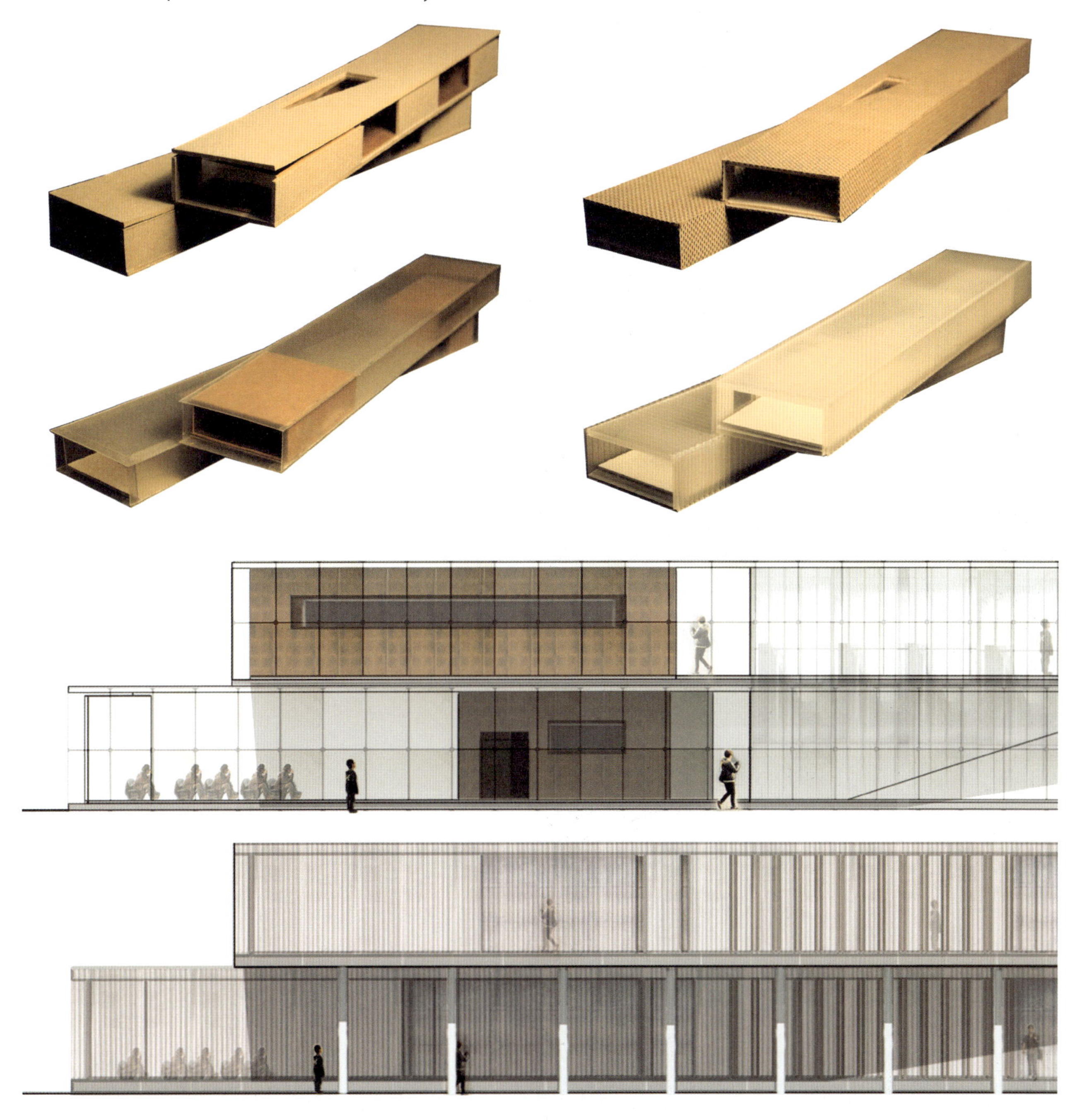

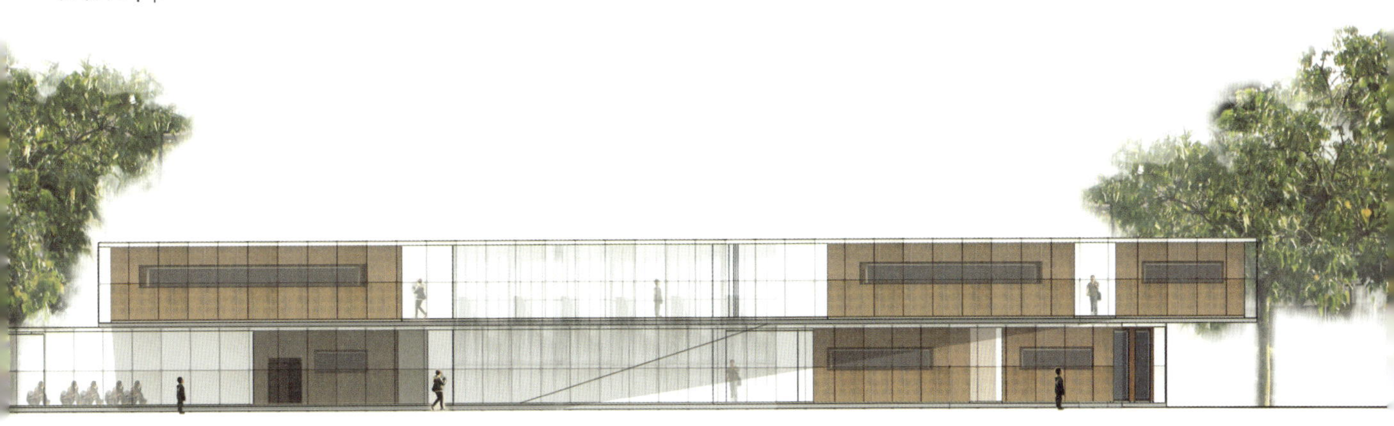

Ground floor

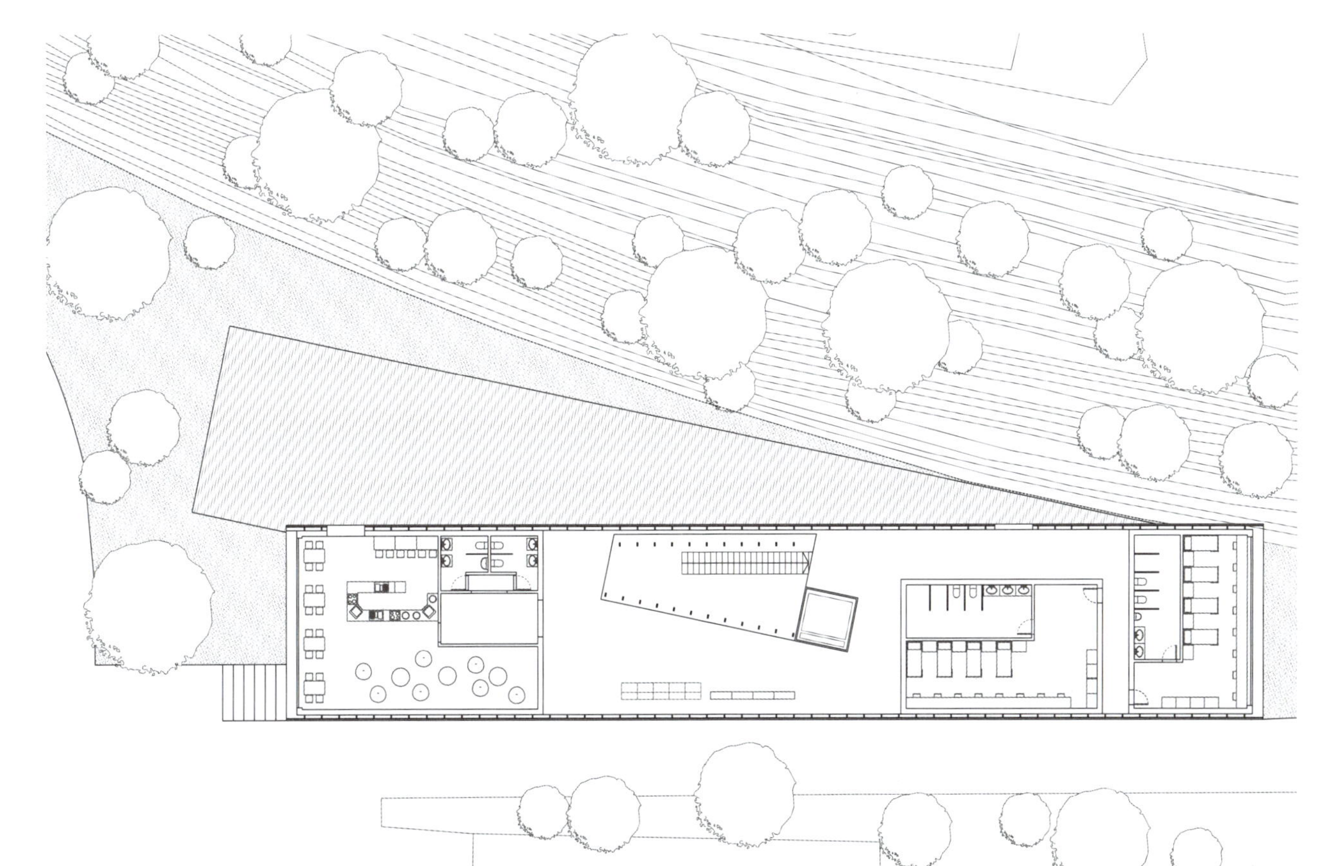

First floor

Performance Space
MUSIC STUDIO
PLACE OF

6 结语 | epilogue

反思与展望 | reflections and prospects

就如同宇宙万物皆有其生命周期一样，一个建筑设计教学体系也有它的生命周期。香港中文大学建筑学系的主题工作室制度从2001年开始实行，到2009年结束，前后运作了8个学年，这也就是建构实验课程和建构工作室的生命周期。一件事情的开始有它的原因，其结束也一定有它的原因。借着这本书完成的时机，我们不妨对这些年的工作作一反思：主题设计工作室的本质究竟是什么，建构实验课程取得哪些成果，以及这个设计研究将来有哪些发展的可能性？

1. 主题工作室制度的再思考

主题工作室制度的终结有多种原因。其中的一个理由是经过若干年的实践，大家发现学生在四个主题工作室的先后顺序的选择方面呈现一个比较明显的倾向（本科二、三年级的学生需要自行选择主题工作室），即建构、人居、技术和城市的顺序。具体来解读，建构工作室的建构实验课程着重基本设计方法的训练，人居工作室着重解决建筑的使用功能的训练，技术工作室着重建筑的建造和环境技术综合，而城市工作室则因为解决问题的规模和复杂程度成为殿后。这样的分析，得到一个学生学习设计的一般规律，即从基础到运用，从简单到复杂，以及从小规模到大规模，有其合理性。其结果就是学系的设计教学制度从垂直式的教学组织回复到水平式的传统教学组织。本科三年的6个学期的顺序为：基础、建构、人居、技术、城市和综合。这大概规定了每个学期的设计教学的重点。设计教学组的名称也改为中性的单元1至6。这里，我们需要进一步思考的是，“垂直”和“水平”这两个教学组织的概念是否就是原先的设计教学体系的本质呢？

正如我们在第一章“教案”中已经强调的，主题工作室制度的本质在于“主题”，而非“垂直”。我们认为主题工作室制度的意义在于：一，一个明确界定的学术研究方向；二，几个学术兴趣和观点相同的同事组成一个教学团队；三，一种将设计教学和学术研究相结合的工作室运作机制。我们认为这是本书所呈现的建构实验能够完成的重要条件。其实这也是任何学术研究的基本条件，即你需要有一个明确的研究课题，需要研究的团队，以及需要一个合适的运作机制。总之，主题工作室的实质是强调设计教学的学术性。

但是，这样的一个学术的设计教学观与传统的强调教师的个性和设计教学的经验性的设计教学观相冲突。这才是主题工作室制度最后终结的深层原因。一个学术的设计教学观，以及由此建立的主题工作室制度在现今的以学术研究为惟一目标的大学体制中却难以持续发展，这多少有点讽刺的意味，这也折

Like any other thing in the universe, a studio programme seems to have its own life cycle too. The thematic studio system at the Department of Architecture of CUHK was introduced in 2001 and ended in 2009. The life cycle of the Tectonic Studio was therefore eight years. A thing starts with a reason and ends also with a reason. At the end of this book, it is time for us to take a moment for reflection: what is the essence of the thematic studio system, what have we achieved through the Tectonic Lab and what are the future possibilities of this design experiment?

1 Reflections on the thematic studio system

There are various reasons for the termination of the thematic studio system. One of them is the acceptance of the perception that the students prefer the sequence Tectonics-Habitation-Technics-Urbanisation. In reality for most of the students the sequence will be different, as in each term a quarter of the students work in one of the studios, resulting in many permutations of the sequence. The reason for the preferred sequence could be that the development of the design ability from basic methods to applications, from the simple to the complex, and from the small to the large is easier to understand than the ideas suggested by the thematic studio system. In 2009, the studio system reverted from the vertical studio structure to the traditional horizontal structure by year and term. The above sequence was kept as keywords, but the studios were named Units 1 to 6. Does this change really represent a reaction to the essence of the previous studio system?

As we have already mentioned in Chapter 1 – Programme, we insist on our belief that the essence of the previous studio system is its thematic nature, not its vertical organisation. Its significance lies first, in a clearly defined direction for design research, second, in a team of collaborating teachers who share the same interest in architectural design, and third, in an operational mechanism to combine design teaching with design research. The vertical organisation allows this approach to be applied to the range from basic to advanced. These are the key elements of academic research in any other context. However, this idea of thematic studio seems to be in conflict with the tradition of experience and personality-based design teaching. This might be the real reason for the ending of the thematic studio. It is ironic that such a research oriented studio system has difficulties to survive in a so-called research university context. It might further reveal the confusion in de-

射出现今建筑教育在确定其学术性时的迷茫。

2. 建构实验课程的收获和意义

把设计教学当成学术研究之一种，是一个新的概念。而看一个设计教学是不是带有学术研究的成分，最终的评价标准是看该教学研究是否对推动建筑设计和建筑教育作出贡献。以这样的观点来反思建构实验课程，我们总结有以下四个方面的收获：

1. 发展了一套讨论空间、建构和设计的语言系统
2. 开展了一个以体块、板片和杆件为线索的空间研究
3. 重新界定了模型、透视图和建筑图在设计中的作用
4. 发展了一个以模型的操作为主要手段的设计方法

如果要对这个课程的本质作更进一步的概括，那么，我们认为这个课程提出了一个建筑设计造型的主张：即如何生成一个空间的概念、如何将这个概念发展成为一个建筑，以及如何在建筑的造型中来表达这个空间的概念。通过一个严谨的设计教学计划，这个设计方法变的可以传授。而且，这个方法并不是一开始就已经形成的，它开始于一些对建筑设计方法的思考，一个关于设计方法的假设，借助于设计教学的手段来研究，并经过很多年的反思、肯定和调整才最终形成的。最后，在本书中所呈现的设计方法不应该被视为一个终极成果，而是一个持续设计研究过程中的一个阶段性成果。

我们在建构实验课程中所提出的一系列问题有其特定的学术背景，一方面来自于对中国内地建筑设计和建筑教育的历史和现状的观察，另一方面来自于对我们共同的学习和教学背景，即对苏黎世联邦理工学院设计教学方法的反思。

我们一直以为这个在香港发生的设计课程学术上与当今中国内地的建筑设计和教育的发展有着密切的联系。这一关联性主要体现在两个方面。首先，我们试图通过提出一个替代的设计方法来填补因为长期影响中国的“布杂”设计方法的衰退而形成的学术真空。不同于“布杂”的过多地关注于建筑的表面形式的设计方法，我们则试图从建筑的空间和建造问题，以及设计的模型操作来寻求设计的想法。此外，我们把这个以学校的设计课堂作为基地的设计实验与那些中国建筑师通过实际的建筑工程来进行的设计实验看成是一个并行和互补的关系。中国当前的经济发展和城市化的进程确实为很多的外国的和中国的建筑师提供了进行建筑设计新尝试的绝佳机会。这些在中国建造的新颖的建筑成为推动中国的建筑设计发展的强大动力。但是，一个健康的建筑学发展有赖于实践和教育的良性互动，不但建筑实践影响教育，同时建筑教育通过设计研究和人才培

fining the essence of scholarship in architecture today.

2 The gains and significance of the Tectonic Lab

It is a rather new concept to consider studio teaching as a form of academic research. However, whether studio teaching can be regarded as an academic research should not be based on its manifesto but by its contribution to the advancement of the body of design knowledge. In this respect, what we have achieved through the Tectonic Lab can be summarised by the following four points:

The Tectonic Lab

1 developed a set of terminologies on space, tectonics and design
2 conducted an investigation of space based on three elements
3 redefined the role of model and graphics in design
4 devised a design method based on model operation

Formulated in a different way, the Tectonic Lab proposes a method to deal with architectural form; not by borrowing the form from somewhere, but by arriving at a form through a process starting from a conceptual form and ending with a built form, with mechanisms to clarify and strengthen the idea when dealing with more issues. The programme is a contribution to make this method teachable.

The Tectonic Lab was developed over a period of time, starting with a hypothesis of a method of work, exploring and testing it through the studio teaching. It took the present form after cycles of reflection, confirmation and revision. The development with its changes show that it is neither meant to be final nor dogmatic.

Apart from what we hope is a contribution to the field of architecture, the Tectonic Lab has specific but very different relationships to three places: Switzerland, mainland China and Hong Kong.

The relationship to Switzerland comes from our shared learning and teaching background at the ETH-Z. Part of our inspiration for the programme comes from critical reflections on the design pedagogy at the ETH-Z since the 1960s which also dealt with the formation of architectural space. The design method proposed by Prof. Berhard Hoesli emphasised the logical relationship between use, space and construction. But it dealt essentially only with one type of space – continuous space. Based on this, Prof. Herbert Kramel later developed his own structured and integrated design programme. However, we recognised in the work of contemporary architects an interest in other types

养影响建筑实践。而中国的现状是建筑教育的发展远远落后于建筑实践。我们希望通过这个教学研究来证明建筑教育可以和建筑实践有一个不同的关系，不是跟在后面，而是平行发展，甚至可以走在前面。

我们的学术思考的另外一个重要的来源是对苏黎世联邦理工学院自1960年代以来的空间设计方法的反思。勃那德·赫伊斯利提出的建筑设计方法基于现代主义建筑的连续空间概念，强调空间和结构的逻辑关系。其后又由赫伯特·克莱默进一步发展为一个结构有序的设计教学体系。这一设计方法曾经在1990年代影响到国内的一些院校。我们认为当代的建筑设计的发展出现了一些根本性的变化，如空间的概念更加丰富多样，建筑作品中更多地表现出模型操作的影响，而不单纯是结构的逻辑导致的形式，等等。这些变化提示我们应该发展新的设计方法以及相应的教学方法。从某种意义上来说，我们试图延续和发展这一空间和建构的课题。

最后，我们还必须要讨论一下这个关于空间和建构的设计研究与香港本土的关系，毕竟这是一个发生在香港的事情。相比较前面提到的两个重要的参照系，这个课程和香港当前的建筑设计文化的相关性就显得比较弱。通过对香港1950年代前后的现代建筑设计成就的研究，我们反而觉得我们与那个时代的建筑设计有一种延续性。

3. 从建构工作室到工作室U2

原先的建构工作室在新的学系教学体系中成为一年级第二学期的设计基础课程，即设计工作室单元二。在建构工作室之前，本书的两位作者曾经主持一年级的建筑设计基础课程多年。从建筑设计入门课程到建构工作室，这是设计教学研究的两个不同的阶段。两者之间既有连续，又有所区别。

那个一年级的设计基础课程是受到我们对香港的集装箱建筑研究的启发。以集装箱建筑的单元组合设计原理为主要线索，我们设计了一个包含了坐具、单元、场所、中心和亭子几个设计的课程，每个设计由一系列清楚界定的练习来展开。我们的目的是设计一个具有鲜明地方特点的建筑设计基础课程。这一对建筑设计基本问题的关注在建构实验课程中得到进一步的发展。但是在后者，它成为一个大的设计方法研究的计划的一个部分，它的目标针对的是当代建筑设计的问题，因而具有更加广泛的意义。借助这个新的平台，我们得以更加深入地来探讨我们对设计方法和设计教学法的一些思考。

从一年级的集装箱基础课程，到建构工作室的建构实验课程，再回归到一年级的设计基础教育，形成一个设计研究的轮

of spaces and influences that working with models has on the creation of space and form. We tried to extend the interest in space formation by articulating a new working method with its respective design pedagogy.

The relationship to the contemporary development of architectural design and education in mainland China has two aspects. One relates to the void left by the decline of the long-term domination of the Beaux-Arts design and teaching methods, another to one part of contemporary Chinese design and building practice – not the fundamental methods made the Beaux-Arts approach inappropriate, but its formal preferences. Although there were sporadic attempts to reform architectural education, and there are good individual teachers, no alternative programme has evolved yet. China's recent economic growth and rapid urbanisation provide unique opportunities for Chinese and foreign architects, and lead to a vivid contemporary architectural scene. The resulting new forms of architecture have a great impact on architectural schools, without provoking new teaching methods. Instead of this one-sided influence, a mutual relationship between education and practice would be healthier. But at present, the development of architectural education seems far behind that of architectural practice. We hope that the Tectonic Lab can make a contribution to these two issues.

Although we have developed this programme in Hong Kong and its effect can be clearly seen in the student work, it is rather weakly linked to this context. This might partly be because contemporary practice in Hong Kong has very different interests. But what we are doing has a strong affinity for Hong Kong's design culture demonstrated by the work of modern architects in the 1930s and its further development from the 1950s to the 1970s.

3 From Tectonic Studio to Studio Unit 2

In the present studio sequence, the people who had constituted the Tectonic Studio became responsible for the second term of the first year design studio, called Unit 2. Interestingly, the two authors had taught the first year course Introduction to Architectural Design I & II for several years beforehand. From Introduction to Architectural Design to the Tectonic Studio, there is a continuity as well as a distinction.

In the previous foundation course, we tried to develop a programme based on the study of Hong Kong's container architecture.

回。很显然，这个回归不应该是将建构实验课程的练习简单地照搬到一年级。在建构实验课程的基础上，现在我们可以从一个不同的角度来重新思考建筑设计的基础训练问题。建构实验课程中关于三种空间界定要素的研究，关于以模型为作业手段的设计方法的研究，以及从构思形式到建造形式的转换的研究，这些均会对正在发展的一个新的建筑设计基础课程产生深远的影响。

4. 针对中国内地的一些教学实验

建构实验课程虽然只是中文大学建筑学系设计教学体系中的一个环节，但是中国当代的建筑设计和建筑教育才是这个建构设计研究的真正目的所在。在本书的最后，我们有必要再次回归到这个真正的主题。

2005年春在南京东南大学建筑系的二年级进行的联合教学是最先在国内的建筑院校尝试建构实验课程的一个尝试。两个月的课程完成了从构思到建造的四个练习。学生的反馈以他们先前的学习经验和对国内设计教学的认知为基础，特别表达了对该教案关注于空间和建构问题、方法的可教性和可学性，以及设计研究的探索性等方面的认同。通过这个教学尝试，我们清楚地认识到建构实验课程试图填补国内的建筑教育在“布杂”的形式主义方法式微之后形成的设计方法的遗缺。

2008年夏天我们与深圳市建筑设计研究总院合作进行了一次以培训建筑师为目的的“空间、建构与设计工作坊”，14位建筑师在香港中文大学建筑学系进行了为期两周的封闭式集中训练，除了理论授课和参观访问外，主要完成了构思、抽象和材料三个练习（限于时间原因，略去了第四个练习）。大学的学术机构和建筑设计院所进行合作应该是很常见的，但是合作进行以设计的基本概念和方法的培训为目的的工作坊确实是一个大胆的尝试。最后的结果说明这样的设计基础训练对于实践建筑师来说也是非常有意义的。而与实践建筑师的互动也使我们从一个新的角度来认识建构实验课程的潜能。

2010年夏天我们与南京东南大学建筑学院合作举办了以培训设计教师为目的的“空间、建构与设计工作坊”。10位青年教师在香港进行了为期两周的封闭式集中训练，完成了构思、抽象和材料三个练习。与实践建筑师不同，设计教师在练习的过程中更多从“教”与“学”的角度来探讨教案。与设计教师的互动使的我们认识到这个建构实验课程在设计教师培训方面的潜能，要提高设计教学的水平，首先是提高设计教师的教学能力；要改变中国建筑教育的现状，首先要改变设计教师的设计观念。

The principle of unit organisation demonstrated in container buildings gave us the inspiration to design a sequence of projects: object, unit, place, complex and pavilion, each of which unfolded in a series of well-defined exercises. Both the study of container architecture and the teaching programme were documented and published by China Architecture & Building Press. Our intention was to develop a design foundation course with strong local characteristics. We carried our primary interest in basic design issues forward into the Tectonic Lab where it became an integral part of a larger project of design research. Through the, at that time, new platform of the thematic studio, we could further advance our thoughts about both design method and design pedagogy.

However, from the foundation course to the Tectonic Studio then back to the Studio Unit 2, we don't want to simply transplant the Tectonic Lab exercises to the first year class. Instead, we use this opportunity once more to formulate a new year one studio programme, by letting the studies of the Tectonic Lab flow into a new series of exercises, like studies on the three space defining elements, the role of model operation for generating a space concept and the transition from conceptual form to built form.

4 Outreach to mainland China

Although the Tectonic Lab was a studio course in the previous Department of Architecture at CUHK, in its background there was always a view on the architectural development in mainland China. At the end of this book, we would like to mention three encouraging experiences.

The first attempt to test the Tectonic Lab programme in mainland China was a second year studio course at the Southeast University in Nanjing. Within about two months' time students completed four exercises from concept to construction. In contrast to their previous learning experiences, students in their feedback particularly expressed their appreciation of the focus on space and tectonics, the teachable nature of the programme, and the encouragement of design exploration. From the experience of this studio, we felt confirmed in our hope that our approach could become a component in a proposal to fill the academic gap left by the decline of the Beaux-Arts method in China.

The second attempt was a summer design workshop at CUHK for 14 young architects from the Shenzhen General Institute of Architectural Design and Research. Within a two-week intensive course, the participants completed the first three exercises. It is unique

这三次活动分别从学习、教学和设计三个方面就建构实验课程对中国的建筑教育和建筑设计的可能的影响给于我们不少的启示。今后，我们必然会继续开展这类学术活动。毫无疑问，这些学术交流的目的在于推广建构设计方法。但是，我们要强调的是本书所呈现的建构实验课程并非一个惟一的设计方法，它只是针对特定的设计问题提出的一个可能的答案。所以，我们期望将来会有更多的设计基础研究成果出现，从而推进中国建筑设计和建筑教育的发展。

for a large design firm to be interested in such a training workshop for the purpose of refreshing their young architects' design capability. It was inspiring and encouraging for us. From this workshop, we recognised that the potential of this programme is not limited to the education sector.

In the summer of 2010, a similar workshop was organised at CUHK for 10 young design teachers from Southeast University. Unlike those practising architects, the participants of this workshop were more interested in pedagogic issues behind the programme – not the results but the teaching method which transforms an initial concept into an end product. From this workshop, we recognised that this programme could also serve for teachers' training.

These three attempts tested the applicability of the Tectonic Lab programme to mainland China from three different aspects: learning, designing and teaching. We are expecting to further strengthen our linkage to mainland China through similar activities in the future. Of course, with these activities we promote a method of work in China. But we don't consider this the only method. Rather, we would like to present it as a case of design research. For Chinese architectural education to flourish, more such studies have to emerge in the future.

7 附录 | appendix

致谢 | credits

中文大学建筑学院的主题工作室制度于2001年由当时的系主任白思德所创立，到2009年结束。相应的，建构工作室的运作迭前后达8年之久。本书所呈现的建筑设计研究绝非个人之力所能完成，必有赖于一个教学团队的精诚合作。除了本书的两位作者外，工作室的成员期间经历了几次转变，林云峰和Tim Nutt两位同事离开后，朱竞翔和谭善隆两位新人先后加入。其他参与短期教学的有：黄永发、吴钢和刘珩。

作者要特别感谢瑞士的建筑师马库斯·卢契尔先生，他不但给建构实验课程一个充满希望的开始，其后又数次来中大参与教学。我们还要特别感谢苏黎世联邦理工学院的赫伯·特克莱默教授，他不但是本书两位作者和马库斯·卢契尔的共同渊源所在，还数年担任工作室的荣誉教授，亲临香港给予指导。

香港本地、国内和国际的学者和建筑师通过工作室的客座讲座和期末评图对建构实验课程做出贡献。他们是：张永和、严迅奇、陈丙骅、龚维敏、张雷、刘家琨、René Furer、贾倍思、单皓、赵辰、艾未未、Lisa Ehrensperger、Roland Frei、刘晓都、孟岩、Renate Oehalf、Heinrich Degelo、Reto Pfenniger、Stephan Rutz、王澍、丁沃沃、Dieter Jüngling、王昀、Pia Simmendinger、Meta4 Design、王方戟（以参与时间为序）。这个名单还不包括参加期末评图的很多本地的建筑师和学者。

此外，通过东南大学建筑学院的王建国和龚恺两位教授、建构工作室与东大建立长期的合作关系，2005年在东大的二年级进行了建构实验课程的教学尝试，特别是2010年夏天与东大合作举办了青年教师工作坊。我们与深圳建筑设计研究总院的孟建民先生合作于2008年举办了青年建筑师的工作坊。深圳大学建筑学院的饶小军教授也在工作室与深大建筑系的学生交流等方面给予诸多协助。对于他们对建构工作室的支持，我们表示感谢。

东南大学先后有屠苏南、夏兵、徐小东和张嵩四位青年教师在建构工作室作学术访问，参与了工作室的设计教学和研究工作，我们对此特表感谢。

本书的写作经历了一个几乎与课程的历史一样长的过程，期间很多学生参与的学生作业的资料收集和文本整理，在此不能一一罗列姓名。作者要特别感谢夏兵先生，他协助将相关的资料整合成本书的初稿。

最后作者要特别感谢在这8年的时间里先后在建构工作室学习的学生们，他们对学习设计的热诚是我们坚持设计研究的持续动力。

As the thematic studio structure was introduced in 2001 by the former Department Chairman, Prof. Essy Baniassad, and ended in 2009, the tectonics studio lasted for eight years. The work presented in this book could never have been done without the close collaboration of a group of dedicated colleagues. Besides the two authors, the studio team underwent several changes over time. Zhu Jingxiang and Nelson Tam joined the studio after Bernard Lim and Tim Nutt joined another studio. Wu Gang, as a practising architect, made an important contribution to the studio as studio critic and guest professor. Johnny Wong and Doreen Liu taught in the Tectonic Studio part-time.

We owe special thanks to Markus Lüscher for his initial workshop which gave a good start to the Tectonic Lab and his continuous support through several visits in the following years. We also thank Prof. Herbert Kramel of the ETH-Z. Not only is our work rooted in his teaching at ETH-Z, but he also served as the honorary professor of the studio for several years.

We are grateful to scholars and architects from Hong Kong, mainland China, and other countries, who contributed to the studio through delivering guest lectures and attending final reviews. They are Chang Yungho, Nelson Chen, Gong Weimin, Zhang Lei, Liu Jiakun, René Furer, Jia Beisi, Shan Hao, Zhao Chen, Ai Weiwei, Lisa Ehrensperger, Ronland Frei, Liu Xiaodu, Meng Yan, Renate Oelhaf, Heinrich Degelo, Reto Pfenninger, Stephan Rutz, Wang Shu, Ding Wowo, Dieter Jüngling, Wang Yun, Pia Simmendinger, Meta4 Design and Wang Fangji. We would also like to express our gratitude to the many local architects and scholars who served as guest critics in the studio reviews.

In addition, we would like to thank Wang Jianguo and Gong Kai of Southeast University for the collaboration through joint studios, especially the workshop for young design teachers this sum- mer. For a similar reason, another successful workshop for young architects in 2008, the authors would like to thank Meng Jianmin of the Shenzhen General Institute of Architectural Design and Research. We would also like to thank Rao Xiaojun of Shenzhen University for his assistance with student activities between the two schools.

We also thank Tu Sunan, Xia Bing, Xu Xiaodong and Zhang Song of Southeast University. They contributed to the teaching and research of the Tectonic Studio as visiting scholars.

The writing of this book has been as long a journey as the studio itself. Many students helped with the documentation of student works. We specially thank Xia Bing for his help with an early version of this book.

Finally, we should express our sincere thanks to those students who have gone through the programme in past years. Their enthusiasm and curiosity in design has always been the primary impetus behind this design study.

成员 | studio members and visitors

2001-02

Members
- Vito Bertin
- Gu Daqing 顾大庆
- Bernard Lim 林云峰
- Tim Nutt

Guests:
- Chang Yungho 张永和
- Herbert Kramel
- Markus Lüscher
- Rocco Yim 严迅奇
- Zhu Jingxiang 朱竞翔

2002-03

Members:
- Vito Bertin
- Gu Daqing 顾大庆
- Bernard Lim 林云峰
- Tim Nutt

Guests:
- Nelson Chen 陈丙骅
- Gong Weimin 龚维敏
- Zhang Lei 张雷
- Liu Jiakun 刘家琨
- Markus Lüscher

2003-04

Members:
- Vito Bertin
- Gu Daqing 顾大庆
- Johnny Wong 黄永发 (t1)
- Zhu Jingxiang 朱竞翔 (t2)

Guests:
- René Furer
- Jia Beisi 贾倍思
- Markus Lüscher
- Shan Hao 单皓
- Zhao Chen 赵辰

Trainees:
- Wang Yi feng
- Zhou Chao 周超

2004-05

Members:
- Vito Bertin
- Gu Daqing 顾大庆
- Zhu Jingxiang 朱竞翔

Guests:
- Ai Weiwei 艾未未
- Lisa Ehrensperger
- Roland Frei
- Liu Xiaodu 刘晓都
- Meng Yan 孟岩
- Renate Oelhaf
- Wu Gang 吴钢

Visiting scholar:
- Tu Sunan 屠苏南

2005-06

Members:

- Vito Bertin
- Gu Daqing 顾大庆
- Nelson Tam 谭善隆
- Wu Gang 吴钢 (t2)
- Zhu Jingxiang 朱竞翔

Guests:

- Heinrich Degelo
- Meng Yan 孟岩
- Reto Pfenninger

Visiting scholar:

- Xia Bing 夏兵

2006-07

Members:

- Vito Bertin
- Gu Daqing 顾大庆
- Nelson Tam 谭善隆
- Zhu Jingxiang 朱竞翔

Guests:

- Heinrich Degelo
- Markus Lüscher
- Stephan Rutz
- Wang Shu 王澍

Visiting scholar:

- Xu Xiaodong 徐小东

2007-08

Members:

- Vito Bertin
- Gu Daqing 顾大庆
- Nelson Tam 谭善隆
- Zhu Jingxiang 朱竞翔

Guests:

- Ding Wowo 丁沃沃
- Dieter Jüngling
- Wang Yun 王昀
- Pia Simmendinger

Visiting scholar:

- Zhang Song 张嵩

2008-09

Members:

- Vito Bertin
- Gu Daqing 顾大庆
- Doreen Liu 刘珩 (t2)
- Nelson Tam 谭善隆
- Wu Gang 吴钢 (t1)
- Zhu Jingxiang 朱竞翔

Guests:

- Markus Lüscher
- Meta4 Design
- Wang Fangji 王方戟
- Wu Gang 吴钢

顾大庆 | Gu Daqing

在南京工学院（今东南大学）接受基本的建筑教育，在硕士研究阶段开始对建筑教育产生兴趣，后赴苏黎世联邦理工学院接受作为设计教师的基本训练，他的博士论文研究方向为设计工作室制度及其教学法的演变。他于1994年秋来到香港中文大学建筑学系之前曾于南京和苏黎世任教。

他的主要研究兴趣有如下几个方面：1）建筑教育，包括中国建筑教育的历史和当前的问题，设计工作室制度，建筑教育的不同模式；2）设计教学法与设计课程的设计；3）设计理论，包括“布杂”的构图理论，现代艺术和建筑的关系，视觉教育，建构方法；4）香港现代建筑和集装箱建筑。

作为建构工作室的负责人，他的贡献主要在于建立一个以大纲为中心，以研究为基础，以团队方式运作的设计工作室，以及发展一个可以系统传授的设计方法。建构实验中有关空间问题的研究也得益于他的其他课程的教学，如视觉训练，剖碎与透明，以及绘画与建筑。

He completed his basic architectural education at the Nanjing Institute of Technology (now Southeast University), became interested in architectural education during his master's study and later went to the ETH to receive his further training as a design teacher. His doctoral thesis was about the formation of the design studio and the evolution of design pedagogy. Before arriving at CUHK in 1994, he taught in Nanjing and Zurich.

His main research interests are: 1) architectural education including different models of architectural education and the history, and contemporary issues of Chinese architectural education; 2) studio method and programme design; 3) design theory including Beaux-Arts composition, relationship between modern art and architecture, visual studies, and tectonics; and 4) the design achievements of Hong Kong's modern architecture.

As the coordinator of the Tectonic Studio, his main contribution was to establish a programme-centred, research-based team teaching, and to develop a tectonic approach, which can be taught in a systematic manner. The essential interest in space also benefited from his other required and elective courses such as visual study, poché and transparency, and painting toward architecture.

http://www.arch.cuhk.edu.hk/server1/staff1/gu

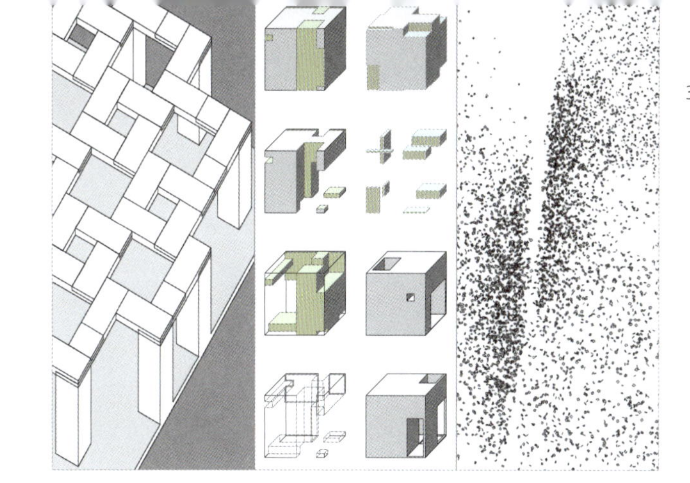

柏庭卫 | Vito Bertin

先在苏黎世联邦理工学院完成建筑教育，后及在京都大学进修，毕业后在ETH做过助教，以及在瑞士和津巴布韦从事设计实践，他于1996年来香港中文大学建筑学系之前分别在南京、苏黎世和神户三地执教，2009年7月从中文大学退休。

他与同事的合作研究项目包括香港集装箱建筑、香港现代建筑，以及建构设计方法。他个人的研究兴趣包括杠杆梁结构、形式和空间的参数描述、互动和动画式的图示表达，以及当代建筑的图示分析。

建构实验课程吸收了他以前在选修课中关于当代瑞士建筑的研究，主要是有关包裹性空间的研究。另外一些与建构方法相关的课题，如体积和空间的连续性，作为透明性的延伸的半透明，以及颗粒所生成的空间，则被结合进他的一个有关计算机媒介的课程。他也在建构工作室的研究生课程中教授参数化设计。

After studying in Zurich and Kyoto, working as an assistant in Zurich, and gaining some practical experience in Switzerland and Zimbabwe, he taught in Nanjing, Zurich, and Kobe, before coming to Hong Kong in 1996 to work at the Department of Architecture of The Chinese University of Hong Kong, which retired him at the end of July 2009.

His research, in co-operation with colleagues is on Hong Kong container architecture, Hong Kong modern architecture and teaching with a tectonics approach. His individual research interests are lever-beam structures, parametric description of form and space, interactive and animated drawings, and graphic analysis of contemporary architecture.

Insights from his earlier electives with studies of contemporary Swiss architecture, mainly under the aspect of enveloped space, have been absorbed by the tectonics lab. He further explored a group of specific aspects relating to the tectonics lab – the continuity of mass and space, translucency as an expansion of transparency, and granularity as particles in space – in the computer course, and explored methods of parametric space definition and organisation within the Tectonics Studio with graduate students.

http://web.me.com/vito.bertin

朱竞翔 | Zhu Jingxiang

他在南京东南大学接受建筑教育，并获得建筑学学士、硕士和博士学位。1997年期间曾在苏黎世联邦理工学院学习。他于2000年开始在南京大学建筑任教，并在盐城市设计建造了多座公共建筑。他于2004年加入香港中文大学，教授建筑设计以及材料和建造课程。他的研究兴趣主要在几个领域：设计方法、结构与空间，以及乡土建筑。

他在建构工作室的教学主要集中于建筑学硕士一年级的高级建构课程。最近几年，该课程的研究方向为新的结构概念所产生的空间组织的可能性。这主要受到欧洲和日本近年来由建筑师和结构工程师密切合作所产生的一些新建筑的启发。

在过去两年，他开始研发一种基于在中国的临时建筑中普遍采用的建造构件的建造体系，并先后在四川省的下寺村和达祖村建造了两所小学。通过这个项目他证明了如何从设计的角度将一种实用技术转化为建筑设计的有利手段。

He studied architecture at Southeast University in Nanjing and graduated with BArch, MArch and PhD degrees. He studied at the ETH during 1997. He started teaching at Nanjing University in 2000 and designed and constructed a number of public buildings in the city of Yancheng. Since 2004, he has taught architectural design, materials and construction at The Chinese University of Hong Kong. His research is in the areas of design methodology, new articulation of structure and space, and the study of vernacular architecture.

In the first year of the master's programme, as a contribution to tectonics studies and teaching on another level, he introduced the students to new spatial possibilities based on structural ideas, referring to recent developments in Europe and Japan as a result of close cooperation between architects and engineers.

In the last two years he has developed a building system based on components commonly used for temporary buildings in China, with which he could build a primary school in Xia Si Village and another in Da Zu Village, Sizhuan province. With this he demonstrated how interest and design sense can transform a pragmatic technique into an architectural contribution.

http://www.arch.cuhk.edu.hk/server1/staff1/zhu

谭善隆 | Nelson Tam

他在香港中文大学接受建筑学教育，并获得建筑学硕士学位。在香港工作两年后，他到了苏黎世联邦理工学院建筑系的巴塞尔工作室任助理，并同时于当地工作，服务过的事务所包括 Diener & Diener，Meili & Peter 及 AGPS。返回香港后他先是在一家本地事务所工作，于2005年开始在香港中文大学兼职任教设计课程，及后于2008年至2010年期间转为全职任教。

他是建构工作室最早的学生之一，接受了建构实验课程的训练，并有出色的表现。任教期间他对建构实验课程的贡献主要在“设计项目习作化”的实施。

教学的同时，他也开始了有关空间组织类型和建筑平面秩序的研究。他亦曾经协助Vito Bertin的杠杆结构研究，去年他参与了朱竞翔的设计团队,设计和建造了四川达祖村小学。他现于北京的维思平建筑设计事务所工作。

He studied architecture at The Chinese University of Hong Kong and graduated with a MArch degree. After practising in Hong Kong for two years, he spent a term as an assistant in the Studio Basel of the ETH and practice parallel. Offices he worked with include Diener & Diener, Meili & Peter and AGPS. After returning to Hong Kong, he joined a private practice, and started teaching from 2005 part-time and until 2008 full-time as a Professional Consultant at the Department of Architecture of CUHK.

Having been a student in the Tectonics Studio himself, he successfully worked with the Studio programme and achieved a high level of work in the process and its results. As a teaching member, he contributed to the development of the programme in which exercises are embedded in a design project.

Along with teaching, he had a research interest in the typology of space organisation and the order of floor plans, and he also supported Vito Bertin's research on leverbeam structures on several occasions. Last year he joined Zhu Jingxiang for the design and construction of the school in Da Zu Village, Sichuan. Currently he works in the architectural office WSP in Beijing.

http://neltam.wordpress.com/tag/architecture/

讲座 | lectures

工作室讲座 | studio lectures

除了与设计练习直接相关的讲座外，我们还安排了由教师主讲，面向建构工作室所有学生的工作室讲座。它们构成建构工作室的理论基础。在此列出的工作室讲座为历年讲座的小结，每学期的讲座并不相同，一般4～5个讲座。客座讲座由外面邀请的建筑师和学者主讲，面向全系师生。一般一个学期安排一个客座讲座。

Except lectures directly related to studio exercises, we also arrange each term a series of studio lectures for the whole studio. These lectures form the theoretic basis of the Tectonic Studio. The list of studio lectures here is a summary over the years. The topic of lectures offered each term varied – usually 4 to 5 lectures. Guest lectures are offered by invited architects and scholars, and are open to the whole department. Usually, there is one guest lecture each term.

Tectonic studio (gdq)

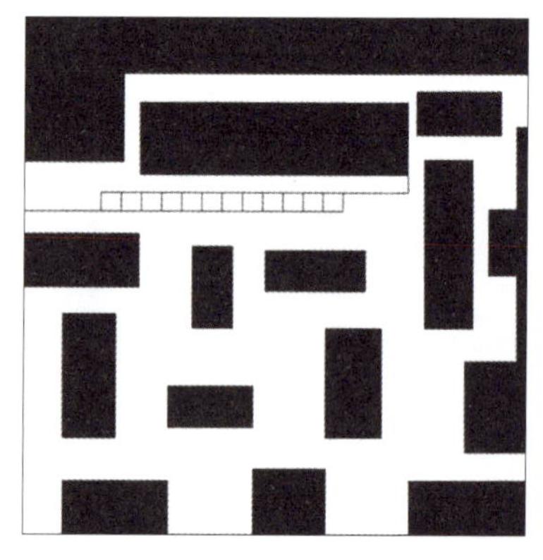

Composition (vb)

Swiss reference (vb)

Structure (zjx)

Practice (zjx)

Articulation (vb)

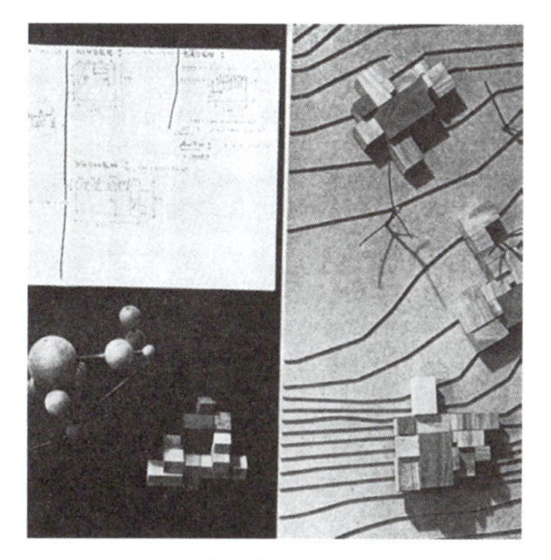

Design method (gdq)

Hong Kong reference (gdq)

Practice (nt)

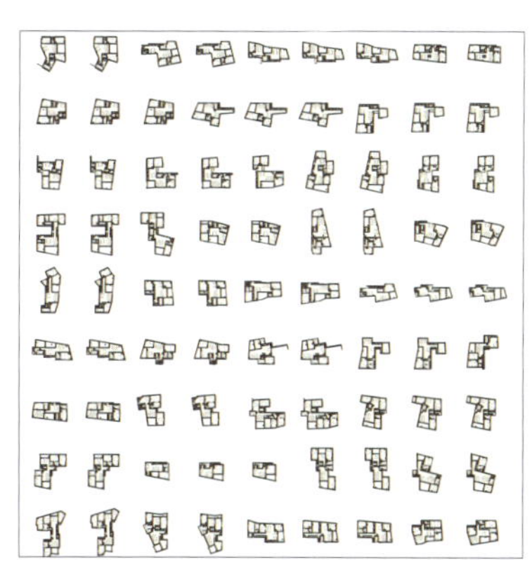

Plan orders (nt)

客座讲座 | guest lectures

26.09.2001 Markus Lüscher

16.10.2002 Nelson Chen

05.03.2003 Liu Jiakun

4.03.2002 Chang Young Ho

05.03.2003 Gong Weimin

05.03.2003 Zhang Lei

16.10.2003 Jia Beisi

04.11.2004 Ai Weiwei

20..10.2005 Meng Yan

00.11.2003 Zhao Chen

23.03.2005 Roland Frei, Lisa Ehrensperger

29.03.2006 Wu Gang

13.11.2006 Heinrich Degelo

01.11.2007 Wang Yun

28.11.2008 Wang Fangji

26.03.2007 Wang Shu

27.02.2008 Dieter Jüngling

18.04.2009 Meta4

引用 | references

书中的插图除了以下注明出处的外，均来自于作者和学生作业。

Except those listed below, all the illustrations in this book are provided by the authors and from students' work.

p. 5 Diagram: Essy Baniassad

p. 22 Course documentation cover: Markus Lüscher

p. 32 Middle and bottom photograpg: Zhu Lei

p. 37 Photograph: Essy Baniassad

p. 38 *The work of G. Rietveld Architect*, A. W. Bruna & Zoom Utrecht, the Netherlands, 1958, p. 128-129

p. 39 Up: *The Architecture of the Ecole des Beaux-Arts, the Museum of Modern Art*, New York, 1977, p. 91; bottom: *Mies van der Rohe Continuing the Chicago School of Architecture*, Birkhäuser Verlag, 1981, p. 20-21

p. 41 *EL Croquis, In Progress II*, vol. 106/107, p. 189-209

p. 45 *EL Croquis, In Progress 1999-2002*, vol. 96/97 + 106/107p. 266-269

p. 47 Bottom: *A+U*, *Peter Zumthor*, 1998.2. extra edition, p. 80

p. 48 Up: *Le Corbusier Oeuvre complète de 1929-34*, Zurich: Éditions d'Architecture Erlenbach, 1948, p. 28 and 26; middle: *Mies van der Rohe : the villas and country houses*, New York: Museum of Modern Art, 1985, plate 10.11, 10.19; bottom: *Frank Lloyd Wright selected houses*, Tokyo: A.D.A. Edita, 1989, p. 25 and 39

p.101 Left: *EL Croquis, Kazuyo Sejima & Ryue Nishizawa 1995-2000*, vol. 99, p. 214; middle: *The Japan Architect: Yoshio Taniguchi*, vol 21, 1996 Spring; bottom: *EL Croquis, Annette Gigon & Mike Guyer 1989-2000*, vol. 102, p. 247

图书在版编目(CIP)数据

空间、建构与设计/顾大庆，柏庭卫著．－ 北京：中国建筑工业出版社，2011.1（2023.3 重印）
ISBN 978-7-112-12780-1

I. ①空… II. ①顾… ②柏… III. ①空间设计 IV. ①TU206

中国版本图书馆CIP数据核字（2010）第254895号

责任编辑：徐纺　滕云飞

空间、建构与设计
顾大庆・柏庭卫　著
*
中国建筑工业出版社出版、发行（北京西郊百万庄）
各地新华书店、建筑书店经销
北京利丰雅高长城印刷有限公司制版
北京利丰雅高长城印刷有限公司印刷
*
开本：965×1270毫米　1/20　印张：16.5　字数：550千字
2011年4月第一版　2023年3月第十二次印刷
定价：150.00元
ISBN 978-7-112-12780-1
（20047）

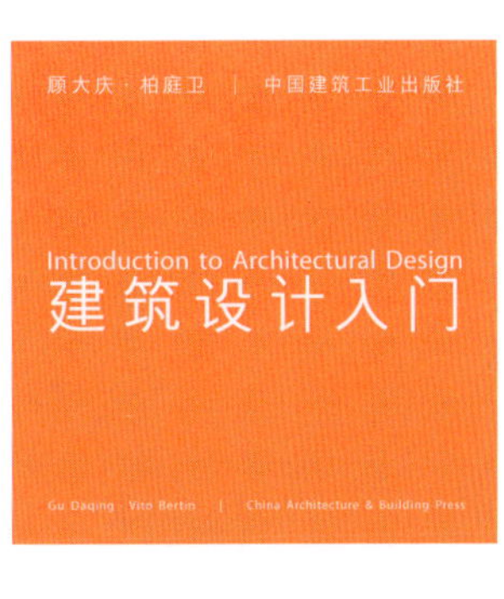

顾大庆 · 柏庭卫 著 | 中国建筑工业出版社

Space, Tectonics and Design

空间、建构与设计

Gu Daqing · Vito Bertin | China Architecture & Building Press

EXIT 出口
TECTONICS STUDIO
CENTRE FOR WORKSHOP AND TRAINING
HO HANG CHUN
CUHK ARC 4130 TECTONIC STUDIO YEAR 2